Atlas of Nevada Conifers

DAVID ALAN CHARLET

Atlas of Nevada Conifers

A Phytogeographic Reference

Foreword by W. D. Billings

Illustrations by Bridget Keimel

University of Nevada Press Reno Las Vegas

University of Nevada Press,
Reno, Nevada 89557 USA

Book design by Kaelin Chappell
Printed in the United States of America

The paper used in this book meets the requirements of American National Standard for Information Sciences—Permanence of Paper for Printed Library Materials, ANSI Z39.48-1984. Binding materials were selected for strength and durability.

This book has been reproduced as a digital reprint.

09 08 07 06 05 5 4 3 2

Research and book production for *Atlas of Nevada Conifers* has been financed in part with funds from the Nevada Biodiversity Initiative.

Library of Congress
Cataloging-in-Publication Data
Charlet, David Alan, 1953–
Atlas of Nevada conifers : a phytogeographic reference / David Alan Charlet; foreword by W. D. Billings; illustrations by Bridget Keimel.
p. cm.
Includes bibliographical references.
ISBN 0-87417-265-9 (paperback : alk. paper)
1. Conifers—Nevada—Identification. 2. Conifers—Nevada—Geographical distribution. I. Title.
QK494.C48 1995
585'.2'09793—dc20 95-9493
CIP

Contents

Maps and Illustrations

Illustrations (*continued*)

Foreword

Spreading eastward over the Great Basin in the lee of the Sierra Nevada, and in its rainshadow, the state of Nevada is characterized by desert and semidesert ecosystems whose vegetation is dominated by light-colored low shrubs and grasses, some native and some not. Nevada is also very mountainous and in no place is close to sea level: even the valleys are high. It is these high mountains—such as the Rubies, the Snakes, and the Toiyabe, Wassuk, and Toquima Ranges—trending north to south, that cool the water vapor of the rising Pacific air masses and catch and preserve the resulting winter snowfall in great drifts that last into the following summer. The meltwater from these long-lasting snows is not only the aquatic life blood of the high-mountain ecosystems themselves, it is also the source of stream flow, irrigation water, and urban water supplies in the areas below.

In the Great Basin, forests dominate ecosystemic vegetation only where water is abundant enough during the summer in the soil-root zone to support a favorable water balance between soil moisture, groundwater, and the transpirational loss of water vapor from the foliage. Most of these forests are confined to the mountains and are dominated by cone-bearing evergreen trees (conifers) whose water source is winter snowfall. Such trees are important not only for timber but also for watershed and runoff protection. Only during the last few decades of this century has any intensive research been accomplished on these very important evergreen conifers in regard to their distribution and ecology throughout Nevada and the Great Basin.

A favorable water balance occurs in three main forest ecosystem types in the basin. The first grows along the main rivers and streams in the lowland valleys and is typified by "galeria forests" of cottonwood trees and willows whose water supply comes from the high groundwater table. Such trees are phreatophytes—literally, from the Greek: "well plants."

The second and third types of forest ecosystems consist of evergreen woodlands and subalpine forests high in the mountains above the sagebrush ecosystems and extending up to timberline. These forests depend largely on winter snowfall and its meltwater. In the eastern Great Basin, some water is supplied as rain by summer thunderstorms, but these are rare in the western basin. Both types of forests are dominated by cone-bearing evergreen trees with small needle-shaped or scalelike leaves. Groves of aspens, deciduous trees related to the cottonwoods and willows of the lower elevations, indicate sites of favorable water balance in the mountains where snow meltwater or springs are abundant. These groves are relics of moister times of centuries ago and survive only by vegetative reproduction where there is enough water. Aspen seedlings are rare and cannot survive even a few days of drought.

The lowest forest ecosystem zone is characterized by single-needle pinyon and junipers (the pinyon-juniper woodland), the most drought-resistant of the Nevada forests. Between the upper limits of this woodland and the true subalpine forest is a "mountain brush" zone of shrubs, grasses, and small trees of mountain mahogany extending up to about 9500 ft (2900 m). Mountains whose uppermost elevations lie between 9000 and 10,000 ft (2700–3050 m) often are capped by this mountain brush vegetational zone and from a distance appear to be "bald" because they have no forests on their summits. There may be as many as a hundred or more of such bare mountains throughout Nevada. In western Nevada, they occur in the Virginia Mountains and the Pine Nut Range.

Mountains higher than 9000–10,000 ft (2750–3050 meters) in elevation support rather open subalpine forest of various pine species, spruce, and fir. Some of the oldest individual trees on earth occur in this zone, particularly in the White Mountains of the western Great Basin of Nevada and eastern California, and also in the Snake Range of eastern Nevada. These trees range from 3500 to more than 5000 years in age. They belong to *Pinus longaeva,* the Great Basin form of the bristlecone pines. Associated with these old trees are individual trees of *Pinus flexilis,* limber pine, that are several centuries old themselves. Both species extend to alpine timberline on the higher peaks as "island forests" often separated from similar subalpine forests by great distances.

The montane forest patterns that now exist have not always been this way. Tree species migrate in both space and time as climates, glaciers, and lakes change. The wet, cold climates of the Pleistocene brought large glaciers to the eastern Sierra Nevada and a number of the higher mountains of the Great Basin. When warmer, drier climates returned during the Holocene, 10,000–15,000 years ago, the glacial ice melted and formed many lakes. Some of these lakes in the Great Basin were very large, such as Bonneville in the eastern Basin and Lahontan in western Nevada, of which Pyramid Lake is but a small remnant. During the thousands of years of full glaciation in the mountains, many of the Sierran conifer populations migrated downward and eastward into the marginally warmer Great Basin and its lower mountains. The evidence for these migrations and their eastern limits comes from several sources. Macrofossils of *Pinus albicaulis,* whitebark pine, the present timberline tree of the Sierra Nevada whose normal elevational limits in the Sierra southwest of Reno are between about 2500 and 3200 m above sea level, have been found by Nowak et al. (1994) embedded in packrat middens near Painted Hills at an elevation of only 1380 m, and only 7.7 km from the western shores of Pyramid Lake. The radiocarbon date at this macrofossil site is 29,630 B.P. The present vegetation around this midden site is dry desert shrub, and the nearest whitebark pines today are about 50 km to the west in the Sierra Nevada. While whitebark pine is extinct around the Painted Hills and has migrated back to the Sierra and up Mount Rose to its timberline at an elevation of about 3150 m, another important Sierran conifer, *Pinus ponderosa,* is still alive and well on Pah-Rah Mountain less than 15 km to the east. This relict stand of ponderosa pine grows at an elevation of only 1600 m, only 3 km from the shore of Pyramid Lake, on a substratum rock of hydrothermally altered andesite (Billings 1950). Ponderosa pine in the region north and east of Reno grows only on such very acid rocks, and the populations of this species have certainly been there overlooking the large glacial meltwater

Lake Lahontan since the ice began to melt in the Sierra Nevada to the southwest some 25,000 to 15,000 years ago.

This book by David Alan Charlet, *Atlas of Nevada Conifers: A Phytogeographic Reference,* fills an important ecologic gap in our biogeographic knowledge of the conifer forests of the Great Basin and Nevada. It is the product of intensive and difficult field exploration and solid voucher documentation by Dr. Charlet over the last decade. It will be the standard reference for vegetational and ecological information on the conifers of the Great Basin far into the twenty-first century.

W. D. Billings
James B. Duke Professor of Ecology
Duke University

Acknowledgments

I have room here to thank only a few of the key people who contributed to this project. My packers (Agee Smith, Dean Hagen, and Lowell Prunty) were essential for my explorations of the Jarbidge Mountains from 1988 to 1994. I am grateful to John Jackson of the YP-Petan Ranch for allowing me to use the roads on his property in the Bull Run Mountains. Lee and Jo Greenstreet, Darrell Lemaire, Betty Lamb, Terri Charlet, Walter and Carrol Cripps, Jon Carpenter, Jack, Nora, and John Conlan, Hugh Marsters, Don Zerga, Betty Vaughn, and Alice Winzeler gave me many kinds of field support. Dr. Bruce Dow kindly gave me emergency medical care in Hawthorne. My major professor, Richard W. Rust, provided reconnaissance and collections for southern and eastern Nevada, and companionship during many of my explorations. I am forever indebted to Rich for putting up with my mood swings, financial depression, and irritating perseverence. The interest, support, and encouragement of Peter Brussard was especially important for the completion of this work.

From the Humboldt National Forest, I thank Steve Anderson, Chrys Olson, and Forest Supervisor John Inman. From the Toiyabe National Forest, I thank Forest Supervisor Jim Nelson, Dave Weixelmann, Desi Zamudio, and Kerry Heise. I greatly appreciate the assistance of John Brack in the Tonopah District and Roland Shaw in the Carson District. I thank Jeanne Chambers, Craig Biggart, Cheryl Nowak, and especially Robin Tausch of the Intermountain Research Station, Reno. From the United States Forest Service Intermountain Region in Ogden, I thank Dean Knighton, Mont E. Lewis, and Cindy and David Chojnacky.

Jim Jeffress of the Nevada Division of Wildlife guided me to several juniper stands in Humboldt and Washoe Counties, and Dan Marlette of the United States Department of the Interior Bureau of Land Management Eagle Lakes Resource Area, Susanville District, directed me to the remarkable "dunes pines" in eastern Lassen County, California. Ranger Bill Champion of Lake Tahoe State Park gave me access to restricted areas. I thank Jeff Knight and Scott Cichowlaz of the Nevada Department of Agriculture for reconnaissance in southern Nevada.

My thanks extend to Peter Wigand and Alvin McLane of the Desert Research Institute, and, from other academic institutions, to Barbara Ertter, John T. Kartesz, and Frank C. Vasek. I am especially grateful to W. Dwight Billings for confirming my yellow pine identifications and compiling the label data from the Nevada conifer collections at Duke University, for his continual encouragement, and for his thoughtful review of this book. Ronald M. Lanner and an anonymous reviewer offered valuable criticisms that substantially improved the text. For their invaluable assistance I thank the directors and staff of all the herbaria I visited, and especially Andrew Sanders at the University of California, Riverside. Tender guidance in the preparation of the text was generously provided by Heather

Goulding, Sara Vélez Mallea, Cam Sutherland, and Melinda Conner of the University of Nevada Press. I thank Kathy Bishop and Lynn Riley of the University of Nevada's Biological Resources Research Center for technical assistance in the preparation of the maps. Nevada Biodiversity Initiative provided field support for two years and paid for my trips to western herbaria. The University of Nevada Program in Ecology, Evolution, and Conservation Biology provided important ancillary support for this work. Finally, I thank my parents, Dale and Jeanette Charlet, for the great genes that gave me the stamina to complete this book, and for their continuous support, encouragement, and, most important, their love.

Introduction

The mission of this book is to document the distributions of all the conifer species found in the state of Nevada. As the most mountainous of all the United States, with 314 discrete mountain ranges recognized by McLane (1978), Nevada exhibits tremendous environmental diversity, from barren playas rising through a series of ecological zones to high mountains that are ecological islands of alpine and subalpine communities.

Biologists have studied the biogeography and ecology of plants and animals in Nevada in the context of their evolution and migration as a response to climatic change, yet the distribution of the organisms most important in shaping community structure in Nevada, the conifers, is not very well known. The most recent treatment of the forests of the United States, *Wild Forests: Conservation Biology and Public Policy* (Alverson et al. 1994), includes Greeley's (1925) three maps showing the distribution of forests in the United States in 1620, 1760, and 1920, but only the 1920 map shows any forests in Nevada, and these only at the north shore of Lake Tahoe and in the Clover Mountains of Lincoln County. However, small forests of montane species occur in 15 of the 17 counties in the state today, and they were certainly better developed in 1620. I hope, in this book, to correct the deficiency in our knowledge of the conifers of Nevada by documenting the distributions of the 22 species native to the state according to rigid standards, and by mapping these distributions at high resolution.

This text simply documents the known distributions of conifers in Nevada; the reader is referred to the following texts for information about other topics relevant to the conifers of Nevada. Ecological zonation in Nevada is treated in Billings (1951) and Cronquist et al. (1972). *Nevada Trees* (Billings 1954) is the only volume that exclusively treats the trees of Nevada. Dichotomous keys for Nevada conifers can be found in that work and in Cronquist et al. (1972), Kartesz (1988), Little (1956), and Morin (1993).

Trees of the Great Basin (Lanner 1984) includes a good account of the natural history of Nevada conifers. A wealth of technical information on conifer seeds and their germination is presented in the two editions of *Seeds of the Woody Plants in the United States* (Schopmeyer 1974; Young & Young 1992), and conifer silvics is treated in Burns and Honkala (1990). The effects of geology on the distribution of conifers in Nevada is discussed by Billings (1950, 1992). The fossil history of conifers in Nevada is available in the monographs by Axelrod (1976, 1995). Wells (1983) provides a thorough treatment of the late Pleistocene and Holocene conifer paleobotany of Nevada, and Critchfield (1984a) discusses the Pleistocene history of Pinaceae in the Great Basin. A recent multidisciplinary contribution to the natural history of Nevada is *The Desert's Past* (Grayson 1993). All terms relating to

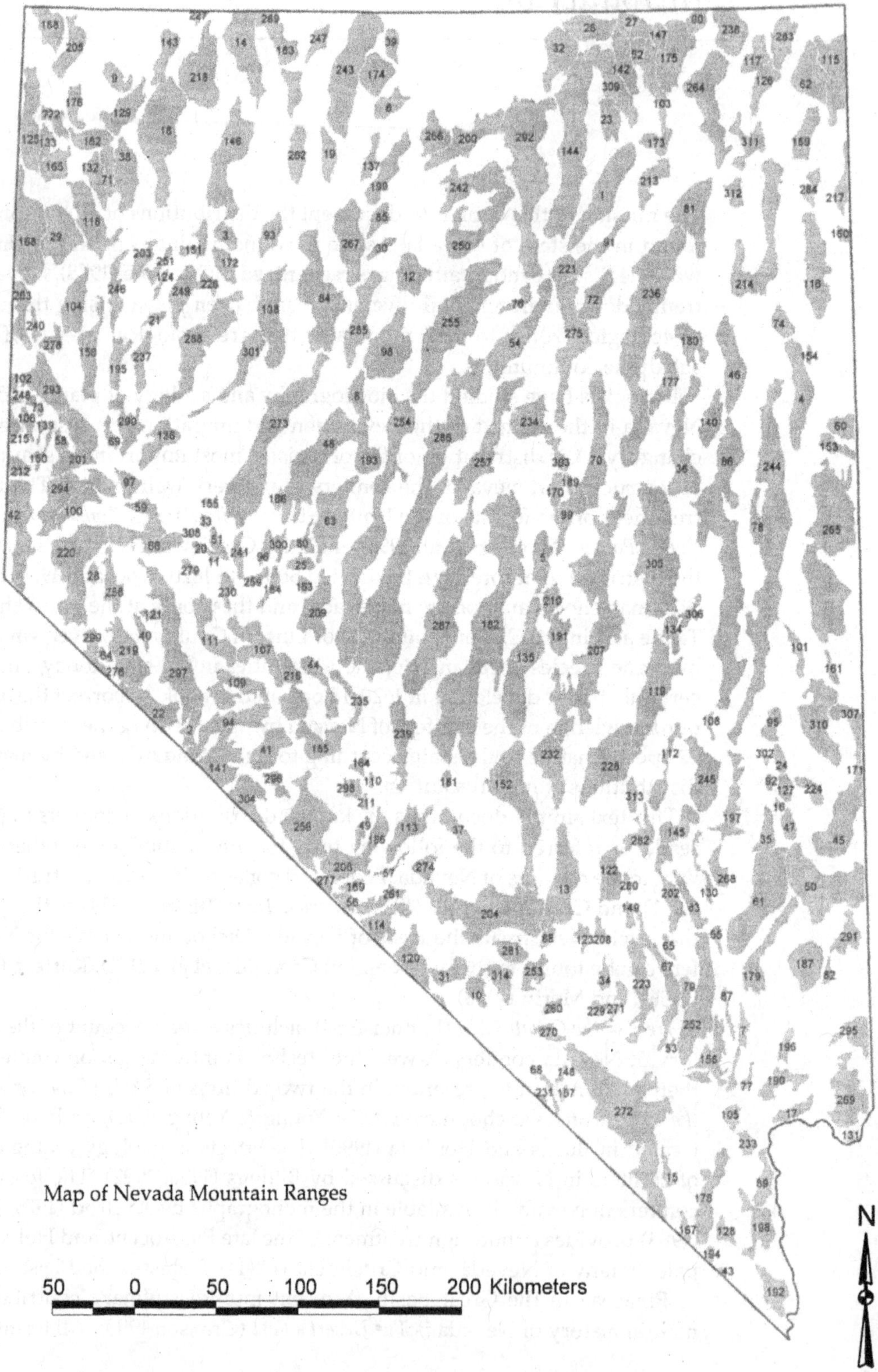

Map of Nevada Mountain Ranges

Table 1.
Herbaria with Nevada conifer collections visited by the author (8 November 1993–1 August 1994).

Acronym	Institution	Location
AUSTIN	USDA Forest Service, Austin District	Austin, NV
BRY	Brigham Young University	Provo, UT
CAS	California Academy of Sciences	San Francisco, CA
DUKE	Duke University	Durham, NC
IFGP	USDA Forest Service, Institute of Forest Genetics	Placerville, CA
IRSR	USDA Forest Service, Intermountain Research Station	Reno, NV
NSM	Nevada State Museum	Carson City, NV
OGDF	USDA Forest Service, Intermountain Region	Ogden, UT
RENO	University of Nevada	Reno, NV
RSA	Rancho Santa Ana	Claremont, CA
UC	University of California	Berkeley, CA
UCR	University of California	Riverside, CA
UNLV	University of Nevada	Las Vegas, NV
UT	University of Utah	Salt Lake City, UT
UTC	Utah State University	Logan, UT

the physiographic regions of the United States are from Hunt's *Physiography of the United States* (1967).

This book is designed to allow future investigators to add newly found distributions, and to plot extirpations should they occur. In addition, the maps can be easily adjusted if taxonomic revisions make that necessary.

Methods

DATA COLLECTION

Data were collected from three sources: the literature, fieldwork, and botanical collections in 15 herbaria in the United States (Table 1). I surveyed the literature that deals with conifer distributions in Nevada (Alverson et al. 1994; Beatley 1976; Billings 1950, 1951, 1954, 1992; Critchfield 1984a, 1984b; Critchfield & Allenbaugh 1969; Cronquist et al. 1972; Goodrich 1981; Grayson 1993; Harper et al. 1978; Kartesz 1988; Lanner 1984; Lavin 1981; Lewis 1971, 1973, 1974, 1975a, 1975b, 1976; Linsdale et al. 1952; Little 1956, 1971; Lloyd & Mitchell 1973; Loope 1969; Lyman & Sanders 1993; Nachlinger et al. 1992a, 1992b; Reveal 1979; Vasek 1966; Wells 1983) for statements on conifer distributions on specific physiographic features. These were recorded as "reports" for the feature by that author.

I visited 202 mountain ranges in Nevada, made 621 conifer collections in 118 of these ranges, and noted species distributions on 1:100,000-scale U.S. Geological Survey topographic maps. I failed to visit only 50 of the 207 mountain ranges in

Table 2.
Abbreviations for Nevada counties.

Abbreviation	County
CC	Carson City (formerly Ormsby County)
CH	Churchill
CL	Clark
DO	Douglas
EL	Elko
ES	Esmeralda
EU	Eureka
HU	Humboldt
LA	Lander
LI	Lincoln
LY	Lyon
MI	Mineral
NY	Nye
PE	Pershing
ST	Storey
WA	Washoe
WP	White Pine

Nevada known to have conifers. As a result of my surveys, I contributed a total of 1616 herbarium sheets to DUKE, UCR, RENO, UC, IRSR, BRY, and UNLV. Complete sets of my juniper collections are at UCR, IRSR, and RENO. Nearly complete sets of my Pinaceae collections are at DUKE, RENO, and UC; and other duplicates have been variously distributed to the other herbaria.

Each of my collections from a specific geographic site was recorded as a "record" for that site, while my observations of living plants (not collected) were recorded as "reports." I examined all the Nevada conifer collections in 15 herbaria in the United States (Table 1) and transcribed the species name, collector, and geographic data from the labels.

I evaluated all the geographic data from the herbarium labels to determine the county (Table 2), physiographic feature, Merriam's Life Zone (Merriam 1898) (Table 3), and the township, range, and section where the collection was made. I then recorded these data as distribution records for that feature and county. Including my own, I examined a total of 1683 collections, which are presented in this volume.

MAP PREPARATION

Using Alvin McLane's working map (unpublished data, 1978) as a guide, I circumscribed the 314 mountain ranges of Nevada on Mylar overlaying a 1:1,000,000 base map (Nevada Bureau of Mines and Geology 1972).

I plotted all reports and records judged to be valid accounts of plants native to the range, and for which a township and range were found, as circles on Mylar

Table 3.
The correspondence of Merriam's (1898) Life Zones to the vegetational zonation in the Great Basin as described by Billings (1951).

Merriam's Life Zone	Billings's Vegetation Zone
Arctic-Alpine	Alpine tundra
Hudsonian	Whitebark pine, Spruce–alpine fir, Limber pine–bristlecone pine, Lodgepole pine–mountain hemlock
Canadian	Red fir, White fir–Douglas-fir, Upper Sagebrush–grass
Arid Transition	Sagebrush-grass, Oak-maple, Yellow pine–white fir
Upper Sonoran	Pinyon-juniper, Sagebrush-grass, Shadscale
Lower Sonoran	Creosote-bush

overlaid on a 1:1,000,000-scale map of Nevada (Nevada Bureau of Mines and Geology, University of Nevada, Reno, 1972). Records were plotted as closed circles; reports were plotted as open circles. The circles were digitized as points and the mountain range circumscriptions were digitized as polygons into the Nevada Biodiversity Initiative Geographic Information System using the ARC-INFO software program. I produced composite maps of species distributions on mountain ranges by assigning a dark gray tone to mountain range polygons and overlaying a species coverage of records and reports. The maps were printed on a Hewlett-Packard plotter.

The Conifers of Nevada

Data format

This atlas presents all the known reports and records (collections) of species belonging to the families Cupressaceae and Pinaceae within the state of Nevada. The list is ordered alphabetically by species within the two families, beginning with Cupressaceae. Records and reports for each species are listed alphabetically by mountain range. The common name of the species, minimum and maximum elevations of collection sites, life zones, and numbers of records and reports are reported immediately under the species name, followed by a brief introduction. The nomenclature follows Little (1971), except for *Pinus longaeva* and *Juniperus occidentalis,* which follow Cronquist et al. (1972).

A block of data is displayed for each mountain range (*sensu* McLane 1978) in which each species occurs. The block begins with a header line, which is in bold print if there is a valid report for the species in that range. Under the name of each mountain range is the block of records and reports for that range.

The number in parentheses after the name of the mountain range refers to the following categories of reports for the range:

1. *Record* Collection made or seen for the mountain range, deposited in a herbarium.
2. *Living plant observed, no record* Observed by me (DAC) and/or RWR (Richard W. Rust) in the mountain range, but no herbarium record.
3. *Other's observations, no record* Reported in a checklist, map, or book for that mountain range, but no material seen in herbaria. Unverified by me or RWR in the field, but no reason to doubt the report. The species almost certainly occurs in the range.
4. *May yet be found* The species occurs within the range but outside Nevada, or in a nearby range. The range in question has suitable habitat, and there is reason to believe the species is present. The species may yet be found in the range.
5. *Possibly occurs; reported, but no successful verification* The species was reported in a checklist, map, or book as occurring in the range; however, no material was seen in herbaria, and all attempts to verify were unsuccessful. The species possibly occurs in the range.
6. *Possibly locally extinct* A valid collection was seen, an unsuccessful attempt to

verify the occurrence was made, and there is reason to believe that the species is now locally extinct.

7. *Nonnative occurrence* The species has been introduced (and hence is nonnative) to the area.

After the mountain range and its report category code, the minimum and maximum elevations of the records in the range are reported. Next is the total number of records for the mountain range and the county (or counties, see Table 2) where records were collected is reported and the first line ends with the total number of records from the mountain range.

Records (if any) and reports of the species in the mountain range are displayed in columns. Most records begin with the township, range, and section (T, R, S) of the record, followed by a brief location and habitat description, elevation, collector, and collection number. It was not always possible to determine the TRS values from the information recorded with herbarium specimens, and only collections with known TRS values were mapped. The record line ends with the acronyms for the herbaria where the collections were seen. All reports of additional locations within the mountain range are listed in the line following the final record for the range. These reports, in brackets, include the name of the reporter and the township and range (TR) where the species was observed (e.g., [Little: T20N R36E, T21N R36E]). All reports and records of native occurrences that can be resolved to a TR square have been plotted on the accompanying distribution maps.

Whether or not the species is recorded on Little's (1971) distribution maps and in other literature relevant to that mountain range is noted in source notes after each list. Ranges visited by DAC and/or RWR are denoted by their initials and observations in footnotes (the years in which our observations were made follow in parentheses). If I was unable to find a collection or report for a species in a mountain range (or county) but I observed the species in that mountain range (or county), that observation is scored as a "new report." This new report is also a "new record" if I deposited a collection at a herbarium. These new reports and records are tallied under the heading for each species. New reports of a species' occurrence in a mountain range are identified by a double asterisk preceding the name of the mountain range on the header line.

Cupressaceae: Cypress Family

Six species of the cypress family and 2 genera (*Calocedrus* and *Juniperus*) are native to Nevada. In addition, *Cupressus arizonica* var. *glabra* (Arizona smooth cypress) has apparently naturalized at one location in extreme southeastern Nevada (Virgin Mountains, Clark County). Only *Juniperus* is widespread in Nevada. The single native species of *Calocedrus* (*C. decurrens*) is nearly restricted to the Carson Range of the Sierra Nevada. *Juniperus osteosperma,* the most widely distributed conifer species in Nevada, is found in all 17 counties, while *J. occidentalis* occurs in 12 counties in western, northern, and central Nevada. *Juniperus scopulorum* is present throughout eastern and southern Nevada and at isolated locations in central Nevada. The shrubby, circumboreal *J. communis* occurs at generally high elevations in eastern Nevada.

Calocedrus decurrens (Torrey) Florin

common name: incense-cedar
elevation: 4500–7800 ft (1370–2375 m)
zones: Canadian, Arid Transition
24 Records for 3 Nevada counties (CC, DO, WA) in 2 mountain ranges

The incense-cedar, once known by the name *Libocedrus decurrens* Torrey, is an important commercial tree of California and Oregon. It occurs in mesic situations in the Arid Transition Zone forests throughout the western slopes of the Sierra Nevada. Incense-cedar appears on the eastern slopes of the Sierra from Lake Tahoe north through the Middle and South Cascades of California and Oregon. Its distribution extends west to near the Pacific Coast in northern California and southern Oregon and is spotty throughout inland portions of the California Coast Ranges south into the mountains of Baja California. The species is not known from north of the Columbia River.

In Nevada, incense-cedar occurs only in 3 far western Nevada counties (Washoe, Carson City, and Douglas) and is restricted to Peavine Mountain and the Carson Range of the Sierra Nevada. The species is not dominant anywhere in Nevada, and its presence in the Carson Range is primarily restricted to canyons, hollows, and near the shores of Lake Tahoe, where stately individuals attain great size. The elevation range over which the species grows in Nevada is 3300 ft (1005 m).

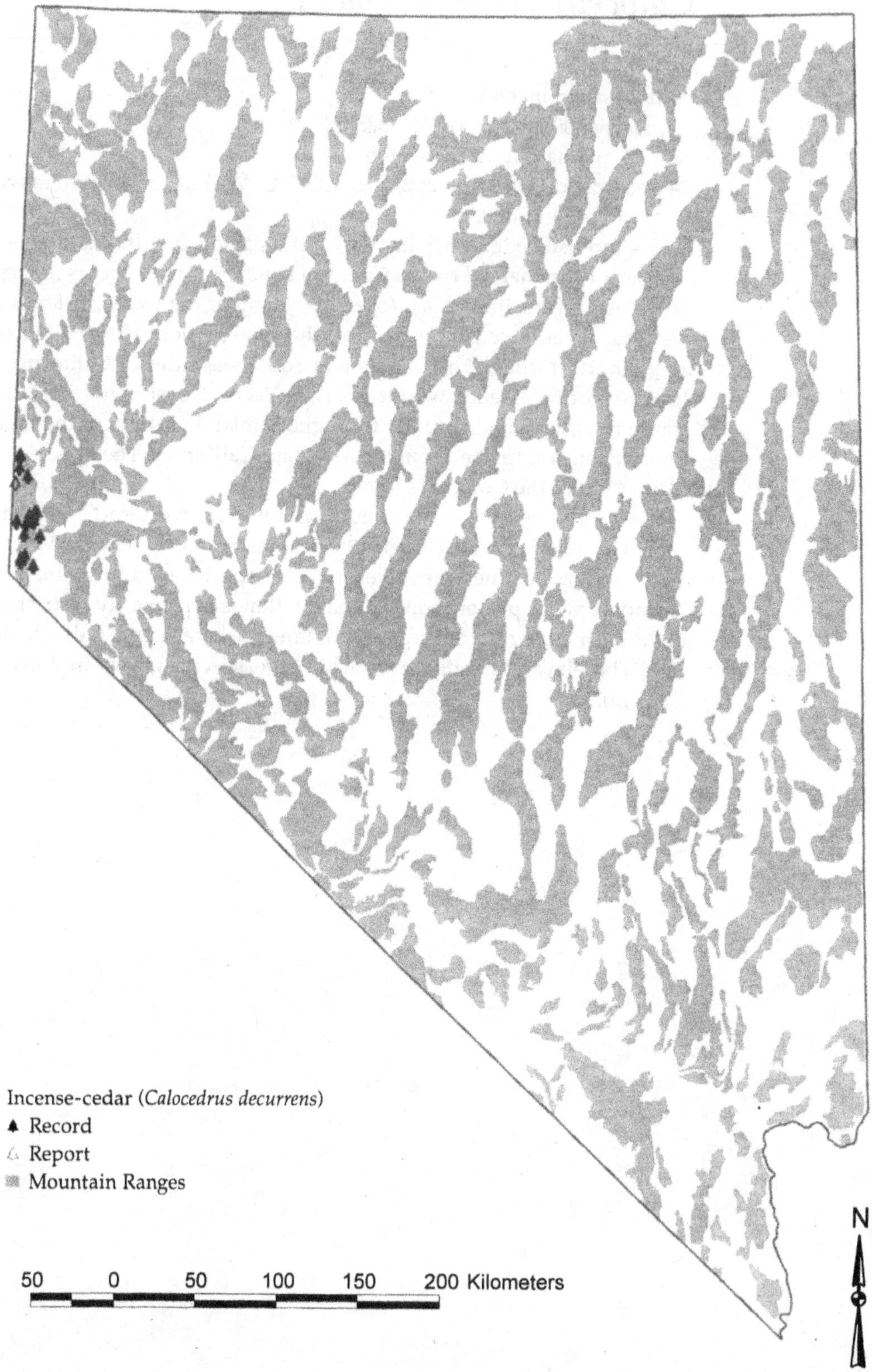
Incense-cedar (*Calocedrus decurrens*)
Record
Report
Mountain Ranges
50 0 50 100 150 200 Kilometers
N

Carson Range (1)
elevation record: 4500–7800 ft
19 records in 3 counties (Carson City, Douglas, and Washoe)

Towns/Range/Section	Location—Habitat	County	Elevation	Collector	Herbarium/a
T13N R19E S21 SW/4	Haines Canyon	DO	6150 ft	*Ann Pinzl 8926*	UNLV, UTC
T14N R18E S15	Logan Shoals near Lake Tahoe	DO	6500 ft	*Ann Pinzl 6001*	NSM, BRY
T14N R18E S22	1 mi N of Cave Rock, with *Pseudotsuga*	DO	6240 ft	*Charlet 2057 & T Charlet*	DUKE, RENO, UC, BRY, UNLV
T14N R18E S27	0.5 mi SE of Cave Rock	DO	6600 ft	*John Darling 81*	RENO
T14N R18E S27	steep slope above Tahoe, just N of Cave Rock	DO	—	*Billings 1641*	RENO
T14N R19E S6	Spooner Summit	DO	7800 ft	*Richard Holland in 1949*	RENO
T15N R18E S2	above E shore Lake Tahoe	DO	6240 ft	*Charlet 1922 & T Charlet*	DUKE, RENO, UC, BRY, UNLV
T15N R19E S23	Kings Canyon	CC	5800 ft	*P Herlan in 1971*	NSM
T16N R18E S18	3 mi E of Brockway	WA	6300 ft	*Nichols & Lund 732*	RENO
T16N R18E	below Mount Rose, near lakeshore, 3 mi N of town [Incline Village]	WA	6200 ft	*Leila Shultz & J Shultz*	UTC
T16N R18E	meadow edge near Incline Village	WA	—	*TC Frantz in 1951*	RENO
T16N R18E	E [NE] shore Lake Tahoe, near Kings Beach	WA	6200 ft	*L Sloan 3*	RENO
T16N R19E S10	Franktown	WA	—	*ME Jones in 1892*	RSA
T16N R19E S10	Franktown	WA	—	*PA Lehenbauer 1865*	RENO
T16N R19E	W of Franktown	WA	—	*NF Petersen 474*	RENO
T17N R19E S27	above Bowers Mansion	WA	—	*PA Lehenbauer 828*	RENO
T17N R19E S32	Prices Lake	WA	7200 ft	*Breena 67 & Sampson*	RENO
T17N R19E S26	0.75 mi N of Bowers Mansion; low & E margins of distribution	WA	4500 ft	*Warren Sandow in 1953*	RENO
T19N R19E	canyon W of Hunter Creek, near old lumber camp	WA	—	*NF Petersen 162*	RENO
[DAC: T18N R18E (above Truckee River)]					

Reported as present in the range by Little (1971).
DAC: common in Truckee Canyon with Jeffrey pine. Enormous trees are rather common in the Tahoe basin; occasional small stands elsewhere. Occasionally locally dominant (i.e., stands of 30–200 trees), but never dominant over large areas (1971–1995).

Peavine Mountain (1)
elevation record: 5400–6550 ft
5 records in 1 county (Washoe)

Towns/Range/Section	Location—Habitat	County	Elevation	Collector	Herbarium/a
T20N R18E	S side Peavine	WA	5600 ft	*M Williams 72-4*	RENO
T20N R18E	SE of Peavine Peak	WA	6550 ft	*M Williams 74-34-4*	RENO
T20N R18E	altered ridge, NE of Mine Spring; 3 mi W of Highland Reservoir	WA	5400 ft	*Billings 2060*	RENO
T19N R18E S7	road between Verdi & Dog Valley	WA	—	*PT Sonne in 1889*	UC
T19N R18E S7	Dog Valley grade	WA	—	*PA Lehenbauer 1317*	RENO

Reported as present in the range by Little (1971), and as present on the northwest, southeast, and southwest sides of the mountain by Nachlinger et al. (1992b).
DAC: rather common in Jeffrey pine forest (1971–1995).

Cupressus arizonica E. Greene var. *glabra* (Sudworth) Little

common name: Arizona smooth cypress
elevation: 4600–6800 ft (1400–2070 m)
zone: Lower Sonoran
6 Records for 1 county (CL) in 1 mountain range; probably nonnative and naturalized in Nevada

The Arizona smooth cypress is one of 5 varieties (once regarded as species by Sudworth [1915]) of Arizona cypress found in southern California, Arizona, extreme southern New Mexico, Texas, and Mexico. In Nevada, *Cupressus arizonica* is known outside of cultivation only from the Virgin Mountains of Clark County. The species was probably introduced there at a homestead in Cabin Canyon but has apparently naturalized and is spreading. The nearest natural population of the species occurs in central Arizona, 275 km southeast of the Virgin Mountains.

Virgin Mountains (7)
elevation record: 4600–6800 ft
6 records in 1 county (Clark)

Towns/Range/Section	Location—Habitat	County	Elevation	Collector	Herbarium/a
T14S R71E S28	Cabin Canyon	CL	—	*Mozingo 71-3*	RENO
T15S R71E S8	NW of Virgin Peak, S of Cabin Creek	CL	6800 ft	*Mozingo*	RENO
T15S R71E S9	Cabin Canyon, 2 trees seen	CL	4600 ft	*Tiehm 3205 & Mozingo*	RENO
T15S R71E	N slope Cabin Creek, "poss. not native at this locality"	CL	4700 ft	*GW Gullion 529*	UC
T15S R71E	Cabin Canyon	CL	5500 ft	*Peter Herlan in 1971*	NSM
T15S R71E	Cabin Canyon "not sure whether native or not"	CL	4800 ft	*Ann Pinzl in 1974*	NSM

Not reported as present in the range by Little (1971).
DAC: EL Little, Jr., is aware of this occurrence and said in a letter written to Ann Pinzl in 1974 (Ann Pinzl, pers. comm. 1993) that it "is almost certainly based upon plants introduced or escaped from cultivation." I have not visited the Virgin Mountains, but Ann Pinzl found several trees near an old homestead site, associated with nonnative elms (*Ulmus* sp.). GW Guillion reported his collection in a letter to WD Billings in 1954 (Billings, in litt. 1994), and mentioned that the few trees he found were near an *Ulmus,* and so were probably not native. Given that Mozingo found the species in three different locations, it appears that the species has naturalized and is spreading.

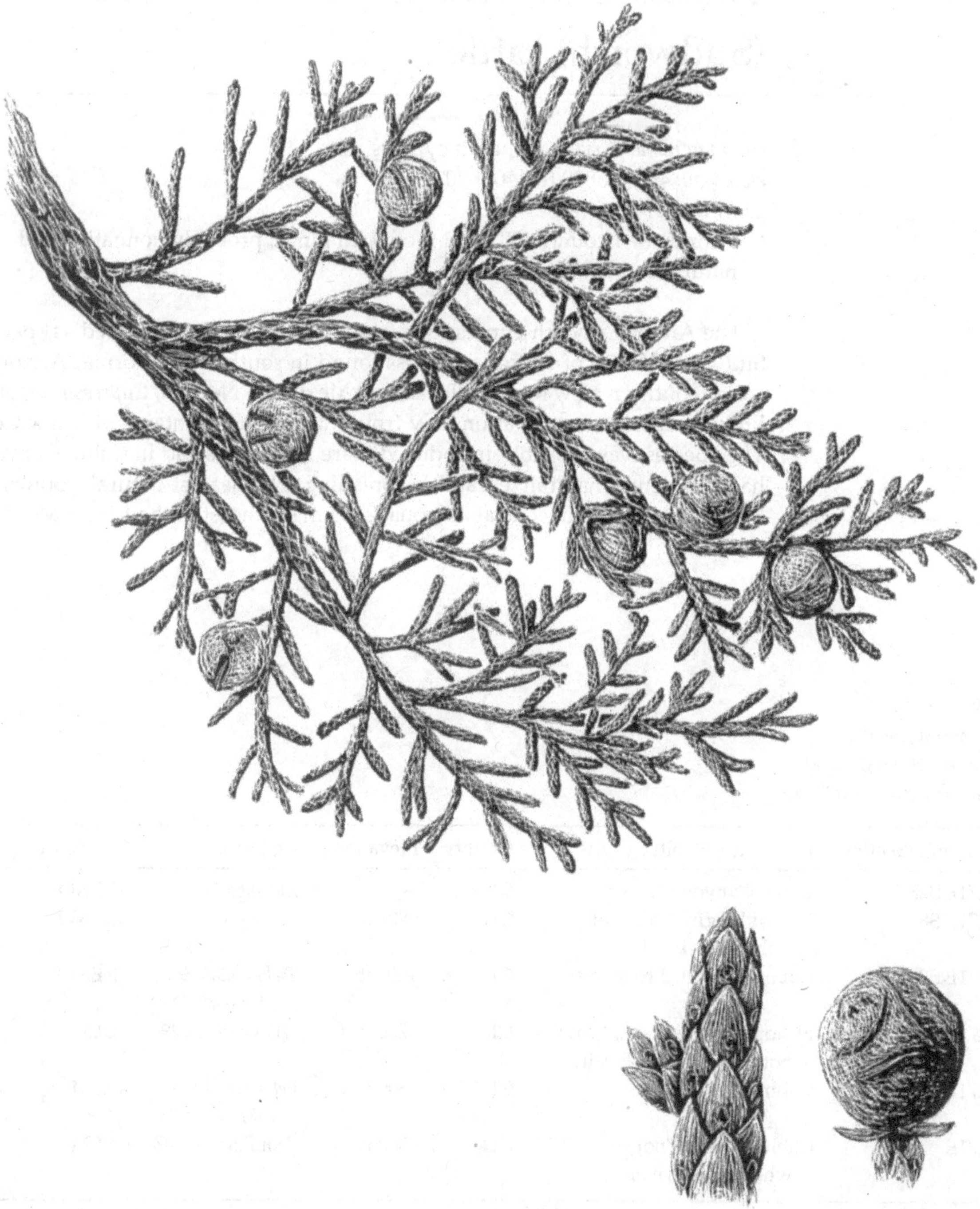

Juniperus californica Carrière

common name: California juniper
elevation: 2500–3600 ft (760–1090 m)
zone: Lower Sonoran
9 Records for 1 Nevada county (CL) in 1 mountain range

California juniper occurs primarily in California and extends into west-central Arizona and Baja California. In Nevada, California juniper is known only from the Newberry Mountains (Clark County) in the extreme southern part of the state. California juniper grows at the lowest elevations of all the Nevada conifers; it occurs only in the Lower Sonoran Zone, and the maximum elevation record is 3600 ft (1090 m). The species may yet be found elsewhere in southern Nevada.

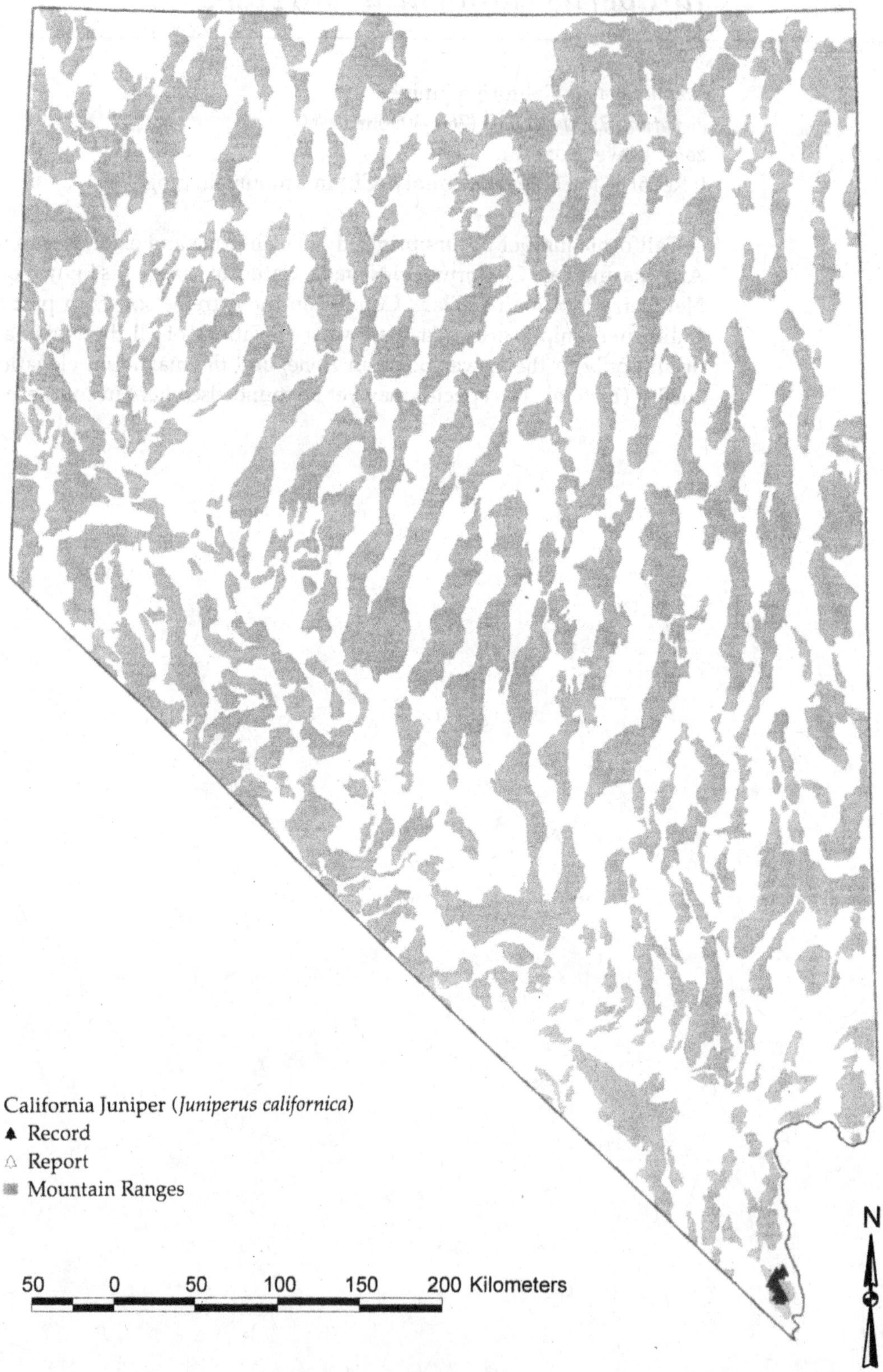
California Juniper (*Juniperus californica*)
Record
Report
Mountain Ranges
50 0 50 100 150 200 Kilometers
N

Newberry Mountains (1)
elevation record: 2500–3600 ft
9 records in 1 county (Clark)

Towns/Range/Section	Location—Habitat	County	Elevation	Collector	Herbarium/a
T30S R65E S34	in canyon due N of Newberry Peak	CL	3400 ft	*Vasek 640131-3*	UCR
T30S R65E S26	in canyon NNE of the high peaks	CL	2950 ft	*Vasek 640131-1*	UCR
T31S R65E S9	1 mi W of Lake Mead NRA boundary, Christmas Tree Pass Rd	CL	3600 ft	*JS Holland 970*	NSM
T31S R65E S29	E of Juniper Mine	CL	3500 ft	*GW Nickell 622*	UNLV, NSM
T31S R65E	(determined JS Holland) [DAC: reported as Dead Mountains, NV]	CL	3000 ft	*WE Niles 1247a*	UNLV
T32S R65E S10	10 mi W of Davis Dam [DAC: reported as Dead Mountains, NV]	CL	—	*FS Stay in 1961*	UNLV
T32S R65E S9	14.6 mi W of Davis Dam on SR 77	CL	2800 ft	*Vasek 621018*	UCR
T32S R65E S2	Bridge Canyon	CL	2500 ft	*JS Holland 1641*	UNLV
T32S R65E S9	S end Newberry Mountains	CL	2900 ft	*Mozingo 74-4*	RENO

Reported as present in the range by Little (1971) and Vasek (1966).

Juniperus communis Linnaeus

common name: common juniper, dwarf juniper
elevation: 6800–11,200 ft (2070–3415 m)
zones: Hudsonian, Canadian, Arid Transition
48 Records for 3 counties (CL, EL, WP), New Reports for 2 additional counties (EU, NY)
Reports for 16 mountain ranges, Records for 14 of these ranges
New Reports for 5 mountain ranges, New Records for 3 of these

Common juniper is a circumboreal, generally shrubby species widespread throughout Canada, northern Europe, and most of the major mountain systems of the Northern Hemisphere. In western North America, it is uncommon outside the Oregon and Washington Cascades and the Rocky Mountains, but common in the northern Midwest east to the Atlantic seaboard. Common juniper is curiously rare in the Sierra Nevada and is easier to find in Nevada than in California.

In Nevada, common juniper always adopts the shrub form. It is reported from 16 mountain ranges in the east and south (Clark, Elko, Eureka, Nye, and White Pine Counties), although Little's (1971) distribution map identifies only 4 locations in the state. New reports for the species are presented here for the Marys River Range (Elko County), White Pine and Egan Ranges (White Pine County), the Monitor Range (Nye County), and Roberts Mountains (Eureka County). The rest of the additions to Little's (1971) map are from extant collections in herbaria.

Common juniper grows over a broad elevation range of 4400 ft (1345 m) in Nevada and usually occurs at high elevations in the subalpine forests and woodlands of the Hudsonian Zone. In the Schell Creek Range, common juniper reaches its maximum elevation in the state, 11,200 ft, higher than any other Nevada juniper. However, in extreme southern Nevada (Spring Mountains, Clark County), it has also been found in Arid Transition Zone forests of ponderosa pine and white fir. Look for common juniper in the Pine Forest Range, the Sweetwater Mountains, the Toiyabe Range, and throughout the Carson Range.

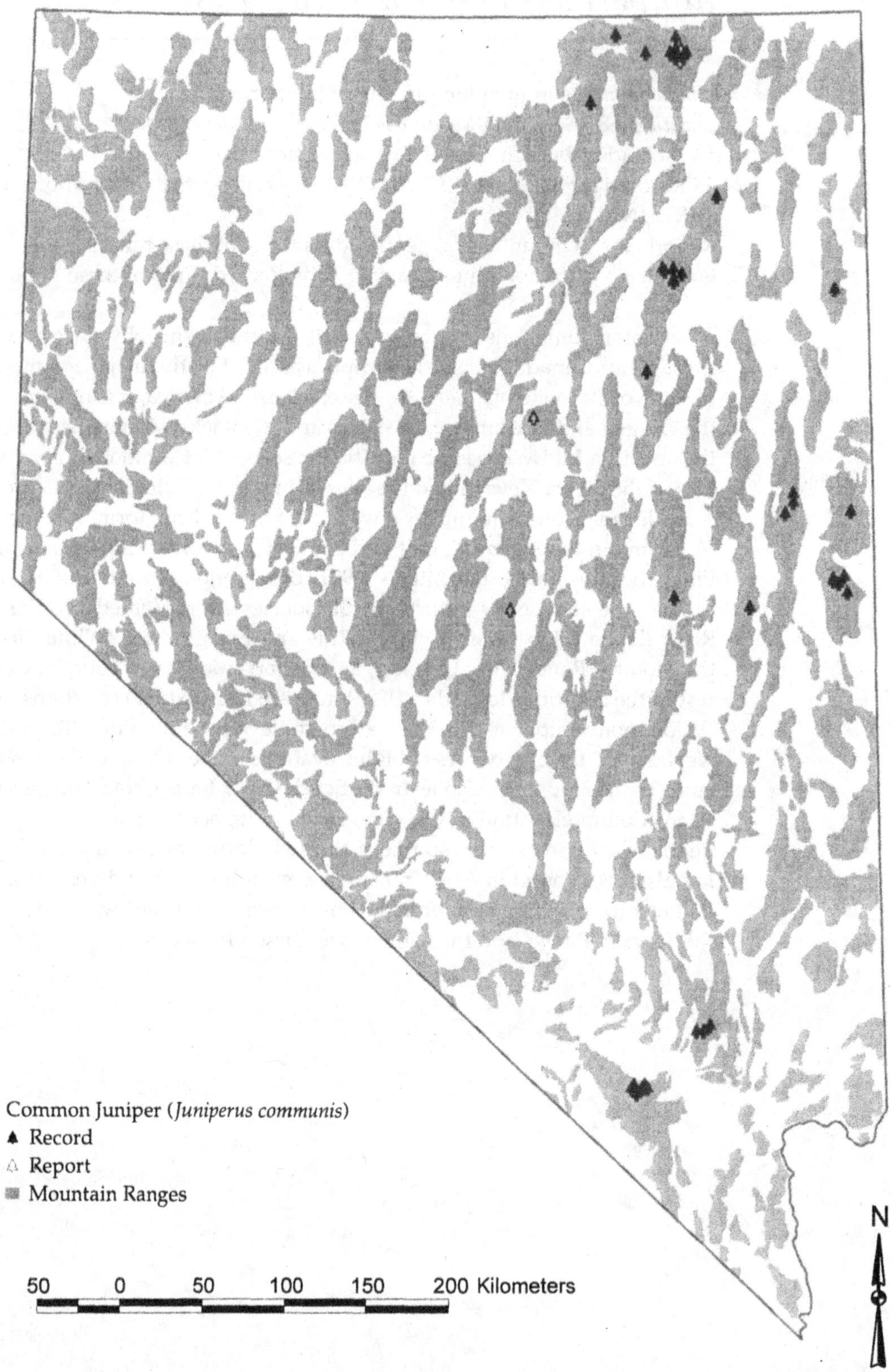
Common Juniper (*Juniperus communis*)
Record
Report
Mountain Ranges
N
50 0 50 100 150 200 Kilometers

Bruneau Range (1)
elevation: 6800 ft
1 record in 1 county (Elko)

Towns/Range/Section	Location—Habitat	County	Elevation	Collector	Herbarium/a
T46N R55E S8	Humboldt NF, Middle Fork Sand Creek, sparse on dry slopes, E aspect	EL	6800 ft	*Royce Hermansen 31H*	UTC, OGDF

Not reported as present in the range by Little (1971).
DAC: this collection is the low-elevation record for the species in Nevada; rare (1989–1994).

Carson Range (4)
Washoe, Douglas, and Carson City Counties

Not reported as present in the range (California or Nevada) by Little (1971).
DAC: *Juniperus communis* is present at numerous locations in the California portion of the Lake Tahoe basin (Smith 1984). Although I have failed to find the species after extensive searching in the Nevada portion of the basin (i.e., the Carson Range), it may yet be found there (1971–1995).

Copper Mountains (1)
elevation record: 8800 ft
1 record in 1 county (Elko)

Towns/Range/Section	Location—Habitat	County	Elevation	Collector	Herbarium/a
T45N R57E	Head of Copper Creek Basin	EL	8800 ft	*Percy Train 654*	RENO

Not reported as present in the range by Little (1971).

East Humboldt Range (1)
elevation record: 8400–9400 ft
2 records in 1 county (Elko)

Towns/Range/Section	Location—Habitat	County	Elevation	Collector	Herbarium/a
T36N R61E S4	East Humboldt, Angel Lake	EL	8400 ft	*PH Raven & Otto Solbrig 13414*	UC
—	Near Deeth	EL	9400 ft	*AA Heller I0227*	RENO

Not reported as present in the range by Little (1971).
DAC: common in the Angel Lake area in the understory of whitebark pine and limber pine woods (1988–1994).

****Egan Range (1)**
elevation record: 7870 ft
1 record in 1 county (White Pine)

Towns/Range/Section	Location Habitat	County	Elevation	Collector	Herbarium/a
T12N R63E S33 SE/4	Lone Pine Swale	WP	7870 ft	*Craig Biggart 9 Sept. 1994*	IRSR

Not reported as present in the range by Little (1971).
New record for the species in the range.

Goshute Mountains (1)
elevation record: 7500 ft
1 record in 1 county (Elko)

Towns/Range/Section	Location—Habitat	County	Elevation	Collector	Herbarium/a
T31N R68E S34	E aspect with *Abies*	EL	7500 ft	*Cooper 142 & Holmgren & Lund*	RENO, UTC

Not reported as present in the range by Little (1971).

Independence Mountains (1)
elevation record: 8790 ft
2 records in 1 county (Elko)

Towns/Range/Section	Location—Habitat	County	Elevation	Collector	Herbarium/a
T42N R53E	Jack Creek headwaters	EL	—	*Bassett Maguire 22469*	UTC
T42N R53E S14	E Fork Mill Creek	EL	8790 ft	*Charlet 1612 & Chrys Olson*	UCR

Not reported as present in the range by Little (1971).
DAC: found as occasional low shrubs under *Pinus albicaulis* on hillsides along both forks of Mill Creek; also occurs under *Abies lasiocarpa* near springs in the cirque basins of Mill Creek (1991–1994).

Jarbidge Mountains (1)
elevation record: 8300–10,100 ft
3 records in 1 county (Elko)

Towns/Range/Section	Location—Habitat	County	Elevation	Collector	Herbarium/a
T46N R58E S13 NW/4	———	EL	8300 ft	*ND Atwood 131;85* [sic]	BRY
T45N R58E S15	Cougar Peak	EL	10,100 ft	*Charlet 70*	UCR
T45N R58E S12 [DAC: T45N R59E]	Matterhorn	EL	9580 ft	*Charlet 453*	UCR

Not reported as present in the range by Little (1971).
Reported as present by Lewis (1975).
DAC: common at high elevations throughout the range. Near the treeline, *J. communis* is sometimes codominant with *Pinus albicaulis* and *Abies lasiocarpa* and up to 2 m tall (1988–1994).

****Marys River Range (1)**
elevation record: 9200–10,200 ft
4 records in 1 county (Elko)

Towns/Range/Section	Location—Habitat	County	Elevation	Collector	Herbarium/a
T45N R59E S15	Gods Pocket Peak	EL	9800 ft	*Charlet 2*	UCR
T45N R59E S21	Divide Peak	EL	9520 ft	*Charlet 308*	UCR
T45N R58E S25	Marys River Peak	EL	10,200 ft	*Charlet 326*	UCR
T45N R58E S25 NE/4	near W Fork Jarbidge River–Marys River divide; Marys River drainage	EL	9200 ft	*Charlet 2141 & T Charlet*	RENO, UC
[DAC: T44N R59E S4 (in rocks in deep portion of Camp Creek Canyon at 7500 ft)]					

Not reported as present in the range by Little (1971).
DAC: common in open situations in whitebark pine and subalpine fir woodlands throughout the higher portions of the range. Some ridgeline stands have individuals 2 m tall (1988–1994).
New record for the species in the range.

****Monitor Range (3)**
**Nye County
[Kerry Heise: T11N R48E N/4]

Not reported as present in the range by Little (1971) or Goodrich (1981).
Kerry Heise (Toiyabe National Forest, Reno) found the species at high elevations in upper Clear Creek (pers. comm. 1995).
New report for the species in the range.
New report for the species in Nye County.

****Roberts Mountains (3)**
**Eureka County
[Peter Wigand: T23N R50E S2]

Not reported as present in the range by Little (1971).
Peter Wigand (Desert Research Institute, Reno) found the species occasional in limber pine woods at high elevations on the ridge north of the highest peak area (1991).
New report for the species in the range.
New report for the species in Eureka County.

Ruby Mountains (1)
elevation record: 7500–9700 ft
8 records in 2 counties (Elko and White Pine)

Towns/Range/Section	Location—Habitat	County	Elevation	Collector	Herbarium/a
T26N R57E S19	Sherman Ridge, head of Sherman Creek Canyon	WP	9100 ft	*N Holmgren 3893 & B Bethers*	UTC
T31N R58E S11	N end Liberty Lakes, 4 mi SE of campground	EL	—	*Ann Pinzl 3624*	NSM
T31N R58E S2	Lamoille Lakes	EL	9000 ft	*Percy Train 434*	NSM
T32N R58E S14	Lamoille Canyon, S slopes	EL	8500 ft	*A Holmgren 848 & N Holmgren*	BRY
T32N R58E	Lamoille Canyon	EL	8650 ft	*Ann Pinzl 2040*	NSM
T32N R58E S20	Thomas Creek Canyon at Thomas Falls	EL	7500 ft	*TL Breene 488*	RENO
T32N R59E S29	glacial basin above Terrace guard station, Lamoille Canyon	EL	8500 ft	*KS Erdman 1696*	BRY
—	Ruby Mountains, Three Lakes	EL	9700 ft	*AE Borell in 1928*	UC

Reported as present in the range by Lewis (1971) and Little (1971).
DAC: common in Lamoille Canyon above campground with *Pinus albicaulis* (1988–1994).

Schell Creek Range (1)
elevation record: 8000–11,200 ft
3 records in 1 county (White Pine)

Towns/Range/Section	Location—Habitat	County	Elevation	Collector	Herbarium/a
T17N R65E S10	N Fork Berry Creek 2.5–4 mi above Berry Creek Ranger Station	WP	8000 ft	*Percy Train 1036*	RENO, RSA
T18N R66E S6	2 km NW of Taft Peak	WP	11,200 ft	*K Bell 76-310*	UNLV
T18N R66E S19	N Schell Peak	WP	10,500 ft	*Guyer 5084 & Nigro*	RENO

Not reported as present in the range by Little (1971).
DAC: *K Bell 76-310* is the high-elevation record for the species in Nevada (1994).

Sheep Range (1)
elevation record: 8000–9200 ft
7 records in 1 county (Clark)

Towns/Range/Section	Location—Habitat	County	Elevation	Collector	Herbarium/a
T15S R60E	Sheep Range, Hidden Forest 36°38'N, 115°11'W	CL	8700 ft	*EC Rockwell in 1977*	IFGP
T15S R60E	above Hidden Forest	CL	9200 ft	*PA Munz 16867*	RSA
T15S R60E	Hidden Canyon [Deadman Canyon]	CL	8000 ft	*LaRivers 613 & Hancock*	UTC
T15S R60E S1	Hayford Peak (determined TL Ackerman)	CL	—	*CG Hansen in 1961*	UNLV
T15S R60E S1	base of Hayford Peak	CL	—	*BH Banta in 1950*	RENO
T15S R61E	saddle SW of Hayford Pk, N slope	CL	9100 ft	*TL Ackerman 30508*	UTC
—	Sheep Range	CL	8500 ft	*Percy Train 1772*	RENO

Reported as present in the range by Little (1971).

Snake Range (1)
elevation record: 7800–10,500 ft
7 records in 1 county (White Pine)

Towns/Range/Section	Location—Habitat	County	Elevation	Collector	Herbarium/a
T12N R69E S9	Snake Creek, E side of range	WP	7900 ft	*Ann Pinzl 1782*	UNLV, NSM
T13N R68E S15	trail to bristlecone pine site, Wheeler Peak area	WP	10,500 ft	*Mozingo 74-41*	RENO
T13N R68E S10	Stella Lake	WP	9000 ft	*Williams 68-WP-2*	RENO
T13N R68E S11	spruce woods, S side Lehman Canyon, Mount Jeff Davis	WP	—	*B Maguire & AG Richards 2476*	UC, UTC
T13N R68E S24	right fork of Baker Creek, 1 mi above forks	WP	9000 ft	*Percy Train 2574*	RENO
T13N R69E S8	Lehman Creek	WP	7800 ft	*Mozingo in 1962*	RENO
T17N R69E S10 SW/4	4.5 mi NW of Moriah Summit	WP	9400 ft	*JC Lyman 248 Phillips & Moyer*	UCR

Not reported as present in the range by Little (1971).
DAC: common in the spruce forests and woodlands of Wheeler Mountain (1994).

Spring Mountains (1)
elevation record: 8700–11,000 ft
7 records in 1 county (Clark)

Towns/Range/Section	Location—Habitat	County	Elevation	Collector	Herbarium/a
T19S R56E S12	Deer Creek, 5 mi NE of Charleston Peak, Canadian Zone	CL	9150 ft	*Ira LaRivers & HF Hancock 490*	UC
T19S R56E S14	NE slope, Mount Charleston	CL	9310 ft	*Richard W. Rust 9*	UCR
T19S R56E S28	Charleston Peak	CL	11,000 ft	*EVA Murphy 671*	RENO
T19S R56E S8	upper Clark Canyon above Adams Ranch, SW slope of north-central Spring Mts	CL	9000 ft	*Beatley 13082*	RSA
T19S R56E	trail to Charleston Peak	CL	9185 ft	*IW Clokey 7822*	UT, UTC, BRY
T19S R57E SW/4	0.5 mi above Boy Scout Camp	CL	8700 ft	*Percy Train 2195*	RENO
—	Spring Mountains	CL	—	*IW Clokey 5388*	UTC

Reported as present in the range by Little (1971).
RWR: common in ponderosa pine–white fir woods above Charleston Lodge, Kyle Canyon.

Toiyabe Range (5)

Not reported as present in the range by Linsdale et al. (1952), Little (1971), or Goodrich (1981).
Reported as present in the range by Wells (1983).
DAC: Although Wells (1983: 372) reports *Juniperus communis* in the Toiyabe Range, he provided no location information nor did he cite specimens. I have seen no material in herbaria, nor have I been able to locate the species on the mountain. Until the species is collected, I consider this to be an erroneous report (1991–1994).

****White Pine Range (1)**
elevation record: 10,600 ft
1 record in 1 county (White Pine)

Towns/Range/Section	Location—Habitat	County	Elevation	Collector	Herbarium/a
T12N R58E S11	Currant Peak	WP	10,600 ft	*Richard W. Rust 21*	UCR

Not reported as present in the range by Little (1971).
RWR: locally common in limber pine and bristlecone pine woodland.
DAC: I have seen no material in herbaria other than *Rust 21*.
New record for the species in the range.

Juniperus occidentalis Hooker

varieties present in Nevada:
J. occidentalis var. *occidentalis*
J. occidentalis var. *australis* (Vasek) A. Holmgren & N. Holmgren
common name: western juniper, Sierra juniper
elevation: 4130–9100 ft (1260–2775 m)
zone: Hudsonian, Canadian, Arid Transition
146 Records for 12 counties (CH, DO, EL, EU, HU, LA, LY, MI, NY, PE, ST, WA)
New Records for 7 counties (CH, EL, EU, HU, LA, NY, PE)
Reports for 54 mountain ranges, Records for 53 of these
New Reports for 44 mountain ranges, New Records for 43 of these

There are 2 recognized varieties of western juniper: the typical one and *J. occidentalis* var. *australis,* or Sierra juniper. The center of distribution of the typical variety is in central and eastern Oregon, on the east slopes of the Cascades and in the southern Columbia Plateau, where it is often dominant in juniper woodlands or occurs as a component of sagebrush steppe.

Typical western juniper extends south into the southern Cascades and Modoc Plateau of northeastern California, north at a few locations in central and western Washington, and along the western border of Idaho. It is known from the Rocky Mountains only in the Seven Devils Mountains, adjacent to the Columbia Plateau in western Idaho. The Sierran variety occurs sporadically in the Sierra Nevada at moderately high elevations, primarily in the central and northern areas, but also reaches the Transverse Ranges of southern California. The Sierran variety is distinguished from the typical variety solely on the basis of the reddish bark of the former and the gray bark of the latter (Vasek 1966; Kartesz 1988; Hickman 1993). Adams (in Morin 1993), in *Flora of North America,* mistakenly reverses the discrimination and calls the red bark forms var. *occidentalis* and the gray bark forms var. *australis.* The Nevada material is confounding because red, brown, and gray bark forms are often found in the same stand. Additionally, there may be a great deal of hybridization (Vasek 1966) in regions where the distributions of Utah juniper and western juniper overlap, which is essentially throughout the Nevada range of western juniper. This species is in need of taxonomic reevaluation in light of the new Nevada collections reported here.

Typical western juniper was previously known in Nevada only from the extreme northwestern part of the state (northern Washoe County), and the Sierran variety was known only from the Sierra Nevada of extreme western Nevada (Douglas and Washoe Counties). I have documented 146 records from 54 mountain ranges in 12 counties (Churchill, Douglas, Elko, Eureka, Humboldt, Lander, Lyon, Mineral, Nye, Pershing, Storey, and Washoe) in western, central, and northern Nevada. The species is apparently absent from Carson City County but may occur there in the vicinity of Ash Canyon in the Carson Range. New records are presented for 43 mountain ranges in 7 counties (Churchill, Elko, Eureka, Humboldt, Lander, Nye, and Pershing), and the known range of the species is extended

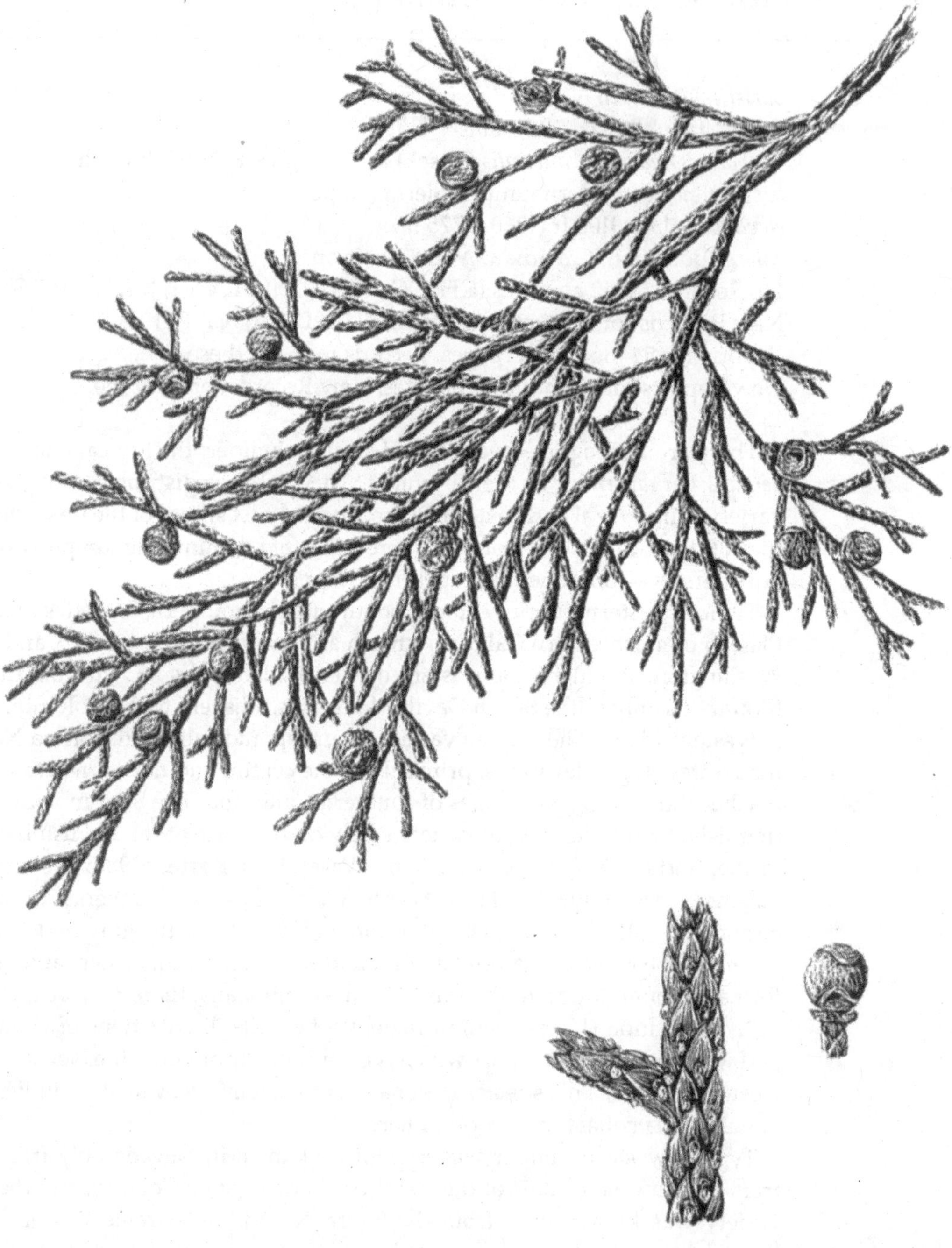

var. *occidentalis*

to the Snake Mountains of northeastern Elko County, less than 70 km from the Utah border.

In Nevada, western juniper is found primarily in high sagebrush steppe or mountain brush communities. When sympatric with Utah juniper, western juniper is often restricted to sites well above the upper limits of the Utah juniper distribution (Pine Nut Range, Douglas County), although it sometimes occurs within the Utah juniper woodlands at lower elevations along streams (Toiyabe Range, Nye and Lander Counties). Often a few individuals will be found in a high-elevation Utah juniper stand; for example, near Tohakum Peak in the Lake Range (Washoe County).

When sympatric with Rocky Mountain juniper, western juniper occurs at lower elevations. I have never seen the ranges of the two species overlap in a mountain range, although western junipers are growing only 460 ft (140 m) below Rocky Mountain junipers on the west slope of the Bull Run Mountains of northeastern Nevada. Western juniper in Nevada grows over an elevation range of 4970 ft (1515 m). The largest specimens are found in the Sierra Nevada at high elevations in rocky openings in the subalpine forests of the Hudsonian Zone. Extensive woodlands of western juniper are found in northern and west-central Washoe County, and the species will probably be found at additional high-elevation sites in the mountains of central Nevada.

****Anchorite Hills (1)**
elevation record: 7620 ft
1 record in 1 county (Mineral)

Towns/Range/Section	Location—Habitat	County	Elevation	Collector	Herbarium/a
T4N R29E S13 SE/4	Anchorite Pass	MI	7620 ft	*Charlet 1963*	UCR, IRSR, RENO

Not reported as present in the range by Little (1971).
DAC: uncommon in the upper elevations of pinyon woodland. This material is var. *occidentalis* (1991–1995).
New record for the species in the range.

****Badger Mountains (1)**
elevation record: 6540 ft
1 record in 1 county (**Humboldt)

Towns/Range/Section	Location—Habitat	County	Elevation	Collector	Herbarium/a
T44N R23E S36	Mahogany Mountain	HU	6540 ft	*Charlet 1853*	UCR, IRSR, RENO, UC

Not reported as present in the range by Little (1971).
DAC: 2 reproductive individuals and 7 saplings are present in low sage (1994).
New record for the species in the range. Not previously known from Humboldt County.

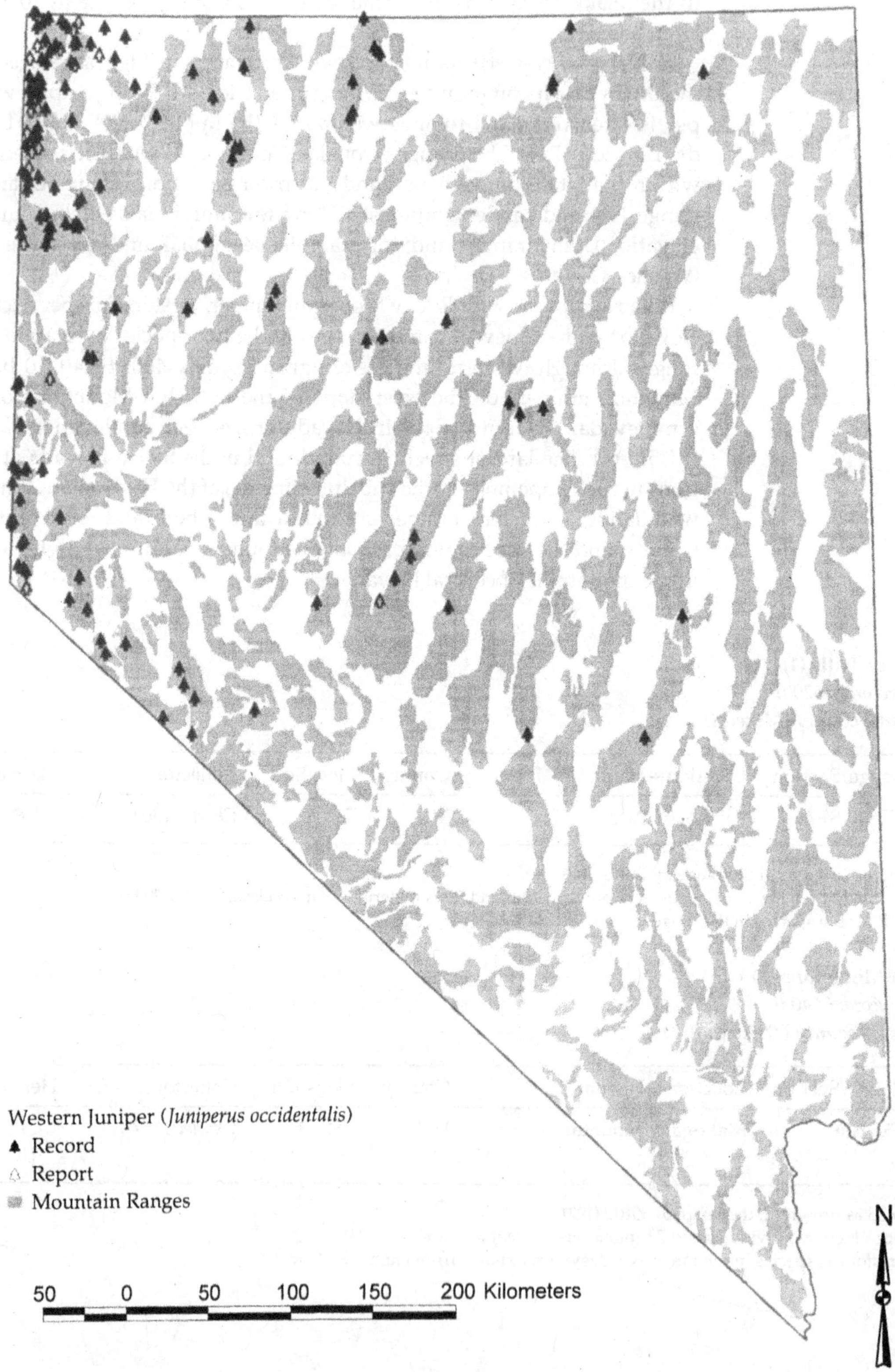
Western Juniper (*Juniperus occidentalis*)
Record
Report
Mountain Ranges
50 0 50 100 150 200 Kilometers
N

****Bilk Creek Mountains (1)**
elevation record: 5560–5790 ft
2 records in 1 county (**Humboldt)

Towns/Range/Section	Location—Habitat	County	Elevation	Collector	Herbarium/a
T44N R32E S5	Table Mountain	HU	5560 ft	*Charlet 2080*	UCR, IRSR, RENO, UC, BRY
T47N R32E S22	Log Cabin Creek	HU	5790 ft	*Charlet 1571 & J Lewis*	UCR, IRSR, RENO

Not reported as present in the range by Little (1971).
DAC: rare; the only conifer individuals seen on the mountain. Based on the bright reddish bark, these collections appear to be var. *australis* (1993–1994).
New record for the species in the range. Not previously known from Humboldt County.

****Black Rock Range (1)**
elevation record: 5900 ft
1 record in 1 county (**Humboldt)

Towns/Range/Section	Location—Habitat	County	Elevation	Collector	Herbarium/a
T41N R26E S6 NW/4	near S shore of Summit Lake	HU	5900 ft	*Charlet 1733 & Alice Winzeler*	UCR, IRSR, RENO, UC

Not reported as present in the range by Little (1971).
DAC: rare; one individual is known from the mountain. A juniper stand spotted with binoculars at 6400–6600 ft, north of Rough Canyon (T41N R27E S9), is likely to be this species. *Charlet 1733 & Winzeler* has the reddish bark of var. *australis* and the branching and internodal lengths of var. *occidentalis* (1993–1994).
New record for the species in the range. Not previously known from Humboldt County.

Bodie Mountains (1)
elevation record: 7500 ft
1 record in 1 county (Mineral)

Towns/Range/Section	Location—Habitat	County	Elevation	Collector	Herbarium/a
T5N R28E	Aurora, in graveyard on hill	MI	7500 ft	*Norman Cooper 2541*	RSA

Reported as present in the range by Little (1971).
DAC: this collection is *J. occidentalis* var. *australis.* Uncommon, particularly in the Nevada portion of the mountain range (1991–1995).

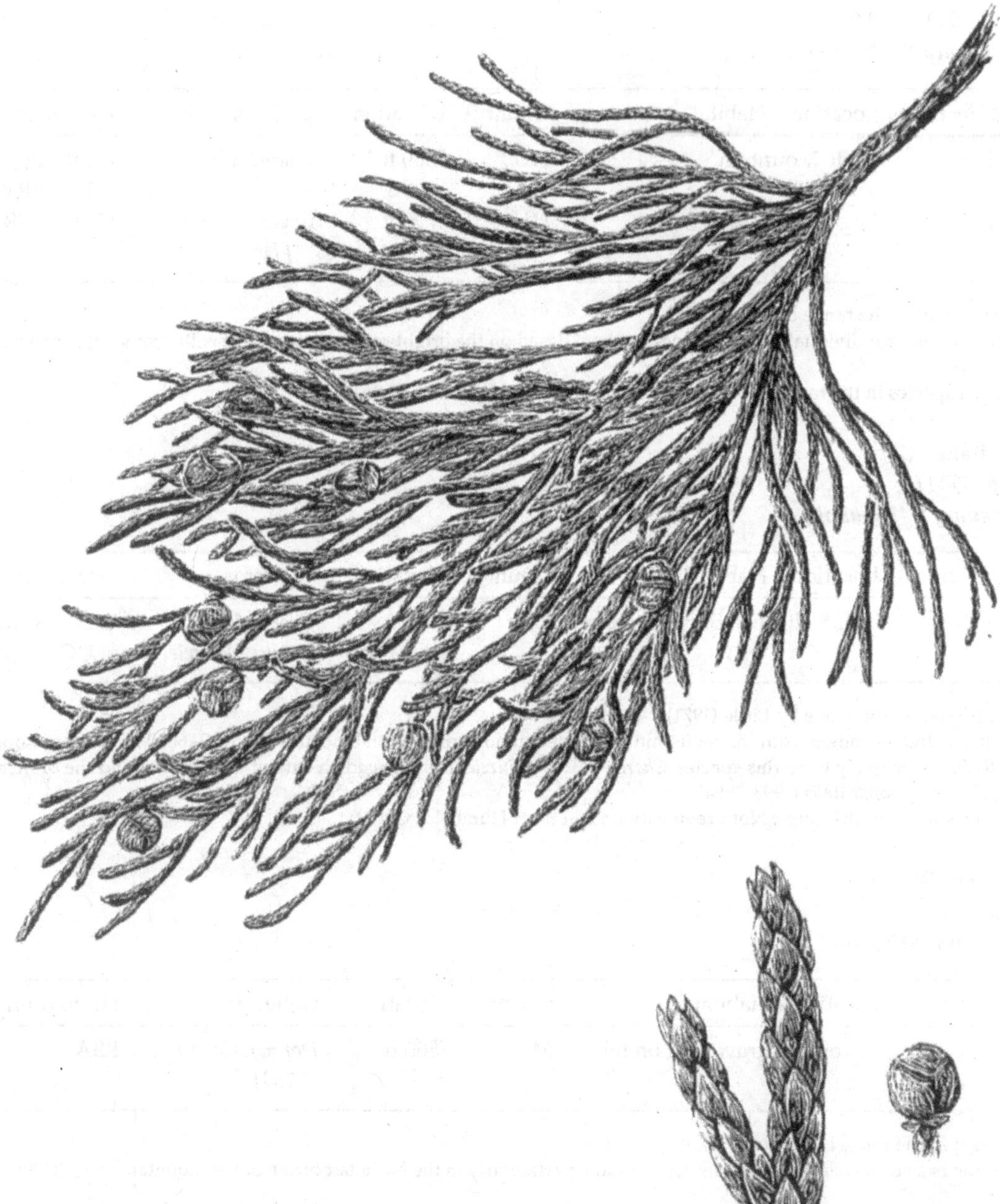

var. *australis*

****Bruneau Range (1)**
elevation record: 5860 ft
1 record in 1 county (**Elko)

Towns/Range/Section	Location—Habitat	County	Elevation	Collector	Herbarium/a
T47N R53E S31	slope above Owyhee River; lone tree w/sagebrush & *Purshia*	EL	5860 ft	*Charlet 2151 & T Charlet*	UCR, IRSR, RENO, UC, BRY

Not reported as present in the range by Little (1971).
DAC: rare; the only individual known from the mountain range. *Juniperus scopulorum* is present nearby along the Owyhee River and in mesic situations on slopes. This collection is var. *occidentalis* (1989–1994).
New record for the species in the range. Not previously known from Elko County.

****Buffalo Hills (1)**
elevation record: 4560–6160 ft
5 records in 1 county (Washoe)

Towns/Range/Section	Location—Habitat	County	Elevation	Collector	Herbarium/a
T35N R21E S27 S/4	just S of pass on SR 447, open juniper woodland	WA	5340 ft	*Charlet 1824*	UCR, IRSR, RENO
T35N R21E S33 W/8	2 trees alone in sagebrush steppe	WA	6160 ft	*Charlet 2072*	UCR, IRSR, RENO
T35N R21E S35	along SR 447, pocket of open woodland with *J. osteosperma*	WA	4560 ft	*Charlet 1823*	UCR, IRSR, RENO, UC
T35N R20E S25	juniper woodland	WA	6100 ft	*Charlet 2077*	UCR, IRSR, RENO
T34N R20E S6 SW/4	NE of the head of N Fork Buffalo Creek	WA	5160 ft	*Charlet 1928*	UCR, IRSR, RENO, UC

Not reported as present in the range by Little (1971).
DAC: these collections (all var. *occidentalis*) are mostly mixed with *J. osteosperma* in *J. osteosperma*-dominated woodlands, but individuals also occur as singletons in sagebrush steppe (1993–1994).
New record for the species in the range.

****Bull Run Mountains (1)**
elevation record: 5855–5970 ft
2 records in 1 county (**Elko)

Towns/Range/Section	Location—Habitat	County	Elevation	Collector	Herbarium/a
T43N R51E S30	2.1 km N of Wilson Peak; sapling	EL	5970 ft	*Charlet 1598*	UCR, IRSR, RENO
T43N R51E S12	3.5 km N of Wilson Peak; tree 8 m tall, 0.5 m dbh	EL	5855 ft	*Charlet 1601*	UCR, IRSR, RENO, UC

Not reported as present in the range by Little (1971).
DAC: on the basis of their bright reddish bark, the above collections appear to be *J. occidentalis* var. *australis*. Rare; 2 individuals seen below *J. scopulorum* stands; a group of 3 individuals thought to be this species was seen with binoculars immediately southwest of Wilson Peak (1990–1994).
New record for the species in the range. Not previously known from Elko County.

Carson Range (1)
elevation record: 6240–9100 ft
13 records in 2 counties (Douglas and Washoe)

Towns/Range/Section	Location—Habitat	County	Elevation	Collector	Herbarium/a
T14N R18E S15	E side of Shakespeare Rock, near summit	DO	7000 ft	*John Darling 20*	RENO
T14N R18E S24	South Camp Peak, W side	DO	8880 ft	*Charlet 2174*	UCR, IRSR, RENO
T14N R18E S24	Logan House Creek, w/aspen	DO	7800 ft	*Charlet 2176*	UCR, IRSR
T14N R19E S30 SE/4	Genoa Peak	DO	8900 ft	*Peter Herlan in 1971*	NSM
T14N R19E S30 SE/4	Genoa Peak	DO	9000 ft	*Margaret Williams 74-D-35*	CAS, RENO
T14N R19E S30 SE/4	Genoa Peak	DO	9100 ft	*Charlet 2172*	UCR, IRSR, RENO
T15N R18E S2 NW/4	Sand Harbor State Park; 1 tree in picnic area w/lodgepole pine	WA	6240 ft	*Charlet 2060 & T Charlet*	UCR, IRSR, RENO, UC
T16N R18E S36 SW/4	———	WA	8750 ft	*Ann Pinzl 8446 & Kimbrough*	UNLV, NSM
T17N R18E S30	NW ridge of Rifle Peak	WA	8500 ft	*Ann Pinzl 1333*	NSM
T17N R18E S31	upper part of Bronco Creek, tall trees, W of Mount Rose	WA	—	*NF Petersen 513*	RENO
T17N R18E S29	W Fork Gray Creek, below Rose knob	WA	8100 ft	*Arnold Tiehm 1457*	RENO
T18N R18E S15	2.5 mi W of Hunter Lake, NW aspect, w/*Cercocarpus* and *P. monticola*	WA	8700 ft	*ND Johnson & JS Spencer in 1954*	DUKE
T19N R18E	4 mi up Hunter Creek	WA	6500 ft	*Kay Johnson in 1951*	RENO

[Little: (DO) T12N R19E]

Reported as present in the range by Little (1971).
DAC: all of the above collections are *J. occidentalis* var. *australis.* Uncommon; scattered throughout the range, generally in rocky, exposed sites. I have been unable to find the species in the Carson City County portion of the range. The species reaches its largest size in Nevada in the Carson Range, and *Charlet 2172* is the high-elevation record for the species in the state (1971–1995).

****Clan Alpine Range (1)**
elevation record: 5800 ft
1 record in 1 county (**Churchill)

Towns/Range/Section	Location—Habitat	County	Elevation	Collector	Herbarium/a
T20N R37E	War Canyon (determined DAC)	CH	5800 ft	*Ann Pinzl in 1974*	NSM

Not reported as present in the range by Little (1971).
DAC: this collection is highly glandular with exudate; nearly all the leaves have distinct glands. This is var. *occidentalis;* originally determined to be *J. osteosperma* by Ann Pinzl. The junipers in the woodlands are primarily *J. osteosperma* (1991–1993).
New record for the species in the range. Not previously known from Churchill County.

****Desert Creek Mountains (1)**
elevation record: 6660 ft
1 record in 1 county (Lyon)

Towns/Range/Section	Location—Habitat	County	Elevation	Collector	Herbarium/a
T9N R24E S34 NW/4	O'Banion Canyon	LY	6660 ft	*Charlet 1947*	UCR, IRSR, RENO

Not reported as present in the range by Little (1971) or Lavin (1981).
DAC: the material is var. *occidentalis,* which is occasional in pinyon woodlands (1987–1994).
New record for the species in the range.

Division Range (1)
elevation record: 5495–5620 ft
2 records in 1 county (Washoe)

Towns/Range/Section	Location—Habitat	County	Elevation	Collector	Herbarium/a
T37N R23E S22 E/4	Leadville Canyon	WA	5495 ft	*Charlet 1700 & Rust*	UCR, IRSR
—	SR 34, in S Willow Creek Canyon, 14.2 mi from Granite Peak turnoff [DAC: this collection is probably from the same tree as *Charlet 1700 & Rust*]	WA	5620 ft	*Vasek 640913-9*	UCR

Not reported as present in the range by Little (1971).
DAC: rare; only 1 individual was seen on the mountain; may occur in the higher portions of the range. Both collections are var. *occidentalis* (1993–1994).

****Dogskin Mountain (1)**
elevation record: 5540–5780 ft
2 records in 1 county (Washoe)

Towns/Range/Section	Location—Habitat	County	Elevation	Collector	Herbarium/a
T24N R19E	just below McKissick Spring	WA	5780 ft	*Charlet 1788 & T Charlet*	IRSR, UCR, RENO, UC, BRY
T24N R19E	McKissick Canyon	WA	5540 ft	*Charlet 1789 & T Charlet*	IRSR, UCR, RENO

Not reported as present in the range by Little (1971).
DAC: var. *australis;* common along McKissick Creek, from the spring to near the mouth of the canyon (1994).
New record for the species in the range.

****Excelsior Mountains (1)**
elevation record: 7215 ft
1 record in 1 county (Mineral)

Towns/Range/Section	Location—Habitat	County	Elevation	Collector	Herbarium/a
T5N R33E S12	pinyon-juniper woods	MI	7215 ft	*Charlet 1315 & Tausch*	IRSR

Not reported as present in the range by Little (1971).
DAC: var. *occidentalis;* this is the only individual of the taxon known from the mountain, although there are likely to be more at the higher elevations; in a large *Pinus monophylla* woodland with some *J. osteosperma,* dominated by pinyon. The *J. osteosperma* populations here have fairly strong glandular expression (1993).
New record for the species in the range.

****Fish Creek Mountains (1)**
elevation record: 5350 ft
5 records in 1 county (**Lander)

Towns/Range/Section	Location—Habitat	County	Elevation	Collector	Herbarium/a
T28N R41E	Daisy Creek	LA	5350 ft	*Charlet 1439-1443 & Tausch*	IRSR

Not reported as present in the range by Little (1971).
DAC: these collections are var. *occidentalis.* Many individuals of this taxon are mixed with somewhat glandular *J. osteosperma* in woodlands in the central portion of the range (1993–1994).
New record for the species in the range. Not previously known from Lander County.

****Fort Sage Mountains (1)**
elevation record: 5200–6660 ft
2 records in 1 county (Washoe)

Towns/Range/Section	Location—Habitat	County	Elevation	Collector	Herbarium/a
T25N R18E S9	NE slope of range, scattered trees in sagebrush steppe	WA	5200 ft	*Charlet 1942*	UCR, IRSR, RENO, UC
T25N R18E S17 NE/4	N of Stateline Peak, near *Pinus jeffreyi*-aspen stand	WA	6660 ft	*Charlet 2161*	UCR, IRSR, RENO, UC

Not reported as present in the range by Little (1971).
DAC: *Charlet 1942* is the lowest of the western junipers on the mountain. Clusters occur along the east and north slopes on Stateline Peak (1993–1994).
New record for the species in the range.

****Granite Range (1)**
elevation record: 5820–7020 ft
2 records in 1 county (Washoe)

Towns/Range/Section	Location—Habitat	County	Elevation	Collector	Herbarium/a
T36N R21E S14 SW/4	Fox Mountain, lone tree in sagebrush steppe	WA	5820 ft	*Charlet 1829*	UCR, IRSR, RENO
T36N R22E S7 W/4	Fox Mountain, above Potato Spring, lone tree in mountain brush	WA	7020 ft	*Charlet 1830*	UCR, IRSR, RENO, UC

Not reported as present in the range by Little (1971).
DAC: these 2 trees were the only ones seen in the north end of the range except for a stand at the head of Negro Creek, which is probably *J. occidentalis.* I have seen only *J. osteosperma* in the lower woodlands of the central and southern Granite Range; however, *J. occidentalis* may occur at the upper elevations there (1974–1994).

****Grant Range (1)**
elevation record: 7100 ft
1 record in 1 county (**Nye)

Towns/Range/Section	Location—Habitat	County	Elevation	Collector	Herbarium/a
T4N R57E S30	Cherry Creek, along creek	NY	7100 ft	*Charlet 2005*	UCR, IRSR, RENO

Not reported as present in the range by Little (1971).
DAC: uncommon along streams in pinyon-juniper-mountain-mahogany woodlands (1994).
New record for the species in the range. Not known previously from Nye County.

Hays Canyon Range (1)
elevation span 5020–6800 ft
15 records in 1 county (Washoe)

Towns/Range/Section	Location—Habitat	County	Elevation	Collector	Herbarium/a
T38N R19E S32	southern limit of juniper distribution in the range (var. *australis*)	WA	5120 ft	*Charlet 1938*	UCR, IRSR, RENO, UC, BRY
T39N R18E S2	Hays Canyon	WA	6060 ft	*Charlet 1936*	UCR, IRSR, RENO
T39N R18E S4	Hays Canyon (var. *australis*)	WA	5280 ft	*Charlet 1937*	UCR, IRSR, RENO
T39N R19E S6	N slope Divine Peak (var. *australis*)	WA	6800 ft	*Charlet 1935*	UCR, IRSR, RENO
T40N R19E S31	below & E of Summit Spring	WA	5880 ft	*Charlet 1835*	UCR, IRSR
T40N R19E S31	juniper woodland near stream, E of Summit Spring	WA	5900 ft	*Charlet 1836*	UCR, IRSR, RENO
T42N R18E S17 E/8	western foothills, along dry wash	WA	5020 ft	*Charlet 1929*	UCR, IRSR, RENO, UC
T42N R19E	W side South Creek, Long Valley	WA	—	*Percy Train 2953*	UC, RENO, RSA, NSM, UTC
T42N R19E	6.1 mi E of NV line on SR 8A	WA	5900 ft	*Vasek 640902-11*	UCR
T42N R19E S9 SE/4	2 km W of Fortynine Lake	WA	5755 ft	*Charlet 1703 & Rust*	UCR, IRSR
T42N R18E	1.5 mi E of NV line	WA	5450 ft	*Vasek 640913-3*	UCR
T42N R18E	1.8 mi E of NV line	WA	5650 ft	*Vasek 640913-5*	UCR
T43N R18E	4.7 mi E of NV line	WA	6350 ft	*Vasek 640913-6*	UCR
T43N R19E S32 S/4	E slope above Vya (var. *australis*)	WA	5700 ft	*Charlet 1930*	UCR, IRSR, RENO, UC
T43N R19E	unnamed range between Long Valley and CA-NV line, 4 mi N of Vya. Vasek: "mostly *J. occ.* (glands), but M/F and broad branch angles like *J. ost.* 2-6-62"	WA	5800 ft	*Percy Train 2795*	UC, UTC

[DAC: T42N R18E, T41N R18E R19E, T40N R18E, T39N R18E R19E, T38N R18E]

Reported as present in the range by Little (1971).
DAC: in extensive juniper woodlands. Most junipers in this range are clearly *J. o.* var. *occidentalis.* A very few are *J. osteosperma,* and some appear to be var. *australis* (e.g., *Charlet 1930, 1935, 1937, 1938*); still others are apparently intermediate. Some collections made from this range are in the "possible hybrids" file at UCR (all det. by FC Vasek; none of these are reported here). Glandular expression tends to weaken from west to east in the range (1993–1994).

****Home Camp Range (1)**
elevation record: 5660 ft
1 record in 1 county (Washoe)

Towns/Range/Section	Location—Habitat	County	Elevation	Collector	Herbarium/a
T40N R19E ns	low edge juniper woodland, N aspect	WA	5660 ft	*Charlet 1833*	UCR, IRSR

Not reported as present in the range by Little (1971).
DAC: a few scattered trees occur in sagebrush steppe (1993–1994).
New record for the species in the range.

****Horse Range (1)**
elevation record: 6300 ft
1 record in 1 county (**Nye)

Towns/Range/Section	Location—Habitat	County	Elevation	Collector	Herbarium/a
T11N R59E S16	NW end of mountain, by stream	NY	6300 ft	*Charlet 1984*	UCR, IRSR, RENO

Not reported as present in the range by Little (1971).
DAC: uncommon along streams in pinyon-juniper woodlands (1994).
New record for the species in the range. Not known previously from Nye County.

****Hot Creek Range (1)**
elevation record: 6120 ft
1 record in 1 county (**Nye)

Towns/Range/Section	Location—Habitat	County	Elevation	Collector	Herbarium/a
T4N R49.5E S24 S/4	S end of mountain range, juniper woodland	NY	6120 ft	*Charlet 1982*	UCR, IRSR, RENO

Not reported as present in the range by Little (1971).
DAC: all of the junipers I encountered at this location were highly glandular var. *occidentalis,* and yet they were not resinous when I collected them (7/1994).
New record for the species in the range. Not known previously from Nye County.

****Humboldt Range (1)**
elevation record: 6790 ft
3 records in 1 county (**Pershing)

Towns/Range/Section	Location—Habitat	County	Elevation	Collector	Herbarium/a
T31N R34E S21	Star Canyon (determined Vasek)	PE	6790 ft	*Charlet 778 & Rust*	UCR, IRSR
T31N R34E S21	Star Canyon (determined Vasek)	PE	6790 ft	*Charlet 779 & Rust*	UCR, IRSR
T30N R34E	Unionville (undetermined *Juniperus* by Jones; determined DAC)	PE	—	*ME Jones in 1901*	RSA

Not reported as present in the range by Little (1971).
DAC: this material is all var. *occidentalis.* Apparently restricted in the Humboldt Range to the Star Canyon and Unionville areas with *J. osteosperma.* Elsewhere, *J. osteosperma* occurs alone (1991–1994).
New record for the species in the range. Not previously known from Pershing County.

Jackson Mountains (1)
elevation record: 4130–6960 ft
4 records in 1 county (**Humboldt)

Towns/Range/Section	Location—Habitat	County	Elevation	Collector	Herbarium/a
T40N R31E S8 NW/4	low edge juniper distribution, on dry wash	HU	4130 ft	*Charlet 1715 & Rust*	UCR, IRSR, RENO
T40N R31E S35	above Jackson Creek, scattered trees in sagebrush steppe	HU	5260 ft	*Charlet 1852*	UCR, IRSR, RENO
T40N R32E S31 SW/4	upper limit juniper woodland, upper Jackson Creek, near Redbird Mine	HU	6960 ft	*Charlet 1851*	UCR, IRSR, RENO
T39N R31E S22 SE/4	lone tree in sagebrush steppe, S aspect	HU	5080 ft	*Charlet 1848*	UCR, IRSR, RENO, UC

Reported as present in the range by Little (1971).
DAC: all this material is var. *occidentalis. Charlet 1715 & Rust* is from the lower end of the juniper distribution, at the end of a dry wash stringer, and is the low-elevation record for the species in Nevada. Others on similar west slope stringers are *J. osteosperma.* Large, open juniper stands occur through much of the northern end of the range, particularly about King Lear Peak and Parrot Peak. I did not visit these higher stands, but the stands at the headwaters of Jackson Creek are principally *J. occidentalis,* and so the species is probably common throughout the northern end of the range. Although *J. occidentalis* is reported as present in the range by Little (1971), I have seen no material for the Jackson Mountains other than my own. JT Kartesz (in litt. 1993) told me that all of Little's Humboldt County reports of the species were based on misidentifications of *J. osteosperma.* The collections reported here are clearly *J. occidentalis.* I have found no records for the species in Humboldt County other than my own (1993–1994).

****Junction House Range (1)**
elevation record: 4860 ft
1 record in 1 county (Washoe)

Towns/Range/Section	Location—Habitat	County	Elevation	Collector	Herbarium/a
T20N R19E S24 N/4	off Quartz Rd., N of radio towers	WA	4860 ft	*Charlet 1797*	UCR, IRSR, RENO, UC

Not reported as present in the range by Little (1971).
DAC: this collection is var. *occidentalis;* uncommon in open juniper woodlands, composed primarily of *J. osteosperma* (1971–1995).
New record for the species in the range.

****Kamma Mountains (1)**
elevation record: 5445–5780 ft
2 records in 1 county (**Pershing)

Towns/Range/Section	Location—Habitat	County	Elevation	Collector	Herbarium/a
T34N R29E S13 W/4	0.7 km S of Rosebud Peak	PE	5780 ft	*Charlet 1584*	UCR, IRSR, RENO
T34N R29E S13 W/4	0.7 km S of Rosebud Peak	PE	5445 ft	*Charlet 1719 & Rust*	UCR, IRSR, RENO

Not reported as present in the range by Little (1971).
DAC: these collections are var. *occidentalis* and are from an isolated stand of 300–400 junipers at an abandoned mine site. *Juniperus osteosperma* is also here, and most have a fairly strong glandular expression (1993).
New record for the species in the range. Not previously known from Pershing County.

****Lake Range (1)**
elevation record: 6400–7060 ft
2 records in 1 county (Washoe)

Towns/Range/Section	Location—Habitat	County	Elevation	Collector	Herbarium/a
T27N R22E S25 E/4	W of Tohakum Peak, juniper woodlands	WA	6400 ft	*Charlet 1739*	UCR, IRSR, RENO, UC, BRY
T27N R22E S27 NE/4	W of Tohakum Peak, juniper woodlands	WA	7060 ft	*Charlet 1741*	UCR, IRSR, RENO, UC, BRY

Not reported as present in the range by Little (1971).
DAC: large juniper stands occur in the vicinity of Tohakum Peak. *Juniperus o.* var. *occidentalis* is present occasionally (< 1%) with eglandular *J. osteosperma* in these juniper woodlands, along with *Holodiscus, Ribes inerme,* and *Artemisia tridentata* (1973–1994).
New record for the species in the range.

****Lost Creek Hills (1)**
elevation record: 5540 ft
1 record in 1 county (Washoe)

Towns/Range/Section	Location—Habitat	County	Elevation	Collector	Herbarium/a
T37N R20E S20 N/4	Juniper Spring; lone tree at vernal spring in sagebrush steppe	WA	5540 ft	*Charlet 1898*	UCR, IRSR, RENO, UC, BRY

Not reported as present in the range by Little (1971).
DAC: rare; only 1 tree found in the range. The red bark of this individual suggests var. *australis* (1994).
New record for the species in the range.

****Madelin Mesa (1)**
elevation record: 4600–6060 ft
12 records in 1 county (Washoe)

Towns/Range/Section	Location—Habitat	County	Elevation	Collector	Herbarium/a
T33N R19E S2 NW/4	near the southern juniper limit in the mountain range	WA	4960 ft	*Charlet 1927*	UCR, IRSR, RENO, UC
T34N R19E S1	head N Fork Buffalo Creek	WA	5120 ft	*Charlet 1925*	UCR, IRSR, RENO, UC
T34N R19E S24	above and W of Buffalo Creek	WA	5140 ft	*Charlet 1926*	UCR, IRSR, RENO, UC
T35N R18E S25 N/4	open juniper woodland	WA	5960 ft	*Charlet 1827*	UCR, IRSR, RENO
T35N R18E S32 NW/4	S end Pilgrim Lake, near CA border, w/mountain-mahogany	WA	6060 ft	*Charlet 1828*	UCR, IRSR, RENO
T35N R19E S11 S/2	alone in sagebrush steppe	WA	5340 ft	*Charlet 1825*	UCR, IRSR, RENO, UC
T35N R19E S16 S/2	juniper woodland	WA	5900 ft	*Charlet 1826*	UCR, IRSR, RENO
T36N R18E S5 SW/4	0.4 km E of California border	WA	5080 ft	*Charlet 2074*	UCR, IRSR, RENO, UC
T37N R18E S2	lone tree in sagebrush steppe, W of SR 447, below juniper woodland	WA	4720 ft	*Charlet 1897*	UCR, IRSR, RENO, UC
T37N R18E S22	Coppersmith Hills, immediately W of Duck Lake	WA	4780 ft	*Charlet 2073*	UCR, IRSR, RENO
T38N R18E S20	2 trees in sagebrush steppe, below juniper woodland, S of SR 447	WA	4600 ft	*Charlet 1896*	UCR, IRSR, RENO, UC
T38N R18E S28 W/4	Coppersmith Hills, lone tree near dry creek w/*Sarcobatus* and sagebrush	WA	4640 ft	*Charlet 1866*	UCR, IRSR, RENO
[DAC: T37N R18E, T36N R18E, T34N R18E]					

Not reported as present in the range by Little (1971).
DAC: the extensive juniper woodlands at the higher elevations of the mesa are exclusively *J. o.* var. *occidentalis;* more common in the north (1994).
New record for the species in the range.

Marys River Range (4)
Elko County

Not reported as present in the range by Loope (1969) or Little (1971).
DAC: found 0.3 km north of the Nevada-Idaho border in Pole Creek, Owyhee County, ID (*Charlet 1604 & Agee Smith* UCR, IRSR). Many individuals (ca. 20) follow the creek, and so the species seems likely to occur within Nevada as well; however, extensive searches in both Pole Creek and Sheep Creek yielded only *J. scopulorum* in the Nevada portion of Pole Creek and neighboring Sheep Creek (1988–1994).

****Massacre Range (1)**
elevation record: 5940–5985 ft
2 records in 1 county (Washoe)

Towns/Range/Section	Location—Habitat	County	Elevation	Collector	Herbarium/a
T41N R21E S32 S/4	Lone Juniper Spring	WA	5940 ft	*Charlet 1701 & Rust*	UCR, IRSR
T41N R21E S33 S/4	Lone Juniper Spring	WA	5985 ft	*Charlet 1702 & Rust*	UCR

Not reported as present in the range by Little (1971).
DAC: a single small stand of perhaps 400 reproductive individuals (var. *occidentalis*) was seen in the area of Lone Juniper Spring; mixed with *J. osteosperma* (population with weak glandular expression), but *J. occidentalis* is more common (1993–1994).
New record for the species in the range.

****Mosquito Mountains (1)**
elevation record: 5500–6840 ft
6 records in 1 county (Washoe)

Towns/Range/Section	Location—Habitat	County	Elevation	Collector	Herbarium/a
T45N R19E S26 W/4	pass from Long Valley to Mosquito Valley, juniper woodland	WA	5520 ft	*Charlet 1839*	UCR, IRSR
T45N R19E S4	juniper-mountain-mahogany woodland, near ridge	WA	6320 ft	*Charlet 1841*	UCR, IRSR, RENO
T46N R19E S28 NW/4	juniper woodland w/mountain-mahogany	WA	6180 ft	*Charlet 1842*	UCR, IRSR, RENO
T46N R20E S11 NE/4	near pass to Coleman Canyon, SW aspect, low end junipers here	WA	5800 ft	*Charlet 1860*	UCR, IRSR, RENO
T47N R19E S27 S/4	4 air miles NNW of Little Coleman Canyon, near upper juniper limit	WA	6840 ft	*Charlet 1863 & John Conlan*	UCR, IRSR, RENO
T47N R20E S33 SE/4	Coleman Canyon, juniper woodland	WA	5500 ft	*Charlet 1861*	UCR, IRSR, RENO
[DAC: T45N R20E, T47N R18E]					

Not reported as present in the range by Little (1971).
DAC: large woodlands of western juniper occur throughout much of the Mosquito Mountains. Ponderosa pine, Washoe pine, and white fir are present, but rare, in the juniper woodlands of the range. The taxon is *J. occidentalis* var. *occidentalis* (1993–1994).
New record for the species in the range.

****Pah Rah Range (1)**
elevation record: 6240 ft
1 record in 1 county (Washoe)

Towns/Range/Section	Location—Habitat	County	Elevation	Collector	Herbarium/a
T21N R23E S30 N/4	Olinghouse Canyon, top of road	WA	6240 ft	*Charlet 1793 & T Charlet*	IRSR, UCR, RENO, UC

Not reported as present in the range by Little (1971).
DAC: uncommon in juniper woodlands at higher elevations (1975–1994).
New record for the species in the range.

Painted Point Range (1)
elevation record: 5440–7020 ft
6 records in 1 county (Washoe)

Towns/Range/Section	Location—Habitat	County	Elevation	Collector	Herbarium/a
T45N R21E S9	Sheldon NWR, 41°50′N, 119°38.5′W, w/*P. ponderosa* (det. Vasek)	WA	6500 ft	*Critchfield & Allenbaugh 2-9/3/65*	IFGP
T45N R21E S9	Bald Mountain headquarters, Sheldon NWR	WA	6500 ft	*Tiehm & Rogers 4709*	CAS, NSM, UTC, RENO
T45N R21E S9	highest juniper on Bald Mountain, in mountain brush	WA	7020 ft	*Charlet 1859*	UCR, IRSR, RENO
T45N R22E S8	widely scattered in low sage	WA	5940 ft	*Charlet 1858*	UCR, IRSR, RENO
T44N R20E S2	E of Middle Lake, low edge juniper woodland	WA	5440 ft	*Charlet 1843*	UCR, IRSR, RENO
T43N R20E S35 SE/4	0.5 km N of Painted Point	WA	5755 ft	*Charlet 1704 & Rust*	UCR, IRSR, RENO, UC
[DAC: T44N R21E, T46N R20E]					

Not reported as present in the range by Little (1971).
DAC: the material is all var. *occidentalis.* Large juniper woodlands composed solely of *J. occidentalis* are found in the higher portions of the range (1993–1994).

****Paradise Range (1)**
elevation record: 6950 ft
1 record in 1 county (**Nye)

Towns/Range/Section	Location—Habitat	County	Elevation	Collector	Herbarium/a
T12N R37E	Green Springs Summit (determined Vasek)	NY	6950 ft	*Charlet 1117*	UCR, IRSR

Not reported as present in the range by Little (1971).
DAC: in extensive pinyon-juniper woodland. This is at the upper end of the juniper distribution, where pinyons are dominant. Most of the summit junipers are *J. o.* var. *occidentalis,* whereas the lower junipers are *J. osteosperma* (1992).
New record for the species in the range. Not previously known from Nye County.

Peavine Mountain (1)
elevation record: 6340–6880 ft
4 records in 1 county (Washoe)

Towns/Range/Section	Location—Habitat	County	Elevation	Collector	Herbarium/a
T20N R18E	Peavine Mountain	WA	—	*Margaret Williams 74-7-3*	CAS
T20N R18E	Peavine Mountain	WA	—	*Margaret Williams & Tiehm 74-54-1*	CAS
T20N R18E S28 SE/4	SW of peak, Bull Creek drainage	WA	6880 ft	*Charlet 2168*	UCR, IRSR, RENO, UC
T20N R18E S24	Near Peaville, E of peak	WA	6340 ft	*Charlet 2165*	UCR, IRSR, RENO

Reported as present in the range by Little (1971).
DAC: the above collections are all *J. occidentalis* var. *australis.* Uncommon; mostly singletons in sagebrush steppe (1971–1995).

****Petersen Mountain (1)**
elevation record: 5315 ft
1 record in 1 county (Washoe)

Towns/Range/Section	Location—Habitat	County	Elevation	Collector	Herbarium/a
T23N R18E S27	singleton in vernal meadow	WA	5315 ft	*Charlet 1738 & T Charlet*	UCR, IRSR, RENO, UC

Not reported as present in the range by Little (1971).
DAC: this singleton (var. *occidentalis*), in a vernal meadow northeast of *Pinus jeffreyi* stands, is the only individual of the species I have seen on the mountain. Well separated from *J. osteosperma* stands on lower sites on this and nearby mountains; *J. occidentalis* is probably present in the higher elevations of the range (1992–1994).
New record for the species in the range.

****Pine Forest Range (1)**
elevation record: 6135–6215 ft
2 records in 1 county (**Humboldt)

Towns/Range/Section	Location—Habitat	County	Elevation	Collector	Herbarium/a
T43N R30E S32	N of Sentinel Peak, SE portion of mountain	HU	6215 ft	*Charlet 1510*	IRSR
T44N R28E S32	between Onion & Knott Creek reservoirs, W side of mountain	HU	6135 ft	*Charlet 1546*	UCR, IRSR, RENO

Possibly reported as present in the range by Little (1971).
DAC: *1546* is an isolated tree in sagebrush steppe; *1510* is part of a mixed stand of *J. osteosperma* (with varying glandular expression) and *J. o.* var. *occidentalis,* about 300 individuals, 15 km distant from *1546* (also var. *occidentalis*); rare. Although Little's distribution map (1971) shows the species as occurring at the north end of the Pine Forest Range (Mahogany Mountain), I know of no juniper from Mahogany Mountain (see comments below under "Pueblo Mountains"). I have seen no material other than my own for the Pine Forest Range, and Vasek (1966) did not report its presence here. Lanner (1984), Cronquist et al. (1972), and Grayson (1993) all visited the range, and all three reported the genus as absent (1990–1994).
New record for the species in the range. I have seen no material from Humboldt County other than my own.

****Pine Grove Hills (1)**
elevation record: 7340 ft
1 record in 1 county (Lyon)

Towns/Range/Section	Location—Habitat	County	Elevation	Collector	Herbarium/a
T 9N R26E S17 NE/4	Rockland Mines, near spring and meadow near head of Rockland Canyon	LY	7340 ft	*Charlet 2206 & T Charlet*	RENO, UC, DUKE

Not reported as present in the range by Little (1971).
DAC: uncommon; only a few trees of the species have been found in the mountain range, and all them are in the upper Rockland Canyon area with northeast aspect (1989–1995).
New record for the species in the range.

Pine Nut Mountains (1)
elevation record: 6200–8510 ft
3 records in 2 counties (Douglas and Lyon)

Towns/Range/Section	Location—Habitat	County	Elevation	Collector	Herbarium/a
T12N R21E	between Pine Nut Creek & E Fork Carson River, 20 ft tall trees JT Howell annotation: "introgression w/ *J. osteosperma?*"	DO	6200 ft	*FW Haasis 117.59*	CAS
T12N R22E S22	1.2 km SE of Mount Siegel, near *Pinus monticola*	DO	8510 ft	*Charlet 705 & Rust*	UCR, IRSR
T14N R22E	Sunrise Pass	LY	6725 ft	*Charlet 1274 & Tausch*	IRSR

Reported as present in the range by Little (1971).
DAC: all of the above appear to be var. *australis.* Found generally separated from and above the extensive pinyon-juniper woodland (1990–1994).

****Pinto Peak Range (1)**
elevation record: 5600 ft
1 record in 1 county (Washoe)

Towns/Range/Section	Location—Habitat	County	Elevation	Collector	Herbarium/a
T41N R19E S10	low end juniper woodland	WA	5600 ft	*Charlet 1934*	UCR, IRSR, RENO

Not reported as present in the range by Little (1971).
DAC: woodlands of the species occur in the northwestern portion of the range, and individuals widely scattered in sagebrush steppe occur elsewhere in the range; all are var. *occidentalis* (1993–1994).
New record for the species in the range.

Pueblo Mountains (5)
Humboldt County

Reported as present in the range by Little (1971).
DAC: I have seen no junipers in the Nevada portion of this mountain range, nor have I seen collections in herbaria. However, I saw a few individuals in this range north of the Oregon border. Little's (1971) map includes the Pueblo Mountains of Nevada and the area of Mahogany Mountain (at the extreme north end of the Pine Forest Range), south of Denio. While *Cercocarpus* is abundant on the northern slope, and at least one *P. flexilis* or *P. albicaulis* occurs near the summit, I have seen no *Juniperus* on Mahogany Mountain. Vasek saw no material; nor did I (1991–1994).

****Roberts Mountains (1)**
elevation record: 6880–7135 ft
2 records in 1 county (**Eureka)

Towns/Range/Section	Location—Habitat	County	Elevation	Collector	Herbarium/a
T23N R49E S3	Tonkin Summit	EU	7135 ft	*Charlet 1390 & Tausch*	IRSR
T24N R50E S36	Dry Creek	EU	6880 ft	*Charlet 2085*	UCR, IRSR, RENO

Not reported as present in the range by Little (1971).
DAC: the material is var. *occidentalis;* uncommon in pinyon-juniper-mountain-mahogany woodlands west of the high peaks of Roberts Mountain, and uncommon in upper elevations of pinyon-juniper woodlands along Dry Creek. *Juniperus osteosperma* is more common, with varying glandular expression (1993–1994).
New record for the species in the range. Not previously known from Eureka County.

****Santa Rosa Range (1)**
elevation record: 5410–7250 ft
6 records in 1 county (**Humboldt)

Towns/Range/Section	Location—Habitat	County	Elevation	Collector	Herbarium/a
T41N R38E S1 NE/4	Morey Creek (determined Vasek)	HU	6330 ft	*Charlet 1265 & Ruth Mazur*	UCR, IRSR, RENO
T42N R39E S30 NE/4	S Fork Hansen Creek (determined Vasek)	HU	6080 ft	*Charlet 1266 & Ruth Mazur*	UCR, IRSR, RENO, UC
T44N R39E S31 SE/4	W of Granite Peak (determined Vasek)	HU	7050 ft	*Charlet 1267 & Ruth Mazur*	UCR, IRSR, RENO
T45N R40E S11 NE/4	lower Long Canyon (determined Vasek)	HU	6450 ft	*Charlet 1270*	UCR, IRSR, RENO
T46N R40E S33 N/2	head, Long Canyon (determined Vasek)	HU	7250 ft	*Charlet 1269*	UCR, IRSR, RENO
T47N R39E S2 S/2	Quinn River (determined Vasek)	HU	5410 ft	*Charlet 1268*	UCR, IRSR, RENO

Not reported as present in the range by Little (1971).
DAC: rare. This is the only conifer species other than *Pinus flexilis* I have seen on the mountain; *1266* was in an area with seedlings and a sapling nearby, all others were singletons, and the only ones of the species observed. Grayson (1993) reported *Juniperus* as absent from the range. All of my material has red bark, suggesting var. *australis,* although Vasek determined all of my material to be the typical variety (1991–1994).
New record for the species in the range. I have seen no material from Humboldt County other than my own.

****Selenite Range (1)**
elevation record: 6365 ft
2 records in 1 county (**Pershing)

Towns/Range/Section	Location—Habitat	County	Elevation	Collector	Herbarium/a
T30N R24E S29 N/4	3 km N of Kumiva Peak	PE	6365 ft	*Charlet 1726 & Rust*	UCR, IRSR, RENO
T30N R24E S29 N/4	3 km N of Kumiva Peak	PE	6365 ft	*Charlet 1727 & Rust*	UCR, IRSR

Not reported as present in the range by Little (1971).
DAC: this is the common species in the higher elevations I visited, in extensive juniper woodlands; *J. osteosperma* is at lower elevations (1993–1994).
New record for the species in the range. Not previously known from Pershing County.

****Seven Troughs Range (1)**
elevation record: 5740 ft
1 record in 1 county (**Pershing)

Towns/Range/Section	Location—Habitat	County	Elevation	Collector	Herbarium/a
T30N R28E S26 N/4	1 km W of Fairview Mine	PE	5740 ft	*Charlet 1724 & Rust*	UCR, IRSR, RENO

Not reported as present in the range by Little (1971).
DAC: var. *occidentalis*; occasional in juniper woodland with *J. osteosperma*, which have varying glandular expression (1993–1994).
New record for the species in the range. Not previously known from Pershing County.

****Shoshone Range (1)**
elevation record: 5890 ft
1 record in 1 county (**Lander)

Towns/Range/Section	Location—Habitat	County	Elevation	Collector	Herbarium/a
T29N R45E	Mill Creek Canyon	LA	5890 ft	*Charlet 1434 & Robin Tausch*	IRSR

Not reported as present in the range by Little (1971).
DAC: var. *occidentalis* is occasional in upper pinyon-juniper woodland on the south face of Mount Lewis; *J. osteosperma* is more common and dominates the lower elevations of the woodlands (1993–1994).
New record for the species in the range. Not previously known from Lander County.

****Simpson Park Range (1)**
elevation record: 6120 ft
1 record in 1 county (**Eureka)

Towns/Range/Section	Location—Habitat	County	Elevation	Collector	Herbarium/a
T24N R48E S13	McClusky Canyon	EU	6120 ft	*Charlet 1394 & Robin Tausch*	IRSR

Not reported as present in the range by Little (1971).
DAC: var. *occidentalis* is occasional along streams in large pinyon-juniper woodlands; with very large female cones. The *J. osteosperma* here has fairly strong glandular expression (1993–1994).
New record for the species in the range. Not previously known from Eureka County.

****Snake Mountains (1)**
elevation record: 6560 ft
1 record in 1 county (**Elko)

Towns/Range/Section	Location—Habitat	County	Elevation	Collector	Herbarium/a
T44N R61E S30	on road to O'Neil Basin	EL	6560 ft	*Charlet 1608 & Dean Hagen*	UCR, IRSR, RENO, UC, DUKE, BRY, UNLV

Not reported as present in the range by Little (1971).
DAC: rare; the only conifer tree I have seen in the northern portion of the mountain range. In the southern part of the range, *J. osteosperma* woodlands are present. *Charlet 1608 & Dean Hagen* has the reddish bark of var. *australis* and the internodal lengths and branching angles of var. *occidentalis.* A duplicate of this collection is at the Ruby Mountain District, Humboldt National Forest, Wells, NV (1988–1994).
New record for the species in the range. Not previously known from Elko County.

Sweetwater Mountains (4)
Lyon County

Reported as present in the range by Little (1971).
Reported as present in the range by Lavin (1981), but he did not find the species in the Nevada part of the range (Matt Lavin, pers. comm. 1994).
DAC: common in the California portion of the range (e.g., *LC Johnson in 1975* IFGP), but I have seen only *J. osteosperma* in Nevada. It is likely to occur in the Nevada portion of the range (1985–1995).

****Tobin Range (1)**
elevation record: 5820 ft
1 record in 1 county (**Pershing)

Towns/Range/Section	Location—Habitat	County	Elevation	Collector	Herbarium/a
T28N R39E S24 NW/4	S of Golconda Pass	PE	5820 ft	*Charlet 1883*	UCR, IRSR, RENO, UC, BRY

Not reported as present in the range by Little (1971).
DAC: uncommon in the upper elevations of the juniper woodlands near Mount Tobin. This individual is var. *occidentalis* (1994).
New record for the species in the range. Not previously known from Pershing County.

****Toiyabe Range (1)**
elevation record: 7000–8200 ft
4 records in 2 counties (**Lander and **Nye)

Towns/Range/Section	Location—Habitat	County	Elevation	Collector	Herbarium/a
T16N R43E	Kingston Creek (determined Vasek)	LA	7345 ft	*Charlet 1052*	UCR, IRSR
T15N R42E S24	Washington Creek	NY	8200 ft	*Tausch 4/21/94*	IRSR, UCR
T15N R42E	Cottonwood Creek	NY	7000 ft	*Scott Fleury in 1995*	DUKE
T14N R41E S36	Marysville Canyon	NY	7480 ft	*Tausch 2-3/29/94*	IRSR, UCR
[Robin Tausch: T12N R41E S6 SE/4 (Stewart Creek)]					

Not reported as present in the range by Little (1971).
DAC: at upper limit of juniper distribution within pinyon-mountain-mahogany woodland, mostly along streams, on east and west sides of the Toiyabe Range. *Juniperus osteosperma* occurs below but is never dominant in pinyon-juniper woodland; *J. scopulorum* may occur nearby (Linsdale et al. 1952), but I have not seen the material (*Linsdale 127*) at AUSTIN, UC, or CAS, where the collection is said to have been deposited. The bark of all these collections is quite red as in var. *australis*, but they have the internodal length and branching angles of var. *occidentalis*. *Charlet 1052* was determined by FC Vasek to be var. *occidentalis*. Robin J. Tausch told me (pers. comm. 1994) that the species is in most of the major drainages in the higher, central, and southern portions of the range (1991–1994).
New records for the range. Not previously known from Lander or Nye County.

****Toquima Range (1)**
elevation record: 7940 ft
1 record in 1 county (**Nye)

Towns/Range/Section	Location—Habitat	County	Elevation	Collector	Herbarium/a
T12N R45E	Moores Creek Summit	NY	7940 ft	*Charlet 1355 & Tausch*	IRSR

Not reported as present in the range by Little (1971).
DAC: var. *occidentalis*, in extensive, open pinyon-juniper woodland; occasional with *J. osteosperma*, which has little glandular expression or is eglandular here (1991–1993).
New record for the species in the range. Not previously known from Nye County.

Trout Creek Mountains (5)
Humboldt County

Reported as present by Little (1971). Species not present in Humboldt County according to Little (1956), Cronquist et al. (1972), and Kartesz (1988).
DAC: I have seen no material in herbaria, nor did I see any conifers on the mountain during my brief visit (7/1993).

****Virginia Mountains (3)**
Washoe County
[Robin Tausch: T25N R20E]

Not reported as present in the range by Little (1971).
Robin J. Tausch (USDA FS, Intermountain Research Station, Reno, pers. comm. 1994) reported the species as occasional (< 1%) in Utah juniper woodland, below Tule Peak.
New report for the species in the range.

Virginia Range (1)
elevation record: 6800 ft
1 record in 1 county (Storey)

Towns/Range/Section	Location—Habitat	County	Elevation	Collector	Herbarium/a
T17N R21E	Geiger Summit	ST	6800 ft	*Peter Raven 21360*	UCR

Not reported as present in the range by Little (1971).
DAC: very rare; I have not seen the species here, but many of the *J. osteosperma* I have seen have some glandular expression (1973–1995).

****Wassuk Range (1)**
elevation record: 8200–8330 ft
3 records in 1 county (Mineral)

Towns/Range/Section	Location—Habitat	County	Elevation	Collector	Herbarium/a
T8N R28E S14	upper Cottonwood Creek, below Mount Grant	MI	8330 ft	*Charlet 907 & Rust*	IRSR
T7N R29E	Lucky Boy Pass (determined DAC)	MI	8200 ft	*Ann Pinzl in 1974*	NSM
T6N R29E S11	ridge N of Buller Mountain	MI	8240 ft	*Charlet 1979*	UCR, IRSR, RENO

Not reported as present in the range by Little (1971).
DAC: *Charlet 907 & Rust* and *Pinzl in 1974* are *J. occidentalis* var. *occidentalis; Charlet 1979* is var. *australis.* The species generally appears above the highest *J. osteosperma* in the range (1990–1995).
New record for the species in the range.

Wellington Hills (1)
elevation record: 5600 ft
1 record in 1 county (Douglas)

Towns/Range/Section	Location—Habitat	County	Elevation	Collector	Herbarium/a
T9N R23E S4	Long Dry Canyon, in pinyon-juniper	DO	5600 ft	*Matt Lavin SW112*	RENO

Reported as present in the range by Little (1971) and Lavin (1981).
DAC: this material is var. *australis.*

White Mountains (4)
Mineral and Esmeralda Counties

Not reported as present in the range by Little (1971).
According to Vasek (1966), var. *australis* is present in the White Mountains of California, and possible hybrids are present on Mustang Mountain, Mineral County, NV.
DAC: one visit to Montgomery Pass did not yield the species, but it may yet be found in the range (12/1993).

OTHER FEATURES IN NEVADA, RECORDS (6)

Humboldt County
elevation record: 6140 ft
1 record

Towns/Range/Section	Location—Habitat	Elevation	Collector	Herbarium/a
T46N R24E S31	Gooch Table, N of Echo Canyon	6140 ft	*Charlet 1856*	UCR, IRSR, RENO

Washoe County
elevation record: 5400–6240 ft
5 records

Towns/Range/Section	Location—Habitat	Elevation	Collector	Herbarium/a
T44N R19E S8 NE/4	W side Long Valley, at Mosquito Valley Ranch	5400 ft	*Charlet 1838*	UCR, IRSR, RENO
T44N R19E S20	Long Valley, W of Alkali Lake	5560 ft	*Charlet 1931*	UCR, IRSR, RENO
T44N R23E S2	Fish Creek Mountain	6240 ft	*Charlet 1855*	UCR, IRSR, RENO, UC, BRY
T46N R23E S7 SE/4	Catnip Mountain, lower slopes	6060 ft	*Charlet 1857*	UCR, IRSR
T46N R18E S20	1 air mile ENE of Barrel Springs	5520 ft	*Charlet 1862*	UCR, IRSR, RENO

[DAC: T434N R18E, T44N R18E, T45N R18E R22E (Bitner Butte), T47N R18E R21E]

Juniperus osteosperma (Torrey) Little

common name: Utah juniper
elevation: 2500–9200 ft (760–2805 m)
zones: Canadian, Arid Transition, Upper Sonoran
533 Records for all 17 Nevada counties
Reports for 178 mountain ranges, Records for 130 of these
New Reports for 33 mountain ranges, New Records for 24 of these

The center of distribution for Utah juniper, formerly known as *Juniperus utahensis* Torrey, is the state of Utah. All the states that border Utah have the species, and only California and Montana have the species but do not share a border with Utah. Utah juniper is a dominant plant throughout much of the Great Basin, and it can be found in the mountains as well as on high plateaus and in valleys.

Utah juniper is the most widespread of all Nevada conifers. It occurs in all 17 counties and is known from 178 of the state's 314 mountain ranges. New reports for the species are made from 33 mountain ranges, 24 of which have records. Little's (1971) distribution map shows the species as absent from Humboldt County, but I present new records for 4 mountain ranges and 1 other feature in the county. Similarly, Little's (1971) map shows Utah juniper as present in only 4 mountain ranges in Pershing County, but I report records for 10 additional mountain ranges in the county.

The 6700-ft (2045-m) elevation range of Utah juniper is the widest of all Nevada conifers. The species generally occurs in Arid Transition and Upper Sonoran woodlands, usually associated with singleleaf pinyon, and typically begins and ends at lower elevations than pinyon. It reaches the Canadian Zone in central Nevada (Toquima Range), where it occurs in mountain brush communities well above pinyon-juniper woodlands, and in northeastern Nevada (Goose Creek Mountains), where it occurs with aspen (*Populus tremuloides*). Utah juniper has a plastic physiognomy in Nevada; mature single-bole trees in excess of 10 m tall and mature, rounded shrubs no taller than 1.5 m commonly occur in the same stand. The species is absent from most of the mountain ranges in northwestern and north-central Nevada.

Adobe Range (1)
elevation record: 5100 ft
2 records in 1 county (Elko)

Towns/Range/Section	Location—Habitat	County	Elevation	Collector	Herbarium/a
T33N R53E S21 W/4	Carlin Canyon, above the Humboldt River	EL	5100 ft	*Charlet 1887*	UCR, IRSR, RENO, UC
T34N R54E	Arid hills 10 mi W of Elko, Humboldt River	EL	—	*Elmer Applegate 8367*	CAS
[DAC: T34N R53E]					

Reported as present in the range by Little (1971).
DAC: dominant in open juniper woodlands (1988–1994).

Anchorite Hills (1)
elevation record: 6320 ft
1 record in 1 county (Mineral)

Towns/Range/Section	Location—Habitat	County	Elevation	Collector	Herbarium/a
T5N R30E S2 W/8	pinyon-dominated pinyon-juniper woodlands	MI	6320 ft	*Charlet 1972*	UCR, IRSR, RENO
[DAC: T5N R30E, T4N R29E R30E, T3N R29E]					

Reported as present in the range by Little (1971).
DAC: in pinyon-dominated woodlands throughout the middle elevations (1993–1994).

Antelope Range (1)
elevation record: 6440–7700 ft
4 records in 1 county (White Pine) and reports in another (Elko)

Towns/Range/Section	Location—Habitat	County	Elevation	Collector	Herbarium/a
T22N R66E S26	4 road miles SE of Stonehouse on road to Tippet Pass	WP	6500 ft	*Ann Pinzl 9340*	UNLV, NSM
T22N R66E S36	Twelvemile Summit	WP	6440 ft	*Charlet 2026*	UCR, IRSR, RENO
T24N R67E S5	Middle Creek	WP	7520 ft	*Charlet 2029*	UCR, IRSR, RENO
T24N R67E	Cliffs N of Antelope Springs	WP	7700 ft	*Cottam 4631*	BRY
[DAC: (WP) T23N R67E, T24N R66E, T25N R67E, T26N R67E]					
[RWR: (EL) T27N R66E R67E, T26N R67E]					

Reported as present in the range by Little (1971).
DAC: in pinyon-juniper woodlands throughout the lower and middle elevations (1994).

Antelope Range (3)
Reports in 2 counties (Nye and Eureka)
[Little: (NY) T15N R51E, T14N R51E, T13N R51E; (EU) T15N R51E]

Reported as present in the range by Little (1971).

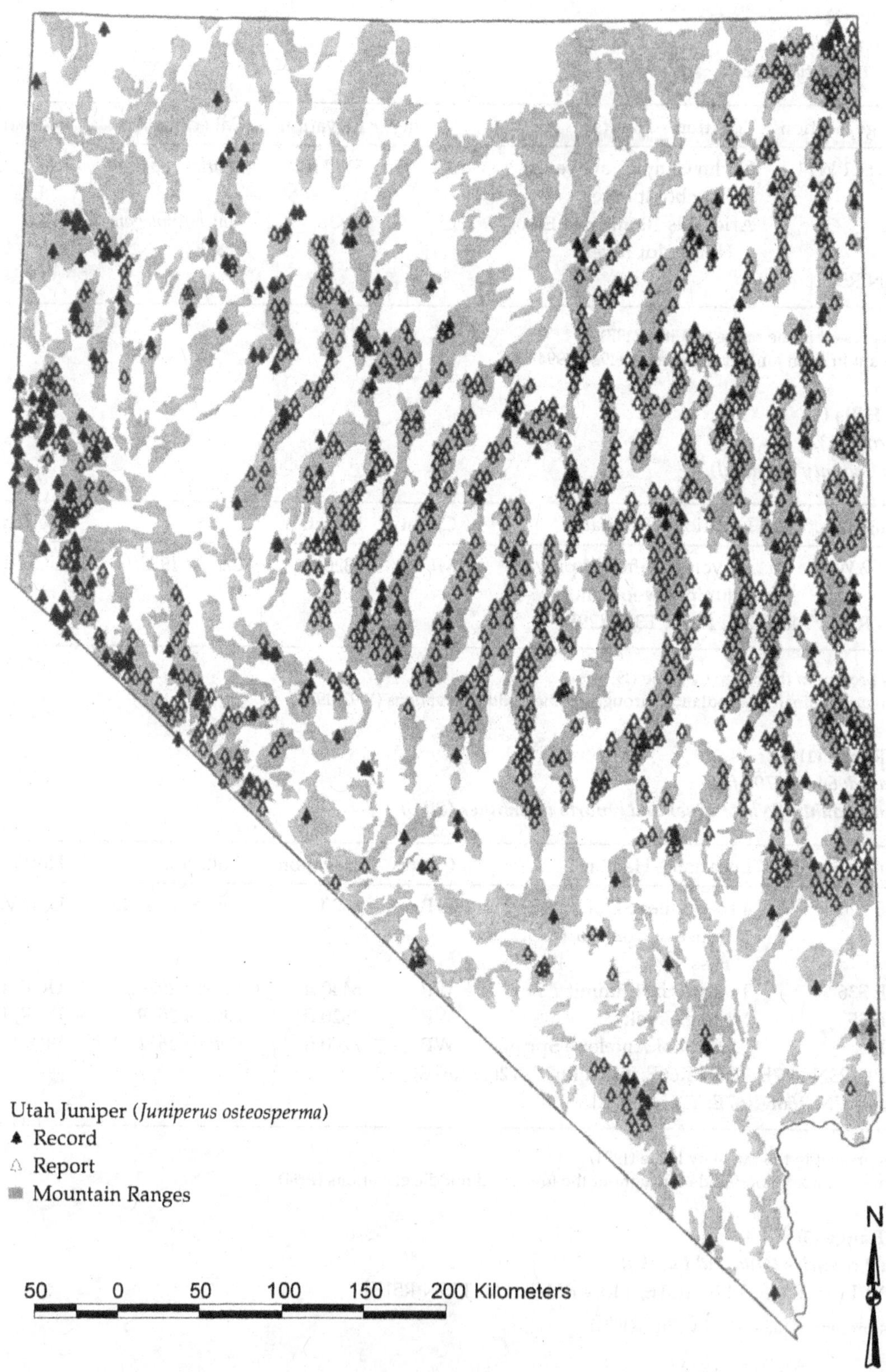
Utah Juniper (*Juniperus osteosperma*)
Record
Report
Mountain Ranges
50 0 50 100 150 200 Kilometers
N

****Augusta Mountains (1)**
elevation record: 5600 ft
1 record in 1 county (Lander) and reports in another (Pershing)

Towns/Range/Section	Location—Habitat	County	Elevation	Collector	Herbarium/a
T25N R40E S8 W/4	lower end juniper woodland	LA	5600 ft	*Charlet 1892*	UCR, IRSR, RENO, UC, BRY
[DAC: (PE) T25N R39E R40E]					

Not reported as present in the range by Little (1971).
DAC: common and widespread throughout the range; forming large juniper woodlands. Pinyon will likely be found with the species from Cain Mountain southward (1993–1994).
New record for the species in the range.

Badger Mountains (5)
Washoe County

Reported as present in the range by Little (1971).
DAC: I have seen no material, and the occurrence of this species was not reported by Vasek (1966). The only juniper species I found on the mountain was *J. occidentalis*; it is highly doubtful that *J. osteosperma* is here (1994).

****Battle Mountain (1)**
elevation record: 5500 ft
1 record in 1 county (Lander)

Towns/Range/Section	Location—Habitat	County	Elevation	Collector	Herbarium/a
T31N R43E S1	Little Cottonwood Canyon, lower end juniper woodland	LA	5500 ft	*Charlet 1886*	UCR, IRSR, RENO

Not reported as present in the range by Little (1971).
DAC: a large juniper woodland occurs in the Little Cottonwood Canyon area. I have seen no other stands on the mountain, but the species is likely to occur in Timber Canyon (1994).
New record for the species in the range.

Belted Range (1)
elevation record: 5800–6725 ft
3 records in 1 county (Nye)

Towns/Range/Section	Location—Habitat	County	Elevation	Collector	Herbarium/a
T8S R52E	Rainier Mesa foothills, below SE end of Mesa, NW Yucca drainage	NY	6500 ft	*J Beatley 3742*	CAS, RENO
T8S R52E	below E rim of Rainier Mesa; lower limit	NY	5800 ft	*J Beatley 4263*	RENO
T8S R52E S18	Gold Meadows, Nevada Test Site	NY	6725 ft	*PJ Leary 3 (4513) et al.*	UNLV

Not reported as present in the range by Little (1971).
Reported as present by Beatley (1976).

Bird Spring Range (1)
elevation record: 4700 ft
1 record in 1 county (Clark)

Towns/Range/Section	Location—Habitat	County	Elevation	Collector	Herbarium/a
T23S R58E S11	———	CL	4700 ft	*Ann Pinzl 2694*	NSM

Not reported as present in the range by Little (1971).

Black Canyon Range (3)
Lincoln County
[Little: T2S R65E]

Reported as present in the range by Little (1971).

Bodie Mountains (2)
Mineral County
[DAC: T6N R26E R27E R28E R29E, T5N R27E R28E R29E, T4N R28E]

Reported as present in the range by Little (1971).
DAC: in extensive pinyon woodlands; the common juniper on the mountain (1991–1995).

Bristol Range (1)
elevation record: 6140 ft
2 records in 1 county (Lincoln)

Towns/Range/Section	Location—Habitat	County	Elevation	Collector	Herbarium/a
T3N R66E	10 mi N of Pioche	LI	—	*V Kay Johnson in 1951*	RENO
T3N R65E S13	Bristol Pass, juniper-dominated pinyon-juniper woodland	LI	6140 ft	*Charlet 1993*	UCR, IRSR, RENO
[DAC: T2N R66E]					

Reported as present in the range by Little (1971).
DAC: abundant in pinyon-juniper woodlands throughout the range (1994).

Broken Hills (3)
Churchill County
[Little: T14N R36E R37E]

Reported as present in the range by Little (1971).

Buckskin Range (2)
Douglas County
[DAC: T14N R23E]

Reported as present in the range by Little (1971).
DAC: a few open stands are scattered at the north end of the range (1993).

****Buffalo Hills (1)**
elevation record: 4050–6020 ft
10 records in 1 county (Washoe)

Towns/Range/Section	Location—Habitat	County	Elevation	Collector	Herbarium/a
T32N R21E S3 W/4	near Smoke Creek playa, w/*Purshia* & *Sarcobatus*	WA	4050 ft	*Charlet 1583*	UCR, IRSR, RENO, UC
T33N R19E S24 SW/4	lower Tule Canyon, lower limit of junipers here	WA	4940 ft	*Charlet 1939*	UCR, IRSR, RENO
T33N R19E S24 NE/4	lower juniper woodland limit	WA	5200 ft	*Charlet 1940*	UCR, IRSR, RENO
T34N R21E S2 SE/4	N of Squaw Valley reservoir	WA	4440 ft	*Charlet 1822*	UCR, IRSR, RENO
T34N R21E S6	———	WA	6020 ft	*Charlet 2079*	UCR, IRSR, RENO
T35N R20E S22	above Lower Coyote Spring	WA	5580 ft	*Charlet 2075*	UCR, IRSR, RENO
T35N R20E S22	Upper Coyote Spring	WA	5620 ft	*Charlet 2076*	UCR, IRSR, RENO
T35N R21E S16 SW/4	lone tree in sagebrush steppe	WA	5920 ft	*Charlet 1900*	UCR, IRSR, RENO, UC
T35N R21E S17 W/2	lone tree in sagebrush steppe; S aspect	WA	5720 ft	*Charlet 1899*	UCR, IRSR, RENO, UC
T35N R21E S30 SW/4	Jones Flat; juniper woodland	WA	5940 ft	*Charlet 2078*	UCR, IRSR, RENO

Not reported for the range by Little (1971).
DAC: present in isolated stands from the desert floor to the lower hills. At higher elevations, the woodlands are more or less continuous and the species is well mixed with *J. occidentalis,* except in the southern part of the range, where the woodlands appear to be exclusively *J. osteosperma* (1993–1994).
New records for the species in the range.

Bull Run Mountains (5)
Elko County

Not reported as present in the range by Little (1971) or Lewis (1975a).
Reported as present in the range by Grayson (1993:26).
DAC: I have not found the species in the range, nor have I seen material in herbaria. However, *J. occidentalis* and *J. scopulorum* are both present. The report of Grayson (1993) may be based on a misidentification of *J. occidentalis* (1990–1994).

Burnt Spring Range (1)
elevation record: 6240 ft
1 record for 1 county (Lincoln)

Towns/Range/Section	Location—Habitat	County	Elevation	Collector	Herbarium/a
T4S R65E	Oak Springs Summit W of Caliente	LI	6240 ft	*Vasek 630716-4*	UCR
[Little: T2S R65E, T3S R65E]					

Reported as present in the range by Little (1971).

Butte Mountains (2)
White Pine County
[RWR: T25N R60E]
[DAC: T18N R59E R60E]
[Little: T24N R60E, T23N R59E, T22N R59E, T21N R59E R60E, T20N R59E R60E, T19N R59E R60E]

Reported as present in the range by Little (1971).
DAC: this is the only juniper species I have found in the Butte Mountains. It occurs in the pinyon-juniper woodlands that are scattered throughout the range (1994).

Cactus Range (1)
elevation record: 6700–7300 ft
2 records in 1 county (Nye)

Towns/Range/Section	Location—Habitat	County	Elevation	Collector	Herbarium/a
T2S R46E	Cactus Range	NY	6700 ft	*P Monnett 968*	CAS
T3S R46E	N Cactus Range, above Cactus Spring	NY	7300 ft	*J Beatley 9588*	RENO

Reported as present in the range by Little (1971) and Beatley (1976).

Candelaria Hills (2)
Mineral County
[Little, RWR: T2N R34E, T4N R35E]

Reported as present in the range by Little (1971).

Carson Range (1)
elevation record: 4800–4880 ft
3 records in 1 county (Washoe)

Towns/Range/Section	Location—Habitat	County	Elevation	Collector	Herbarium/a
T19N R18E	high ground, Hunter Creek Canyon	WA	—	*PB Kennedy 873*	RENO
T19N R19E S19	Juniper Hill, w/*Cercocarpus* in sagebrush steppe	WA	4800 ft	*Charlet 1903 & Buffy Charlet*	UCR, IRSR, RENO, UC
T19N R19E S29	Juniper Hill, in juniper woodland	WA	4880 ft	*Charlet 1913*	UCR, IRSR, RENO

Reported as present in the range by Little (1971).
DAC: very uncommon; the northeastern foothills, where these collections are from, is the only location I know for the species in this mountain range (1971–1995).

Cedar Mountains (1)
elevation record: 5500–7215 ft
11 records in 1 county (Mineral) and reports in another (Esmeralda)

Towns/Range/Section	Location—Habitat	County	Elevation	Collector	Herbarium/a
T8N R37E	Finger Rock Wash, 22 mi S of SR 23	MI	5500 ft	*P & H Vreeland II-1*	RENO
T8N R37E	Cedar Summit	MI	7215 ft	*Tausch 092992-1-10*	IRSR
[Little: (MI) T7N R38E, T6N R38E, (ES) T7N R38E]					

Reported as present in the range by Little (1971).

Cedar Range (1)
elevation record: 5600 ft
1 record in 1 county (Lincoln)

Towns/Range/Section	Location—Habitat	County	Elevation	Collector	Herbarium/a
T1S R69E S15 S16	mesas E of Condor Canyon, 3 mi E of Delmues	LI	5600 ft	*Steve Boyd 820503-5*	UCR
[Little: T1S R70E R71E, T2S R70E R71E, T3S R69E R70E, T4S R68E R69E R70E]					

Reported as present in the range by Little (1971).

Cherry Creek Range (1)
elevation record: 6760 ft
1 record in 1 county (Elko) and reports in another (White Pine)

Towns/Range/Section	Location—Habitat	County	Elevation	Collector	Herbarium/a
T25N R62E S25	Paris Creek	EL	6760 ft	*Charlet 2032*	UCR, IRSR, RENO
[Little: (EL) T29N R62E R63E, T28N R62E R63E, T27N R62E R63E, T26N R62E R63E; (WP) T26N R62E R63E R64E, T25N R63E, T24N R63E]					
[RWR: (WP) T24N R62E]					

Reported as present in the range by Little (1971).
DAC: abundant in the pinyon-juniper woodlands at lower elevations (1994).

Chief Range (3)
Lincoln County
[Little: T2S R66E]

Reported as present in the range by Little (1971).

Clan Alpine Mountains (1)
elevation record: 4600–6000 ft
2 records in 1 county (Churchill)

Towns/Range/Section	Location—Habitat	County	Elevation	Collector	Herbarium/a
T20N R37E	22 mi N of East Gate on Edwards Creek Ranch, 9 mi from Alpine	CH	6000 ft	*RA Allen 399*	RENO
T22N R37E	Hoyt Canyon	CH	4600 ft	*S Williams in 1972*	RENO

[DAC: T22N R38E, T21N R37E R38E, T20N R36E, T19N R36E R37E, T18N R36E]

Reported as present in the range by Little (1971).
DAC: abundant in pinyon-juniper woodlands throughout the middle elevations (1991–1994).

Clover Mountains (1)
elevation record: 5340–5900 ft
2 records in 1 county (Lincoln)

Towns/Range/Section	Location—Habitat	County	Elevation	Collector	Herbarium/a
T5S R71E	Beaver Dam State Park	LI	5340 ft	*Robertson BD25-74*	RENO
T5S R67E S23	Cabin Pines campground	LI	5900 ft	*Ann Pinzl 8625*	RSA, NSM

[Little: T4S R67E R71E, T5S R68E, T6S R67E R68E R69E R70E R71E, T7S R67E R68E R69E R70E, T8S R68E]

Reported as present in the range by Little (1971).

Cortez Mountains (1)
elevation record: 5870 ft
10 records in 1 county (Eureka)

Towns/Range/Section	Location—Habitat	County	Elevation	Collector	Herbarium/a
T26N R48E S6	at pass, N end of Grass Valley	EU	5870 ft	*Charlet 1405–1414 & Tausch*	IRSR

[Little: T27N R48E, T26N R49E]

Reported as present in the range by Little (1971).
DAC: codominant with *Pinus monophylla* in pinyon-juniper woodlands, which are currently restricted to the southern tip of the range (1993).

Curnow Range (1)
elevation record: 6500 ft
2 records in 1 county (Washoe)

Towns/Range/Section	Location—Habitat	County	Elevation	Collector	Herbarium/a
T22N R21E	0.2 mi N of jct Winnemucca Ranch Rd & SR 33 [SR 445], on SR 33 [SR 445]	WA	—	*Vasek 630715-9*	UCR
T21N R21E	1.3 mi WSW of Spanish Spring Peak	WA	6500 ft	*Roy W Saarni 106*	UC

Reported as present in the range by Little (1971).
DAC: abundant throughout the range in juniper woodlands (1988–1994).

Deep Creek Mountains (2)
White Pine County
[DAC: T23N R69E R70E, T22N R69E R70E, T21N R69E R70E]

Reported as present in the range by Little (1971).
DAC: abundant throughout the Nevada portion of the range in pinyon-juniper woodlands (1994).

De Lamar Mountains (1)
elevation record: 4200 ft
2 records in 1 county (Lincoln)

Towns/Range/Section	Location—Habitat	County	Elevation	Collector	Herbarium/a
T8S R65E	30 mi SW of Caliente	LI	—	*sn in 1967*	RENO
T5S R66E	Meadow Valley Wash, 8 mi S of Caliente	LI	4200 ft	*Billings 2201*	RENO
[Little: T4S R65E, T5S R65E, T6S R65E R66E, T7S R65E]					

Reported as present in the range by Little (1971).

Delano Mountains (2)
Elko County
[DAC: T44N R68E R69E, T43N R67E R68E R69E, T42N R68E]

Reported as present in the range by Little (1971).
DAC: forms extensive juniper woodlands in the range (1993).

Desatoya Mountains (1)
elevation record: 7480 ft
2 records in 1 county (Churchill) and reports in another (Lander)

Towns/Range/Section	Location—Habitat	County	Elevation	Collector	Herbarium/a
T16N R38E	1-6 mi W of Carroll Summit	CH	—	*Goodner 671 & Henning*	RENO, UTC
T16N R38E	W of Carroll Summit, Vasek annotation (6/93): "prob. *J. ost.*")	CH	7480 ft	*Charlet 1051*	UCR, IRSR

[DAC: (CH) T19N R38E R39E, T18N R37E R38E, T17N R37E, T16N R37E, T15N R37E; (LA) T19N R39E, T18N R38E R39E, T17N R38E R39E, T16N R38E, T15N R38E]

Reported as present in the range by Little (1971).
DAC: codominant with pinyon in extensive woodlands throughout the range. Individuals at higher elevations (e.g., *Charlet 1051*) are highly glandular (1991–1994).

Desert Creek Mountains (1)
elevation record: 6040 ft
1 record in 1 county (Lyon)

Towns/Range/Section	Location—Habitat	County	Elevation	Collector	Herbarium/a
T9N R24E S27	O'Banion Canyon, in pinyon-dominated pinyon-juniper woodland	LY	6040 ft	*Charlet 1948*	UCR, IRSR, RENO

[DAC: T8N R24E R25E]

Reported as present in the range by Little (1971).
DAC: common at lower elevations, particularly on the eastern slope, in pinyon-dominated pinyon-juniper woodland (1987–1994).

Desert Range (1)
elevation record: 6330 ft
1 record in 1 county (Lincoln)

Towns/Range/Section	Location—Habitat	County	Elevation	Collector	Herbarium/a
T11S R58E	N slope of 1935-m peak	LI	6330 ft	*TL Ackerman*	UNLV

Not reported as present in the range by Little (1971).

Diamond Mountains (1)
elevation record: 5880–7500 ft
3 records in 1 county (Eureka) and reports in another (White Pine)

Towns/Range/Section	Location—Habitat	County	Elevation	Collector	Herbarium/a
T18N R54E S17	2.5 mi E of Pinto Summit (8 mi SE of Eureka)	EU	6900 ft	*RC Holmgren 496*	UTC, BRY
T19N R54E S13	Newark Summit	EU	7500 ft	*P & H Vreeland III-56*	RENO
T26N R54E S24 SE/4	Railroad Pass	EU	5880 ft	*Charlet 2082*	UCR, IRSR, RENO, UC

[DAC: (EU) T25N R54E, T24N R54E, T23N R54E, T22N R54E, T21N R54E, T20N R54E; (WP) T25N R55E, T24N R55E, T21N R55E, T20N R55E, T19N R55E, T18N R54E]

Reported as present in the range by Little (1971).
DAC: abundant in the sporadic pinyon-juniper woodlands that grow in the lower and middle elevations (1994).

Dixie Hills (1)
elevation record: 5640 ft
1 record in 1 county (Elko)

Towns/Range/Section	Location—Habitat	County	Elevation	Collector	Herbarium/a
T31N R55E S3	juniper woodlands	EL	5640 ft	*Charlet 2081*	UCR, IRSR, RENO

[DAC: T30N R54E R55E, T31N R54E R55E]

Reported as present in the range by Little (1971).
DAC: dominant in juniper woodlands throughout the range (1994).

Dogskin Mountains (1)
elevation record: 5400–6725 ft
4 records in 1 county (Washoe)

Towns/Range/Section	Location—Habitat	County	Elevation	Collector	Herbarium/a
T23N R19E S3 NE/4	E of Beddell Spring, S side of mountain	WA	5400 ft	*Charlet 1790 & T Charlet*	UCR, IRSR
T23N R20E S18	SW side of mountain	WA	5480 ft	*Charlet 1772*	UCR, IRSR, RENO
T24N R19E S14	lower McKissick Canyon	WA	5380 ft	*Charlet 1792 & T Charlet*	UCR, IRSR, RENO
T24N R19E	hilltop above McKissick Spring	WA	6725 ft	*Charlet 1787 & T Charlet*	UCR, IRSR, RENO

Reported as present in the range by Little (1971).
DAC: dominant throughout most of the range, particularly in the west, in juniper woodlands. *Juniperus occidentalis* occurs in upper McKissick Canyon (1991–1994).

Dolly Varden Mountains (1)
elevation record: 6230 ft
1 record in 1 county (Elko)

Towns/Range/Section	Location—Habitat	County	Elevation	Collector	Herbarium/a
T29N R66E S35	lower Dolly Varden Canyon, 5 mi from Dolly Varden Spring	EL	6230 ft	*Ann Pinzl 9302*	NSM, UTC
[RWR: T28N R66E]					

Reported as present in the range by Little (1971).

Duck Creek Range (1)
elevation record: 7300 ft
1 record in 1 county (White Pine)

Towns/Range/Section	Location—Habitat	County	Elevation	Collector	Herbarium/a
T15N R65E S5	Elk Flat Campground, in pinyon-juniper woodland	WP	7300 ft	*Charlet 2101 & T Charlet*	UCR, IRSR, RENO
[DAC: T19N R64E, T18N R64E, T17N R64E, T16N R64E]					

Reported as present in the range by Little (1971).
DAC: abundant in pinyon-juniper woodlands in the lower elevations (1994).

East Gate Range (1)
elevation record: 6140 ft
1 record in 1 county (Churchill)

Towns/Range/Section	Location—Habitat	County	Elevation	Collector	Herbarium/a
T16N R36E S24	lower end pinyon-juniper woodland	CH	6140 ft	*Charlet 2051*	UCR, IRSR, RENO
[Little: T15N R36E R37E]					

Reported as present in the range by Little (1971).
DAC: abundant in pinyon-juniper woodlands in the central and southern portions of the range (1991–1994).

East Humboldt Range (1)
elevation record: 6345–7345 ft
4 records in 1 county (Elko)

Towns/Range/Section	Location—Habitat	County	Elevation	Collector	Herbarium/a
T33N R61E S16	Outhouse Draw	EL	7345 ft	*Charlet 1611*	UCR, IRSR, RENO
T34N R60E S6	NW of Secret Pass	EL	6345 ft	*Charlet 1609*	UCR, IRSR, RENO
T34N R60E S22	Secret Pass	EL	6610 ft	*Charlet 1610*	UCR, IRSR, RENO
T36N R61E S2	along road to Angel Lake	EL	6700 ft	*Charlet 2122 & T Charlet*	UCR, IRSR

[DAC: T32N R61E, T34N R61E, T35N R61E, T36N R60E]

Reported as present in the range by Little (1971).
DAC: occasional isolated trees and small stands on the west slope; forms extensive woodlands with pinyon on the east slope (1988–1994).

****East Range (1)**
elevation record: 5300–5440 ft
2 records in 1 county (Pershing)

Towns/Range/Section	Location—Habitat	County	Elevation	Collector	Herbarium/a
T27N R37E S11 E/4	0.5 mi N of McKinney Pass, juniper woodland	PE	5440 ft	*Charlet 1870 & Rust & Sanderson*	UCR, IRSR, RENO, UC
T30N R38E S4 N/4	Table Mountain	PE	5300 ft	*Charlet 1879*	UCR, IRSR, RENO

[DAC: T34N R36E R37E, T33N R36E R37E, T32N R36E R37E, T31N R36E R37E R38E, T30N R36E R37E R38E, T29N R37E R38E, T28N R37E R38E]

Not reported as present in the range by Little (1971).
DAC: juniper woodlands are widespread but not continuous in the range; some large gaps apparently due to recent (< 50 yr) fire (1994).
New records for the species in the range.

Egan Range (1)
elevation record: 6500–8135 ft
3 records in 1 county (White Pine) and reports in 2 others (Lincoln and Nye)

Towns/Range/Section	Location—Habitat	County	Elevation	Collector	Herbarium/a
T12N R63E	Sawmill Canyon	WP	8135 ft	*Richard W. Rust 4*	UCR, RENO
T14N R63E	Ward Mountain	WP	7500 ft	*Dimaggio in 1967*	RENO
T16N R63E	US 50 & US 93, 2 mi S of Ely	WP	6500 ft	*Vasek 630715-9*	UCR

[DAC: (WP) T23N R62E, T22N R62E R63E, T21N R62E R63E, T20N R62E R63E, T19N R62E, T18N R61E R62E R63E, T17N R62E, T16N R62E, T15N R62E R63E, T14N R62E, T13N R62E R63E, T12N R62E R63E R64E, T11N R62E R63E, T10N R62E R63E; (LI) T9N R62E R63E, T8N R62E R63E R64E, T7N R62E R63E, T6N R62E R63E; (NY) T9N R62E, T6N R62E]

Reported as present in the range by Little (1971).
DAC: abundant in pinyon-juniper throughout the lower elevations in the north, and the middle elevations in the south (1994).

Elko Mountains (1)
elevation record: 5500–5800 ft
3 records in 1 county (Elko)

Towns/Range/Section	Location—Habitat	County	Elevation	Collector	Herbarium/a
T34N R55E	3.3 mi S of RR tracks, S edge Elko on SR 46	EL	—	*Vasek 640811-2*	UCR
T34N R55E	5 mi SE of Elko, hillside	EL	5800 ft	*C McMillan 245*	UT
—	Elko	EL	5500 ft	*EVA Murphy 241*	UTC, RENO
[DAC: T33N R55E, T34N R56E]					

Reported as present in the range by Little (1971).
DAC: dominant in open juniper woodlands throughout much of the range (1988–1994).

Ely Springs Range (3)
Lincoln County
[Little: T1N R65E]

Reported as present in the range by Little (1971).

****Eugene Mountains (1)**
elevation record: 4680 ft
1 record in 1 county (Humboldt) and reports in another (Pershing)

Towns/Range/Section	Location—Habitat	County	Elevation	Collector	Herbarium/a
T35N R34E S26	NE end of range, below Blackbird Mine, lower end juniper woods	HU	4680 ft	*Charlet 1891*	UCR, IRSR, RENO, UC
[DAC: (PE) T34N R33E NE/4; T34N R34E N/4]					

Not reported as present in the range by Little (1971).
DAC: there are juniper woodlands at the northern end of the range, from just above the valley floor to near the summit (1994).
New record for the species in the range.

Excelsior Mountains (1)
elevation record: 7200–7380 ft
9 records in 1 county (Mineral)

Towns/Range/Section	Location—Habitat	County	Elevation	Collector	Herbarium/a
T5N R33E S12	above Pepper Springs	MI	7200 ft	*Charlet 1308–1314 & Tausch*	IRSR
T5N R33E S12	above Pepper Springs	MI	7250 ft	*Charlet 1316 & Tausch*	IRSR
T5N R33E S12	above Pepper Springs	MI	7380 ft	*Charlet 1317 & Tausch*	IRSR
[DAC: T6N R34E, T5N R31E R32E R34E, T4N R30E R31E R32E, T3N R30E]					

Reported as present in the range by Little (1971).
DAC: present, but rarely common, with *Pinus monophylla* in pinyon woodlands throughout the range, but particularly in the eastern portion (1993).

Fairview Range (1)
elevation record: 5500 ft
1 record in 1 county (Lincoln)

Towns/Range/Section	Location—Habitat	County	Elevation	Collector	Herbarium/a
T5N R66E S6	Pony Spring, 30 mi N of Pioche	LI	5500 ft	*LaRivers & Hancock 786*	RENO, UTC
[DAC: T5N R65E, T4N R65E, T3N R65E]					

Reported as present in the range by Little (1971).
DAC: codominant with pinyon in extensive pinyon-juniper woodlands (1994).

****Fairview Range (1)**
elevation record: 6620–6680 ft
2 records in 1 county (Churchill)

Towns/Range/Section	Location—Habitat	County	Elevation	Collector	Herbarium/a
T16N R34E S33 SE/4	pinyon-juniper woods	CH	6680 ft	*Charlet 2055*	UCR, IRSR
T16N R34E S34 NW/4	on earthquake fault	CH	6620 ft	*Charlet 2053*	UCR, IRSR, RENO
[DAC: T15N R34E]					

Not reported as present in the range by Little (1971).
DAC: abundant throughout all but the lowest elevations in pinyon-juniper woodland. Most of the individuals I have seen here have visible glands (1994).
New record for the species in the range.

****Fish Creek Mountains (1)**
elevation record: 4940–5525 ft
6 records in 2 counties (Lander and Pershing)

Towns/Range/Section	Location—Habitat	County	Elevation	Collector	Herbarium/a
T26N R40E S9 N/4	Cedar Canyon, in juniper woodland	PE	4940 ft	*Charlet 1885*	UCR, IRSR, RENO
T28N R41E	above Dacies Creek, N aspect	LA	5345 ft	*Charlet 1435–1438 & Tausch*	IRSR
T28N R41E	above Dacies Creek, N aspect	LA	5525 ft	*Charlet 1444 & Tausch*	IRSR
[DAC: (LA) T29N R42E, T28N R42E, T27N R40E R41E, T26N R40E R41E R42E]					

Not reported as present in the range by Little (1971).
DAC: abundant in juniper woodlands, with varying glandular expression; *J. occidentalis* is also present in the range (1993–1994).
New record for the species in the range.

Fish Creek Range (1)
elevation record: 8000 ft
1 record in 1 county (Eureka)

Towns/Range/Section	Location—Habitat	County	Elevation	Collector	Herbarium/a
T18N R53E S15	Prospect Mountain, SE of Prospect Peak, w/*Cercocarpus* and limber pine	EU	8000 ft	*Charlet 2039*	UCR, IRSR, RENO
[DAC: T19N R53E, T18N R53E, T17N R52E R53E, T16N R52E, T15N R52E]					

Reported as present in the range by Little (1971).
DAC: abundant in pinyon-juniper woodlands throughout most of the range, extending well into *Cercocarpus* woodlands at higher elevations (1994).

Flowery Range (1)
elevation record: 6000 ft
2 records in 1 county (Storey)

Towns/Range/Section	Location—Habitat	County	Elevation	Collector	Herbarium/a
T17N R21E S28	Sixmile Canyon, 1 mi E of Virginia City	ST	—	*BO Moore 38 & GE Franklin, Jr*	RENO
T17N R21E S26	3 mi E of Virginia City, Sixmile Canyon	ST	6000 ft	*Roy Allen 145*	RENO
[DAC: T17N R22E, T18N R22E R23E]					

Reported as present in the range by Little (1971).
DAC: common throughout the pinyon-juniper woodlands of the range; dominant at the lower elevations, becoming rare with *Pinus monophylla* in the higher-elevation woodlands (1973–1994).

Fortification Range (1)
elevation record: 6000 ft
1 record in 1 county (Lincoln) and report in another (White Pine)

Towns/Range/Section	Location—Habitat	County	Elevation	Collector	Herbarium/a
T9N R67E S15	Indian Springs	LI	6000 ft	*Ann Pinzl 1845*	NSM
[Little: (LI) T8N R67E]					
[DAC: (WP) T10N R66E]					

Reported as present in the range by Little (1971).
DAC: common in open pinyon-juniper woodlands (1994).

Fort Sage Mountains (1)
elevation record: 5000–5140 ft
2 records in 1 county (Washoe)

Towns/Range/Section	Location—Habitat	County	Elevation	Collector	Herbarium/a
T25N R18E	W of Newcombe Dry Lake, SE of Fort Sage Mountain	WA	5000 ft	*Billings 2092*	RENO
T25N R18E S8 NW/4	N slope of range, near ponderosa pine stand	WA	5140 ft	*Charlet 2158*	UCR, IRSR, RENO

Reported as present in the range by Little (1971).
DAC: abundant in the Nevada portion of the range in the south (where *Billings 2092* was collected), but mostly absent in the north and northeast, except for a small stand of about 30 where *Charlet 2158* was collected (1993–1994).

Fox Range (1)
elevation record: 4420–5100 ft
3 records in 1 county (Washoe)

Towns/Range/Section	Location—Habitat	County	Elevation	Collector	Herbarium/a
T29N R21E	Fox Canyon, Corral Springs	WA	5100 ft	*Ann Pinzl in 1975*	NSM
T30N R21E S33 NW/4	Wild Horse Canyon, near lower juniper limit	WA	4420 ft	*Charlet 1943*	UCR, IRSR, RENO, UC
T30N R21E S34 SW/4	Wild Horse Canyon, along creek	WA	4780 ft	*Charlet 1944*	UCR, IRSR, RENO, UC
[Little: T31N R22E, T30N R22E, T28N R21E R22E]					

Reported as present in the range by Little (1971).
DAC: forms the juniper woodlands that occur throughout most of the middle elevations (1988–1994).

French Mountains (1)
elevation record: 5580 ft
1 record in 1 county (Washoe)

Towns/Range/Section	Location—Habitat	County	Elevation	Collector	Herbarium/a
T22N R19E S6 NE/4	Sand Hills area	WA	5580 ft	*Charlet 1771*	UCR, IRSR, RENO, UC
[DAC: T23N R19E]					

Reported as present in the range by Little (1971).
DAC: the only conifer species present, in open juniper woodlands scattered throughout the range (1980–1994).

****Gabbs Valley Range (1)**
elevation record: 6330–6840 ft
12 records in 1 county (Mineral)

Towns/Range/Section	Location—Habitat	County	Elevation	Collector	Herbarium/a
T9N R34E S23	1.6 km ESE of Cliff Spring, on Cliff Spring Rd (determined Vasek)	MI	6840 ft	*Charlet 713 Rust & Tausch*	UCR, IRSR, RENO
T9N R34E S24	1.9 km ESE of Cliff Spring, on Cliff Spring Rd (determined Vasek)	MI	6805 ft	*Charlet 714 Rust & Tausch*	UCR, IRSR
T9N R34E S17	near Middle Spring, E of road to Rhyolite Pass, SE portion of range	MI	6330 ft	*Charlet 1318–1327 & Tausch*	IRSR
[DAC: T10N R33E]					

Not reported as present in the range by Little (1971).
DAC: abundant throughout the pinyon-juniper woodlands. I have seen no material other than my own; *713* & *714* are from a very shrubby population with a strong glandular expression, gray bark, and very large female cones, characteristics of *J. californica; 1318–1327* are typical *J. osteosperma,* although some individuals are somewhat glandular (1991–1993).
New record for the species in the range.

Gap Mountains (2)
Nye County
[RWR: T5N R62E S7, S8]

Reported as present in the range by Little (1971).

****Goldfield Hills (1)**
elevation record: 5700 ft
1 record in 1 county (Esmeralda)

Towns/Range/Section	Location—Habitat	County	Elevation	Collector	Herbarium/a
T3S R42E S36	4 mi S of Goldfield, W of US 95, rocky	ES	5700 ft	*Charlet 2193*	RENO, UC, DUKE

Not reported as present in the range by Little (1971).
DAC: uncommon; in juniper-*Purshia stansburiana–Grayia spinosa* vegetation fragment within open *Yucca brevifolia* woodlands (1985–1995).
New record for the species in the range.

****Golden Gate Range (2)**
Lincoln and Nye Counties
[DAC: T4N R59E, T1S R58E]

Not reported as present in the range by Little (1971).
DAC: present in open juniper woodlands scattered through the range (1994).
New report for the species in the range.

Goose Creek Mountains (2)
Elko County
[Little: T46N R68E R70E, T45N R68E R69E R70E, T44N R69E R70E, T43N R69E R70E, T43N R69E R70E, T42N R69E R70E, T41N R69E R70E]

Reported as present in the range by Little (1971).
DAC: dominant throughout the range; the only conifer species seen on the mountain, in places growing so close together that there is virtually no understory (1993).

Goshute Mountains (1)
elevation record: 6000–6900 ft
3 records in 1 county (Elko)

Towns/Range/Section	Location—Habitat	County	Elevation	Collector	Herbarium/a
T29N R68E	5.6 mi N of White Horse Pass, ca. 24 mi S of Wendover	EL	6000 ft	*Duane Atwood 15226*	BRY
T29N R68E	on US 50A at White Horse Pass	EL	6025 ft	*Vasek 630715-2*	UCR
T30N R68E S32	———	EL	6900 ft	*Cooper 95 & Holmgren & Lund*	BRY

[Little: T35N R68E, T34N R68E, T33N R68E R69E, T32N R68E R69E, T31N R68E, T30N R69E]

Reported as present in the range by Little (1971).

****Granite Range (2)**
Elko County
[DAC: T44N R64E R65E, T45N R65E]

Not reported as present in the range by Little (1971).
DAC: thick stands occur locally at higher elevations (1993).
New report for the species in the range.

Granite Range (1)
elevation record: 4460–5200 ft
5 records in 1 county (Washoe) and report in another (Pershing)

Towns/Range/Section	Location—Habitat	County	Elevation	Collector	Herbarium/a
T33N R23E S30 W/4	8.5 km S of Granite Peak	WA	4460 ft	*Charlet & Rust 1699*	UCR, IRSR, RENO
T33N R23E	between Granite & Mahogany Peaks, on road at bad creek crossing, 5.2 mi from SR 34	WA	—	*Vasek 640814-2*	UCR
T33N R23E	7.2 mi N of Gerlach on SR 34	WA	—	*Vasek 640913-12*	UCR
T33N R23E S22 NW/4	Granite Basin, upper Bowen Canyon	WA	5120 ft	*Douglas Smith & Tristen Faith in 1994*	UCR
T34N R21E S10	23 mi NW of Gerlach, SW side of Granite Mts on SR 447, S aspect	WA	5200 ft	*RC Holmgren 494*	UTC, BRY
[DAC: (WA) T34N R22E R23E, T33N R22E R23E; (PE) T33N R24E]					

Not reported as present in the range by Little (1971).
DAC: in large, mostly open juniper woodlands, at least at the lower elevations. The woodlands are denser on the northern and eastern exposures. *Juniperus occidentalis* is found on Fox Mountain in the northern part of the range and may yet be found at higher elevations in the central and southern parts (1974–1994).

Grant Range (1)
elevation record: 7200–7440 ft
3 records in 1 county (Nye)

Towns/Range/Section	Location—Habitat	County	Elevation	Collector	Herbarium/a
T4N R57E S19 E/4	between Burnt Canyon and Cherry Creek Summit	NY	7200 ft	*Charlet 2004*	UCR, IRSR, RENO
T5N R58E S15	Scofield Canyon	NY	7440 ft	*Charlet 2015*	UCR, IRSR, RENO
T5N R58E S15	Scofield Canyon	NY	7440 ft	*Charlet 2016*	UCR, IRSR, RENO
[DAC: T10N R58E, T9N R58E R59E, T8N R57E R58E, T7N R58E R59E, T6N R57E R58E R59E, T5N R57E R58E, T4N R58E, T3N R57E]					

Reported as present in the range by Little (1971).
DAC: abundant in pinyon-juniper woodlands throughout the lower and middle elevations (1994).

Grapevine Mountains (1)
elevation record: 7800–8600 ft
2 records in 2 counties (Esmeralda and Nye)

Towns/Range/Section	Location—Habitat	County	Elevation	Collector	Herbarium/a
T11S R44E	Grapevine Mts along ridge, head of Phinney Canyon	NY	7800 ft	*Beatley & Reveal 12315*	CAS, RENO, RSA
T11S R43E S22	———	ES	8600 ft	*K Kingsley, M Kurzius 1139*	RSA
[Little: (ES) T10S R43E; (NY) T10S R43E]					

Reported as present in the range by Little (1971).

Groom Range (1)
elevation record: 6200 ft
1 record in 1 county (Lincoln)

Towns/Range/Section	Location—Habitat	County	Elevation	Collector	Herbarium/a
T6S R55.5 S29	upper Groom Mine Rd	LI	6200 ft	*G Marrs-Smith 22 & J Nachlinger*	UNLV

Not reported as present in the range by Little (1971).

Halfpint Range (3)
Nye County
[Beatley: T9S R53E]

Not reported as present in the range by Little (1971).
Reported as present in the range by Beatley (1976).

Hays Canyon Range (1)
elevation record: 6350 ft
1 record in 1 county (Washoe)

Towns/Range/Section	Location—Habitat	County	Elevation	Collector	Herbarium/a
T43N R18E	4.7 mi E of NV line	WA	6350 ft	*Vasek 640913-8*	UCR

Not reported as present in the range by Little (1971).
DAC: occasional on the eastern slope; with varying glandular expression. Found with the much more common *J. occidentalis* in large juniper woodlands (1993–1994).

****HD Range (2)**
Elko County
[DAC: T43N R65E R66E, T44N R66E, T42N R66E]

Not reported as present in the range by Little (1971).
DAC: forms the juniper woodlands of the range (1993).
New report for the species in the range.

Highland Range (1)
elevation record: 6000 ft
2 records in 1 county (Lincoln)

Towns/Range/Section	Location—Habitat	County	Elevation	Collector	Herbarium/a
T1N R66E	12 mi E of Caliente, on Highland Range divide	LI	6000 ft	*LaRivers 758 & Hancock*	RENO
T1S R66E	Comet Peak, Pioche	LI	–	*ME Jones in 1912*	CAS, RSA
[DAC: T1N R66E, T1S R67E]					

Reported as present in the range by Little (1971).
DAC: abundant in pinyon-juniper–*Quercus gambellii* woodlands throughout the lower and middle elevations (1994).

****Hiko Range (2)**
Lincoln County
[RWR: T3S R61E S17]

Not reported as present in the range by Little (1971).
RWR: only 5 trees found at one spot, at the north end of the range (1993).
New report for the species in the range.

Horse Range (2)
Nye County
[DAC: T11N R59E, T10N R59E R60E, T9N R59E]

Reported as present in the range by Little (1971).
DAC: abundant throughout the range in pinyon-juniper woodlands; *J. occidentalis* is also present, at least in the northwest (1994).

Hot Creek Range (2)
Nye County
[Little: T12N R50E R51E, T11N R50E R51E, T10N R50E R51E, T9N R49E R50E R51E, T8N R49E R50E, T7N R49E R50E, T6N R49E R50E, T5N R49E R49.5E, T4N R49E R49.5E]

Reported as present in the range by Little (1971).
DAC: abundant in juniper-dominated pinyon-juniper and juniper woodlands scattered throughout the range; *J. occidentalis* is also present, but rare (1994).

Humboldt Range (1)
elevation record: 4600–6100 ft
6 records in 1 county (Pershing)

Towns/Range/Section	Location—Habitat	County	Elevation	Collector	Herbarium/a
T26N R34E S5 E/4	extreme S end of range	PE	4600 ft	*Charlet 1869 & Rust & Sanderson*	UCR, IRSR
T28N R34E S4	Limerick Pass (determined Vasek)	PE	6100 ft	*Charlet 780 & Rust*	UCR, IRSR
T31N R33E	"W Humboldt Range" Star Peak, W slope	PE	—	*AA Heller*	CAS
T31N R33E S1 NE/4	USGS Imlay, Humboldt Canyon	PE	5800 ft	*Vreeland & Lugaski XIII-5*	RENO
T31N R34E S7	Lakeview Mine	PE	5850 ft	*M Williams 78-221-2 & Tiehm*	RENO
—	Victory Hwy, along the road between Imlay & Lovelock	PE	—	*Alice Eastwood 140 & JT Howell*	CAS

[DAC: T32N R34E, T30N R33E R34E, T29N R33E, T28N R34E, T27N R34E R35E]

Not reported as present in the range by Little (1971).
DAC: common in juniper woodlands throughout most of the lower and middle elevations, particularly in the southern portion; with *J. occidentalis* in Star Canyon and near Unionville (1991–1994).

Hungry Range (1)
elevation record: 5200 ft
1 record in 1 county (Washoe)

Towns/Range/Section	Location—Habitat	County	Elevation	Collector	Herbarium/a
T21N R19E S2 NW/4	open juniper woods	WA	5200 ft	*Charlet 1773*	UCR, IRSR, RENO, UC

[DAC: T22N R19E R20E, T23N R20E]

Reported as present in the range by Little (1971).
DAC: this species forms scattered, open, limited juniper woodlands (1980–1994).

Hunter Point Mountains (1)
elevation record: 6720 ft
1 record in 1 county (White Pine)

Towns/Range/Section	Location—Habitat	County	Elevation	Collector	Herbarium/a
T25N R62E S28	pinyon-juniper woodland	WP	6720 ft	*Charlet 2118 & T Charlet*	UCR, IRSR, RENO

[DAC: T24N R61E R62E, T23N R61E R62E, T22N R61E R62E, T21N R61E]

Reported as present in the range by Little (1971).
DAC: codominant with pinyon in the widespread pinyon-juniper woodlands (1994).

Huntoon Mountains (2)
Mineral County
[Little, DAC: T1N R32E]
[Little: T2N R30E R31E R32E, T1N R31E]

Reported as present in the range by Little (1971).
DAC: common in pinyon woodlands throughout the range (1994).

****Independence Mountains (1)**
elevation record: 5260 ft
1 record in 1 county (Elko)

Towns/Range/Section	Location—Habitat	County	Elevation	Collector	Herbarium/a
T34N R53E S30	Above Susie Creek, extreme S end of range	EL	5260 ft	*Charlet 1890*	UCR, IRSR, RENO, UC

Not reported as present in the range by Little (1971) or Lewis (1975).
Reported as present by Grayson (1993).
DAC: I have verified the presence of the species only in the extreme southern tip of the range, where small woodland stands occur above Susie Creek; however, this occurrence does not explain Grayson's report (1993:26) that "these have Utah juniper woodlands—the Granite Range . . . and the Independence, Bull Run, and Jarbidge mountains to the east," all ranges described as "north of the Humboldt River." I have not seen any material in herbaria. Grayson's (1993) report for the Independence Mountains is probably the result of the use of the name "Independence Mountains" or "East Independence Mountains" on labels for collections made in the Wood Hills (east of Wells, Elko County), which is southeast of the Humboldt River. I have found *J. osteosperma* in the Independence Mountains north of the Humboldt River only at the Susie Creek location. I have not seen the species in the Bull Run or Jarbidge Mountains. The only other juniper woodlands in the Independence Mountains are along the Owyhee River at the extreme northeastern portion of the range, but the species there is *J. scopulorum. Juniperus occidentalis* grows singly on the periphery of the mountain range to the north, east, and west (1988–1994).
New record for the species in the range.

Irish Mountain (1)
elevation record: 5000–6000 ft
1 record in 1 county (Lincoln)

Towns/Range/Section	Location—Habitat	County	Elevation	Collector	Herbarium/a
T3S R59E	Irish Mountain	LI	5000–6000 ft	*Percy Train 2403*	RENO
[RWR: T3S R58E R59E, T4S R59E]					

Not reported as present in the range by Little (1971).

Jackson Mountains (1)
elevation record: 4130–7660 ft
5 records in 1 county (Humboldt)

Towns/Range/Section	Location—Habitat	County	Elevation	Collector	Herbarium/a
T36N R31E S29	southern Jackson Mts, 8.5 mi WNW of Jungo	HU	5400 ft	*N Holmgren 4015 & P Holmgren*	UTC
T39N R30E S21	Jackson Mts, W slope	HU	4200 ft	*N Holmgren 4052 & P Holmgren*	CAS, UTC
T39N R31E S36 NW/4	near Trout Creek–Clover Creek confluence	HU	4800 ft	*Charlet 1847*	UCR, IRSR, RENO
T40N R31E S31 SE/4	9.5 km N of King Lear Peak, W slope	HU	4130 ft	*Charlet 1716 & Rust*	UCR, IRSR, RENO
T40N R32E S31 SE/4	near Redbird Mine, upper Jackson Creek, at upper juniper limit	HU	7660 ft	*Charlet 1849*	UCR, IRSR, RENO, UC

Not reported as present in the county or range by Little (1971).
DAC: the extensive juniper woodlands in the west-central portion of the range, especially below King Lear Peak and Parrot Peak, are primarily *J. occidentalis; J. osteosperma* is more common in the southern and eastern parts of the range. I collected both species (1993–1994).

Jarbidge Mountains (5)
Elko County

Not reported as present in the range by Loope (1969), Little (1971), or Lewis (1975).
Reported as present in the range by Grayson (1993).
DAC: Grayson's report (1993:26) must be in error. I have seen no material in herbaria and have searched extensively for the species for 7 years, but have not found it here or in any of the mountains in the Jarbidge Mountain complex. The nearest *J. osteosperma* to the Jarbidge Mountains are 50 km to the southeast in the Snake Mountains, Elko County. This report may be a misidentification of *J. occidentalis,* which occurs 0.3 km north of the Idaho border along Pole Creek (*Charlet 1604 & Agee Smith* UCR, IRSR, RENO) (1988–1994).

Junction House Range (1)
elevation record: 4600–5700 ft
2 records in 1 county (Washoe)

Towns/Range/Section	Location—Habitat	County	Elevation	Collector	Herbarium/a
T22N R20E	NE side of Hungry Valley	WA	5700 ft	*K Bradshaw 152*	UC
T22N R20E S13 S/4	NE end of the range, along SR 34	WA	4600 ft	*Charlet 1902*	UCR, IRSR, RENO
[Little, DAC: T21N R20E, T20N R19E]					

Reported as present in the range by Little (1971).
DAC: occurs in open juniper woodlands scattered throughout much of the range (1971–1995).

****Kamma Mountains (1)**
elevation record: 5445 ft
2 records in 1 county (Pershing)

Towns/Range/Section	Location—Habitat	County	Elevation	Collector	Herbarium/a
T34N R29E S13 E/4	0.7 km S of Rosebud Peak	PE	5445 ft	*Charlet 1717 & Rust*	UCR, IRSR, RENO
T34N R29E S13 E/4	0.7 km S of Rosebud Peak	PE	5445 ft	*Charlet 1718 & Rust*	UCR, IRSR, RENO

Not reported as present in the range by Little (1971).
DAC: apparently restricted to a population of some 400 individuals in an area of hydrothermal alteration below Rosebud Peak, at an abandoned mine; with *J. occidentalis* (1993).
New record for the species in the range.

Kawich Range (1)
elevation record: 6240 ft
1 record in 1 county (Nye)

Towns/Range/Section	Location—Habitat	County	Elevation	Collector	Herbarium/a
T4N R49.5E S25 S/4	near Warm Springs Summit	NY	6240 ft	*Charlet 1981*	UCR, IRSR, RENO
[Little: T2N R49E R50E, T1N R49E R50E, T1S R49E R50E, T2S R50E R51E, T3S R50E]					
[RWR: T3N R50E R51E]					

Reported as present in the range by Little (1971) and Beatley (1976).
DAC: abundant, at least in the northern portion of the range, and forming juniper woodlands (1994).

Kern Mountains (1)
elevation record: 6840 ft
1 record in 1 county (White Pine)

Towns/Range/Section	Location—Habitat	County	Elevation	Collector	Herbarium/a
T22N R69E S16	Tippett Canyon, pinyon-juniper woodland	WP	6840 ft	*Charlet 2105 & T Charlet*	UCR, IRSR, RENO
[DAC: T21N R68E R69E R70E, T22N R68E]					

Reported as present in the range by Little (1971).
DAC: codominant with pinyon throughout most of the lower and middle elevations (1994).

Kinsley Mountains (2)
Elko County
[RWR: T26N R67E]

Reported as present in the range by Little (1971).

****Lake Range (1)**
elevation record: 7040–7309 ft
2 records in 1 county (Washoe)

Towns/Range/Section	Location—Habitat	County	Elevation	Collector	Herbarium/a
T27N R22E S26 NE/4	ridgetop W of Tohakum Peak, juniper woodland	WA	7309 ft	*Charlet 1740*	UCR, IRSR, RENO, UC
T27N R22E S26 NE/4	E aspect, W of Tohakum Peak, juniper woodland	WA	7040 ft	*Charlet 1742*	UCR, IRSR, RENO, UC, BRY
[DAC: T27N R23E, T26N R22E]					

Not reported as present in the range by Little (1971).
DAC: there are juniper woodlands near Tohakum Peak; *J. osteosperma* occurs with occasional (< 1%) *J. occidentalis* (1973–1994).
New record for the species in the range.

Las Vegas Range (1)
elevation record: 5800 ft
1 record in 1 county (Clark)

Towns/Range/Section	Location—Habitat	County	Elevation	Collector	Herbarium/a
T16S R61E	Peek-a-boo Canyon	CL	5800 ft	*T Ackerman 3503*	UNLV

Not reported as present in the range by Little (1971).

Leach Range (3)
Elko County
[Little: T38N R67E, T39N R67E, T40N R67E R68E, T41N R68E]

Reported as present in the range by Little (1971).

Leppy Range (3)
Elko County
[McLane: T34N R70E]

Not reported as present in the range by Little (1971).
Reported as present in the range by McLane (1978).

Limestone Hills (3)
Lincoln County
[Little: T7N R69E, T8N R69E]

Reported as present in the range by Little (1971).

Lone Mountain (1)
elevation record: 6560–7350 ft
13 records in 1 county (Esmeralda)

Towns/Range/Section	Location—Habitat	County	Elevation	Collector	Herbarium/a
T2N R40E S16	upper Springdale Canyon	ES	7350 ft	*Richard W. Rust 44*	UCR
T2N R40E	Springdale Canyon, lower end of juniper distribution	ES	6560 ft	*Robin Tausch 1-12 11-17-93*	IRSR

Reported as present in the range by Little (1971).

****Mahogany Hills (1)**
elevation record: 6240 ft
1 record in 1 county (Eureka)

Towns/Range/Section	Location—Habitat	County	Elevation	Collector	Herbarium/a
T20N R52E S28	lower end of juniper distribution	EU	6240 ft	*Charlet 2042*	UCR, IRSR, RENO
[DAC: T19N R52E, T18N R52E]					

Not reported as present in the range by Little (1971).
DAC: codominant with pinyon in pinyon-juniper woodlands throughout the range (1994).

Mahogany Mountains (3)
Lincoln County
[Little: T1N R71E, T2N R71E]

Reported as present in the range by Little (1971).

****Majuba Mountains (1)**
elevation record: 5840–5855 ft
4 records in 1 county (Pershing)

Towns/Range/Section	Location—Habitat	County	Elevation	Collector	Herbarium/a
T33N R30E S25 E/4	Antelope Summit	PE	5855 ft	*Charlet 1722 & Rust*	UCR, IRSR, RENO
T33N R30E S25 E/4	Antelope Summit	PE	5855 ft	*Charlet 1723 & Rust*	UCR, IRSR, RENO
T33N R30E S25 N/4	Antelope Summit	PE	5840 ft	*Charlet 1720 & Rust*	UCR, IRSR, RENO
T33N R30E S25 N/4	Antelope Summit	PE	5840 ft	*Charlet 1721 & Rust*	UCR, IRSR, RENO
[DAC: T33N R30E S36, T33N R31E S31 S32, T32N R31E S5 S6]					

Not reported as present in the range by Little (1971).
DAC: there are rather large juniper woodlands at and near Antelope Summit, at the northern end of the range (10/1993).
New record for the species in the range.

Maverick Springs Range (3)
Elko and White Pine Counties
[Little: (EL) T26N R59E; (WP) T25N R59E, T24N R58E R59E, T23N R58E]

Reported as present in the range by Little (1971).

McCullough Range (1)
elevation record: 5000 ft
1 record in 1 county (Clark)

Towns/Range/Section	Location—Habitat	County	Elevation	Collector	Herbarium/a
T28S R61E	15 mi W of Searchlight	CL	5000 ft	*Mozingo in 1960*	RENO

Not reported as present in the range by Little (1971).

Meadow Valley Mountains (1)
elevation record: 3700 ft
1 record in 1 county (Lincoln)

Towns/Range/Section	Location—Habitat	County	Elevation	Collector	Herbarium/a
T7S R67E	Meadow Valley Wash, mile 16	LI	3700 ft	*ME Jones in 1904*	RSA

Not reported as present in the range by Little (1971).

Medicine Range (1)
elevation record: 6480 ft
1 record in 1 county (Elko)

Towns/Range/Section	Location—Habitat	County	Elevation	Collector	Herbarium/a
T29N R60E S32 S/4	Medicine Spring; at spring in juniper woodland	EL	6480 ft	*Charlet 2120 & T Charlet*	UCR, IRSR, RENO
[DAC: T28N R60E R61E, T27N R60E R61E, T26N R61E]					

Reported as present in the range by Little (1971).
DAC: dominant at lower elevations in juniper woodlands, codominant with pinyon in middle elevations in pinyon-juniper woodlands throughout most of the range (1994).

Monitor Range (1)
elevation record: 6280–8040 ft
11 records in 2 counties (Eureka and Nye)

Towns/Range/Section	Location—Habitat	County	Elevation	Collector	Herbarium/a
T3N R45E S24	W of Woodchopper Canyon	NY	6280 ft	*Charlet 1980*	UCR, IRSR
T15N R49E	Rye Grass Summit	EU	8040 ft	*Charlet 1364–1373 & Tausch*	IRSR

[DAC: (NY) T4N R45E R46E, T5N R45E R46E, T6N R45E R46E R47E, T7N R36E R46E R47E, T8N R46E R47E, T9N R47E R48E, T10N R47E R48E, T11N R47E R48E, T12N R47E R48E R49E, T13N R48E R49E, T14N R48E R49E, T15N R48E; (EU) T16N R48E R49E, T17N R48E R49E, T18N R48E R49E]

Reported as present in the range by Little (1971).
DAC: in large pinyon-juniper woodlands throughout the middle elevations; the common juniper of the range; *J. scopulorum* is rare at higher elevations, and *J. occidentalis* may yet be found here (1994).

****Monte Cristo Range (1)**
elevation record: 6070 ft
1 record in 1 county (Esmeralda)

Towns/Range/Section	Location—Habitat	County	Elevation	Collector	Herbarium/a
T4N R38.5E	——	ES	6070 ft	*Richard W. Rust 43*	UCR

Not reported as present in the range by Little (1971).
DAC: I have seen no material from the Monte Cristo Range other than *Rust* 43.
New record for the species in the range.

Mormon Mountains (1)
elevation record: 4800 ft
1 record in 1 county (Lincoln)

Towns/Range/Section	Location—Habitat	County	Elevation	Collector	Herbarium/a
T12S R68E	S end of Mormon Mts	LI	4800 ft	*WE Niles 3674*	UNLV

Not reported as present in the range by Little (1971).

****Mountain Boy Range (1)**
elevation record: 6160 ft
1 record in 1 county (Eureka)

Towns/Range/Section	Location—Habitat	County	Elevation	Collector	Herbarium/a
T20N R52E S26 SW/4	low end of juniper distribution	EU	6160 ft	*Charlet 2043*	UCR, IRSR, RENO
[DAC: T19N R52E R53E]					

Not reported as present in the range by Little (1971).
DAC: abundant in pinyon-juniper woodlands throughout the range (1994).
New record for the species in the range.

Needle Range (1)
elevation record: 6625 ft
10 records in 1 county (Nye)

Towns/Range/Section	Location—Habitat	County	Elevation	Collector	Herbarium/a
T10N R52E	——	NY	6625 ft	*Robin Tausch 073193-1-10*	IRSR
[Little: T9N R52E]					

Reported as present in the range by Little (1971).

Newberry Mountains (1)
elevation record: 2500–5500 ft
4 records in 1 county (Clark)

Towns/Range/Section	Location—Habitat	County	Elevation	Collector	Herbarium/a
—	Newberry Mts	CL	3000 ft	*Beley-Larson 74-671*	CAS
—	"Dead Mts," Davis Dam Rd	CL	2500 ft	*Billings 1783*	RENO
T31S R65E	Summit of Spirit Mountain	CL	5500 ft	*JS Holland 2583 & WE Niles*	UNLV
T31S R65E	Summit of Spirit Mountain	CL	5500 ft	*JS Holland 2588 & WE Niles*	BRY

Not reported as present in the range by Little (1971).
DAC: this is the only mountain range in Nevada where *J. californica* is known to occur. *Billings 1783* is the low-elevation record for *J. osteosperma* in Nevada.

New Pass Range (1)
elevation record: 6380 ft
1 record in 1 county (Churchill) and 1 report in another (Lander)

Towns/Range/Section	Location—Habitat	County	Elevation	Collector	Herbarium/a
T20N R40E S29	lower end of pinyon-juniper woodlands	CH	6380 ft	*Charlet 2049*	UCR, IRSR, RENO

[DAC: (LA) T20N R40E R41E, T21N R40E, T22N R40E; (CH) T22N R40E, T21N R40E]

Reported as present in the range by Little (1971).
DAC: abundant in pinyon-juniper woodlands (1992–1994).

New York Mountains (1)
elevation record: 5100 ft
1 record in 1 county (Clark)

Towns/Range/Section	Location—Habitat	County	Elevation	Collector	Herbarium/a
T28S R61E	Crescent Peak, turquoise mine	CL	5100 ft	*Percy Train 1467*	UC, RENO, UTC

Reported as present in the range by Little (1971).

Nightingill Mountains (3)
Pershing County
[Little: T25N R24E]

Reported as present in the range by Little (1971).
DAC: I have not found any junipers on the mountain (1991–1994).

****North Pahroc Range (2)**
Lincoln County
[RWR: T1N [illegible] T1S R63E, T2S R63E, T3S R63E]

Not reported as present in the range by Little (1971).
New report for the species in the range.

Pah Rah Range (1)
elevation record: 4500–6260 ft
5 records in 1 county (Washoe)

Towns/Range/Section	Location—Habitat	County	Elevation	Collector	Herbarium/a
T21N R23E S2 N/4	Fort Defiance Creek	WA	4720 ft	*Charlet 1795 & T Charlet*	UCR, IRSR, RENO
T21N R23E S30 N/4	upper Olinghouse Canyon	WA	6260 ft	*Charlet 1794 & T Charlet*	UCR, IRSR, RENO, UC
T21N R23E	2 mi S of Olinghouse	WA	5500 ft	*KC Williams 55*	RENO
T22N R22E S20	below Chimney Spring	WA	4960 ft	*Charlet 1796 & T Charlet*	UCR, IRSR, RENO, UC
T23N R22E	1 mi S of Pyramid Lake; NW foothills of Pah Rah Mts, 300 yd up canyon	WA	4500 ft	*Billings 2040*	RENO
[DAC: T22N R23E, T21N R22E]					

Reported as present in the range by Little (1971).
DAC: occasional in juniper woodland pockets at middle elevations on the eastern slope, common at middle elevations on the western slope (1975–1994).

****Pahranagat Range (2)**
Lincoln County
[RWR: T4S R58E R59E]

Not reported as present in the range by Little (1971).
New report for the species in the range.

****Palmetto Mountains (2)**
Esmeralda County
[RWR: T5S R39E]

Not reported as present in the range by Little (1971).
New report for the species in the range.

Pancake Range (1)
elevation record: 6000 ft
1 record in 1 county (Nye) and reports in another (White Pine)

Towns/Range/Section	Location—Habitat	County	Elevation	Collector	Herbarium/a
T7N R51E S36	SW of Iceberg Spring, W slope Palisade Mesa	NY	6000 ft	*Charlet 1983*	UCR, IRSR, RENO
[Little: (NY) T4N R53E, T7N R52E, T9N R55E, T10N R54E R55E R56E, T11N R55E, T12N R55E, T13N R55E, T14N R53E R54E R55E; (WP) T15N R54E, T16N R55E R56E, T17N R55E R56E, T18N R56E]					

Reported as present in the range by Little (1971).

Papoose Range (1)
elevation record: 5200–5800 ft
1 record in 1 county (Lincoln)

Towns/Range/Section	Location—Habitat	County	Elevation	Collector	Herbarium/a
T9S R55E	large white mountain N of Papoose Lake Rd, W of playa	LI	5200–5800 ft	*Beatley & Reveal 12409*	RENO

Not reported as present in the range by Little (1971).

Paradise Range (1)
1 record in 1 county (Nye)

Towns/Range/Section	Location—Habitat	County	Elevation	Collector	Herbarium/a
T14N R37E	2 mi NW of Penelas Mill, 15 mi NW of Ione	NY	—	*KH Beach 1043*	UC, UT
[DAC: T11N R36E R37E, T12N R36E R37E, T13N R37E]					

Reported as present in the range by Little (1971).
DAC: common throughout the range in pinyon-juniper woodlands; with *J. occidentalis* at higher elevations (1991–1993).

Park Range (3)
Nye County
[Little: T12N R52E, T11N R52E]

Reported as present in the range by Little (1971).

Peavine Mountain (1)
elevation record: 6580 ft
2 records in 1 county (Washoe)

Towns/Range/Section	Location—Habitat	County	Elevation	Collector	Herbarium/a
T20N R18E S20	W side of mountain	WA	6580 ft	*M Williams 82-127-1 & Stebbins*	RENO
T19N R19E	0.5 mi W of Highland Reservoir	WA	—	*Billings 1184*	RENO

Reported as present in the range by Little (1971).
DAC: reported as rare by Williams et al. (1992). I have not seen *J. osteosperma* on the mountain (1971–1995).

Pequop Mountains (1)
elevation record: 6000–6200 ft
3 records in 1 county (Elko)

Towns/Range/Section	Location—Habitat	County	Elevation	Collector	Herbarium/a
T30N R63E	Spruce Mountain, N to Warm Springs Ranch along Ely-Wells Hwy	EL	—	*Percy Train 965*	RENO, UTC
T37N R65E	Pequop Summit, 35 mi W of Wendover	EL	6200 ft	*CL Porter & MW Porter 7694*	CAS, RSA
—	Pequop Mts	EL	6000 ft	*JH Thomas 2617*	CAS
[Little: T38N R65E R66E, T37N R66E, T36N R65E R66E, T35N R65E R66E, T34N R63E R65E, T33N R63E R64E R65E, T32N R65E, T31N R63E R64E R65E]					

Reported as present in the range by Little (1971).

Petersen Mountain (1)
elevation record: 4980–6000 ft
4 records in 1 county (Washoe)

Towns/Range/Section	Location—Habitat	County	Elevation	Collector	Herbarium/a
T21N R18E	1.75 mi SW of Cold Springs	WA	6000 ft	*HC Lee 684*	UC
T21N R18E S21	NV side of NV-CA border, along US 395	WA	4980 ft	*Charlet 1782 & T Charlet*	UCR, IRSR, RENO, UC
T22N R18E S2	6 mi NE of Cold Springs	WA	6000 ft	*CA Graham 46*	UC
T22N R18E	8 mi NE of Black Springs on Red Rock Rd	WA	—	*J Henrichs 401*	RENO
[Little: T23N R18E]					

Reported as present in the range by Little (1971).
DAC: in juniper woodlands, which in Nevada are restricted to the south side of the range. These stands survived the June 1994 fire (1988–1994).

Pilot Mountains (1)
elevation record: 6725 ft
10 records in 1 county (Mineral)

Towns/Range/Section	Location—Habitat	County	Elevation	Collector	Herbarium/a
T7N R37E	Graham Spring	MI	6725 ft	*Robin Tausch 093093-1-10*	IRSR
[RWR: T7N R36E, T6N R36E]					

Reported as present in the range by Little (1971).

Pilot Range (1)
elevation record: 4300–6000 ft
3 records in 1 county (Elko)

Towns/Range/Section	Location—Habitat	County	Elevation	Collector	Herbarium/a
T36N R70E	Pilot Peak, road from McCuiston Ranch to Debbs Creek Canyon	EL	6000 ft	*Percy Train 3777*	RENO
T36N R70E	near Danger Cave, E of Pilot Peak, Bonneville level, dry slopes	EL	4300 ft	*WP Cottam 13820*	UT
T36N R70E	canyon on E side of Pilot Mountain	EL	5200 ft	*WP Cottam 12696*	UT
[Little: T37N R70E]					

Reported as present in the range by Little (1971).

****Pine Forest Range (1)**
elevation record: 6250–6500 ft
5 records in 1 county (Humboldt)

Towns/Range/Section	Location—Habitat	County	Elevation	Collector	Herbarium/a
T43N R30E S32	N of Sentinel Peak (determined Vasek)	HU	6500 ft	*Charlet 1049*	UCR
T43N R30E S32	N of Sentinel Peak	HU	6345 ft	*Charlet 1501*	IRSR
T43N R30E S32	N of Sentinel Peak	HU	6345 ft	*Charlet 1502*	IRSR
T43N R30E S32	N of Sentinel Peak	HU	6315 ft	*Charlet 1504*	IRSR
T43N R30E S32	N of Sentinel Peak	HU	6250 ft	*Charlet 1507*	IRSR

Not reported as present in the range by Little (1971). Grayson (1993) states that *Juniperus* is absent.
DAC: the only juniper stand known in this range is north of Sentinel Peak, with *J. occidentalis,* in a woodland fragment containing about 1000–1500 reproductive individuals (1990–1994).
New record for the species in the range.

Pine Grove Hills (1)
elevation record: 6400–8240 ft
8 records in 1 county (Lyon)

Towns/Range/Section	Location—Habitat	County	Elevation	Collector	Herbarium/a
T8N R25E S9	lower Nye Canyon, pinyon woodland	LY	6420 ft	*Charlet 1949*	UCR, IRSR, RENO
T8N R25E S2	upper Nye Canyon, meadow edge	LY	7300 ft	*Charlet 1950*	UCR, IRSR, RENO
T8N R25E S2	upper Nye Canyon, meadow edge	LY	7340 ft	*Charlet 1951*	UCR, IRSR, RENO
T8N R26E S7	N of Bald Mountain, pinyon-juniper upper limit	LY	7900 ft	*Charlet 1953*	UCR, IRSR, RENO
T8N R26E S7	N of Bald Mountain	LY	7880 ft	*Charlet 1954*	UCR, IRSR, RENO
T9N R26E S18	between Dead Ox Spring and Pine Grove Summit	LY	7980 ft	*Charlet 1955*	UCR, IRSR, RENO
T9N R26E S18	between Pine Grove Summit and Rockland Mine	LY	8240 ft	*Charlet 1956*	UCR, IRSR, RENO
—	Pine Grove Hills	LY	6400 ft	*Ann Pinzl 1017*	NSM
[DAC: T10N R25E, T7N R24E R25E]					

Reported as present in the range by Little (1971).
DAC: common in extensive pinyon-dominated pinyon-juniper woodlands (1989–1995).

Pine Nut Mountains (1)
elevation record: 4800–7400 ft
20 records in 3 counties (Carson City, Douglas, and Lyon)

Towns/Range/Section	Location—Habitat	County	Elevation	Collector	Herbarium/a
T16N R21E S36	2.5 mi SSE of Dayton	LY	4900 ft	*V Stackhouse 22109*	UC
T15N R22E	just S of Lyon Peak Pass [DAC: has glands & exudate]	LY	7400 ft	*J Darling 14*	RENO
T15N R22E	1.5 mi SE of Prison Hill	CC	4800 ft	*Don Tillotson 264*	UC
–	Empire City	CC	–	*ME Jones in 1904*	RSA
T14N R22E	in pinyon-juniper	DO	7400 ft	*M Williams 78-194-2 & Tiehm & Wise*	RENO
T14N R22E	0.4 mi NNE of Mineral Peak, SE of Carson City in pinyon-juniper	DO	6700 ft	*Dan Tillotson 276*	UC
T14N R22E	Sunrise Pass	DO	6725 ft	*Charlet 1273 & Tausch*	IRSR
T14N R22E	Sunrise Pass	DO	6725 ft	*Charlet 1275–1282 & Tausch*	IRSR
T13N R22E	near summit of Como Peak	DO	–	*Vasek 640814-3*	UCR
T13N R23E	Como Rd, 5 mi below Como summit	DO	–	*Vasek 640814-2*	UCR
T12N R21E S16	———	DO	5400 ft	*M Williams 82-171-5 & Tiehm*	RENO
T10N R22E	12.4 mi N of Yerington Rd jct on US 395 S of Gardnerville	DO	–	*Vasek 650901-2*	UCR
–	Pine Nut Range	DO	–	*M Williams 82-171-5 & A Tiehm*	CAS

[DAC: (CC) T14N R21E, T15N R20E R21E; (DO) T14N R21E, T13N R22E, T12N R23E, T11N R21E R22E R23E]

Reported as present in the range by Little (1971).
DAC: often codominant with pinyon, but usually subordinate, in vast pinyon-juniper woodlands (1988–1994).

Piñon Range (2)
Elko County
[DAC: T32N R53E R54E]

Reported as present in the range by Little (1971).
DAC: in large pinyon-juniper woodlands (1991–1994).

Pioche Hills (1)
elevation record: 5700 ft
1 record in 1 county (Lincoln)

Towns/Range/Section	Location—Habitat	County	Elevation	Collector	Herbarium/a
T1N R67E [DAC: T1S R68E]	2 mi S of Pioche	LI	5700 ft	*Billings 2200*	RENO

Reported as present in the range by Little (1971).
DAC: abundant in pinyon-juniper-oak (*Quercus gambellii*) and pinyon-juniper woodlands; throughout the range (1994).

Quinn Canyon Range (1)
elevation record: 7000 ft
1 record in 1 county (Nye) and reports in another (Lincoln)

Towns/Range/Section	Location—Habitat	County	Elevation	Collector	Herbarium/a
T3N R57E S6	Cherry Creek Campground	NY	7000 ft	*Ann Pinzl 1898*	NSM
[DAC: (NY) T5N R56E, T4N R55E R56E, T3N R54E R55E, T1N R54E; (LI) T2N R54E R55E R56E, T1N R54E R56E]					

Reported as present in the range by Little (1971).
DAC: *Pinzl 1898* is rather glandular, with thinner than normal branchlets. Abundant in pinyon-juniper woodlands throughout most of the lower and middle elevations (1994).

Reveille Range (3)
Nye County
[Little: T1N R52E, T2N R52E, T3N R51.5E R52E]

Reported as present in the range by Little (1971).

Roberts Mountains (1)
elevation record: 6560 ft
9 records in 1 county (Eureka)

Towns/Range/Section	Location—Habitat	County	Elevation	Collector	Herbarium/a
T23N R49E	Tonkin Summit	EU	6560 ft	*Charlet 1384–1389 & Tausch*	IRSR
T23N R49E	Tonkin Summit	EU	6560 ft	*Charlet 1391–1393 & Tausch*	IRSR
[DAC: T24N R50E R51E, T23N R50E R51E, T22N R49E R50E R51E R52E, T21N R51E]					

Reported as present in the range by Little (1971).
DAC: the common juniper in extensive pinyon-juniper woodlands; *J. scopulorum, J. occidentalis,* and *J. communis* are all present, but rare (1993–1994).

Royston Hills (3)
Nye County
[Little: T7N R38E R39E]

Reported as present in the range by Little (1971).

Ruby Mountains (1)
elevation record: 5600–7000 ft
5 records in 1 county (Elko) and reports for another (White Pine)

Towns/Range/Section	Location—Habitat	County	Elevation	Collector	Herbarium/a
T26N R57E S11	12.7 mi S of Harrison Pass Rd	EL	7000 ft	*Williams 82-139-4 & Tiehm*	RENO
T27N R57E S31	Ruby Lake campground	EL	6150 ft	*Ann Pinzl 6808*	NSM
T28N R57E	E slope, Harrison Pass	EL	6500 ft	*AA Heller 9542*	RENO
T28N R58E S7	Longhair Smith Canyon	EL	6760 ft	*Charlet 2121 & T Charlet*	UCR, IRSR, RENO
T31N R56E	Boyd Ranch NW of Lee	EL	5600 ft	*DJ Vanderwal 244*	RSA

[DAC: (WP) T26N R56E R57E, T25N R56E R57E, T24N R56E R57E, T23N R57E, T22N R56E R57E, T20N R56E R57E, T21N R57E; (EL) T32N R59E, T31N R57E R59E, T30N R57E R58E]

Reported as present in the range by Little (1971).
DAC: common only in parts of the eastern slope and in the south, usually with pinyon (1988–1994).

Sahwave Mountains (1)
elevation record: 5700 ft
1 record in 1 county (Pershing)

Towns/Range/Section	Location—Habitat	County	Elevation	Collector	Herbarium/a
T28N R26E S32	Juniper Spring; juniper woodlands	PE	5700 ft	*Charlet 2070*	UCR, IRSR, RENO, UC

[DAC: T27N R26E]

Reported as present in the range by Little (1971).
DAC: limited juniper woodlands are present, particularly on the western slope. There are no junipers at Juniper Pass, at the extreme north end of the range (1994).

San Antonio Mountains (3)
Nye County
[Little: T6N R42E]

Reported as present in the range by Little (1971).

Schell Creek Range (1)
elevation record: 6000–7700 ft
7 records in 2 counties (White Pine and Lincoln)

Towns/Range/Section	Location—Habitat	County	Elevation	Collector	Herbarium/a
T4N R63E S3	Silver King Pass	LI	6320 ft	*Charlet 1991*	UCR, IRSR, RENO
T9N R65E	Patterson Pass Rd, 1-2 mi E [W] of US 93	LI	—	*George T Austin in 1979*	BRY
T14N R65E S35	Connors Pass	WP	7700 ft	*Olga Reifschneider in 1950*	RENO
T14N R65E S35	Connors Pass	WP	7700 ft	*Billings 2195*	RENO
T17N R66E	Taft Creek, along stream	WP	7300 ft	*W Sandow in 1953*	RENO
T17N R66E S14 SW/4	S Fork Taft Creek	WP	6500 ft	*Charlet 1997*	UCR, IRSR, RENO
T19N R66E	1 mi from Spring Valley Rd, Meadow Canyon	WP	6000 ft	*Irwin Ting in 1959*	RENO

[DAC: (WP) T25N R65E R66E, T24N R64E R65E R66E, T23N R64E R65E, T22N R64E R65E, T21N R64E R65E, T20N R64E R65E R66E, T19N R64E, T18N R65E R66E, T16N R65E R66E, T15N R65E R66E, T14N R66E, T13N R65E R66E, T12N R64E R65E R66E, T11N R65E R66E, T10N R64E R65E; (LI) T9N R64E R65E, T8N R64E R65E, T7N R64E R65E, T6N R64E R65E, T5N R63E R64E, T4N R63E]

Reported as present in the range by Little (1971).
DAC: widespread throughout the length of the range, forming woodlands with pinyon at middle and upper elevations in the south, and lower and middle elevations in the north (1994).

****Seaman Range (2)**
Lincoln County
[RWR: T1N R62E, T2N R62E]

Not reported as present in the range by Little (1971).
New report for the species in the range.

Selenite Range (1)
elevation record: 4680–5580 ft
2 records in 1 county (Pershing)

Towns/Range/Section	Location—Habitat	County	Elevation	Collector	Herbarium/a
T30N R23.5E S6	below Luxor Peak, W slope, in small juniper woodland	PE	4680 ft	*Charlet 1901*	UCR, IRSR, RENO, UC
T30N R24E S32	along creek	PE	5580 ft	*Charlet 2071*	UCR, IRSR

[DAC: T31N R24E, T32N R24E, T30N R24E]

Reported as present in the range by Little (1971).
DAC: in juniper woodlands at lower elevations; *J. occidentalis* is present with *J. osteosperma* in woodlands at higher elevations (1993–1994).

Seven Lakes Mountain (1)
elevation record: 4520 ft
1 record in 1 county (Washoe)

Towns/Range/Section	Location—Habitat	County	Elevation	Collector	Herbarium/a
T24N R18E S32 NE/4	near CA-NV border, near altered area	WA	4520 ft	*Charlet 1770*	UCR, IRSR, RENO, UC
[DAC: T24N R18E]					

Reported as present in the range by Little (1971).
DAC: abundant in juniper woodlands throughout the range; the only juniper species I found on the mountain (1985–1994).

****Seven Troughs Range (1)**
elevation record: 4900–5740 ft
2 records in 1 county (Pershing)

Towns/Range/Section	Location—Habitat	County	Elevation	Collector	Herbarium/a
T30N R28E S26 N/4	upper juniper woods	PE	5740 ft	*Charlet 1725 & Rust*	UCR, IRSR, RENO
T30N R28E S34 S/8	low end of the juniper distribution	PE	4900 ft	*Charlet 2069*	UCR, IRSR, RENO
[DAC: T31N R28E R29E]					

Not reported as present in the range by Little (1971).
DAC: common in juniper woodlands at middle elevations. Junipers are more common toward the north end of the range. I have seen no material in herbaria (1993–1994).
New record for the species in the range.

Sheep Range (1)
elevation record: 6200–8040 ft
3 records in 1 county (Clark)

Towns/Range/Section	Location—Habitat	County	Elevation	Collector	Herbarium/a
T15S R60E S13	Wiregrass Spring	CL	8040 ft	*Richard W. Rust 53*	UCR
T16S R60E S12	Cow Camp Spring	CL	6200 ft	*AM Solomon in 1972*	UNLV
—	Sheep Range	CL	6400 ft	*DA Davidson 218 & SE Meyer*	UNLV
[Little: T13S R61E]					

Reported as present in the range by Little (1971).

****Shoshone Mountain (3).**
Nye County
[Paul Tueller: T11S R51E, T12S R50E R51E]

Not reported as present in the range by Little (1971).
Reported to me as present in the range by Paul Tueller (Dept. of Environmental and Resource Sciences), Univ. Nevada, pers. comm. 1994).
DAC: woodlands are clearly visible on Landsat satellite images of areas throughout the upper elevations, although few botanists have visited this restricted-access range. Tueller has and describes the *J. osteosperma* here as codominant with *Pinus monophylla.* Although Tueller has known of this location for many years, the species has never appeared on a distribution map or in a published report. New report for the species in the range.

Shoshone Mountains (1)
elevation record: 7675–7715 ft
10 records in 1 county (Nye) and reports in another (Lander)

Towns/Range/Section	Location—Habitat	County	Elevation	Collector	Herbarium/a
T12N R40E	Grantsville Summit	NY	7715 ft	*Charlet 1116*	UCR, IRSR
T11N R40E	Indian Valley	NY	7675 ft	*Charlet 1332–1340 & Tausch*	IRSR

[DAC: (NY) T9N R39E, T10N R39E R40E, T11N R39E, T12N R39E, T13N R39E R40E, T14N R39E R40E; (LA) T23N R42E R43E, T22N R42E R43E, T21N R42E, T17N R41E, T16N R39E R40E R41E, T15N R39E R40E R41E, T18N R41E, T19N R41E]

Reported as present in the range by Little (1971).
DAC: *J. osteosperma* is the only juniper species in the extensive pinyon-juniper woodlands (1993).

Shoshone Range (1)
elevation record: 5590–6215 ft
20 records in 1 county (Lander)

Towns/Range/Section	Location—Habitat	County	Elevation	Collector	Herbarium/a
T29N R45E	Mill Creek, below Mount Lewis	LA	5890 ft	*Charlet 1425–1433 & Tausch*	IRSR
T29N R45E	Mill Creek picnic area, 30 mi SE of Battle Mountain	LA	5590 ft	*B Ross in 1959*	RENO
T25N R44E S17	Horseshoe Basin	LA	6215 ft	*Charlet 1415–1424 & Tausch*	IRSR

[DAC: T29N R44E, T28N R44E R45E, T27N R44E, T25N R43E]

Reported as present in the range by Little (1971).
DAC: common in restricted pinyon-juniper woodlands (1993–1994).

Silver Peak Range (1)
elevation record: 6500 ft
1 record in 1 county (Esmeralda)

Towns/Range/Section	Location—Habitat	County	Elevation	Collector	Herbarium/a
T2S R37E	Silver Peak Range, above Cave Spring Canyon	ES	6500 ft	*WA Archer 7276*	CAS, RENO

Not reported as present in the range by Little (1971).
DAC: uncommon in extensive pinyon woodlands.
RWR: about 99% pinyon, 1% *J. osteosperma.*

Simpson Park Range (1)
elevation record: 6115–6820 ft
12 records in 2 counties (Eureka and Lander)

Towns/Range/Section	Location—Habitat	County	Elevation	Collector	Herbarium/a
T19N R46E S36	1.7 mi above Hickison Summit	LA	6820 ft	*Robertson NS 116-74*	RENO
T24N R48E	McClusky Canyon	LA	6115 ft	*Charlet 1395–1404 & Tausch*	IRSR
T24N R48E S27	Gund Ranch, Keystone area	EU	6600 ft	*Rowan Sage in 1980*	RENO
[DAC: (EU) T25N R48.5 R49E R50E, T24N R48.5 R49E, T23N R48E, T21N R48E; (LA) T21N R47E, T20N R46E R47E]					

Reported as present in the range by Little (1971).
DAC: codominant with pinyon in extensive pinyon-juniper woodlands (5/1993).

Singatse Range (1)
elevation record: 5000 ft
1 record in 1 county (Lyon)

Towns/Range/Section	Location—Habitat	County	Elevation	Collector	Herbarium/a
T11N R25E	15 mi S of Yerington	LY	5000 ft	*Koch N-120 & Holstein & Carlson*	RSA

Not reported as present in the range by Little (1971).

Snake Mountains (2)
Elko County
[DAC: T39N R62E, T40N R62E, T41N R62E]

Reported as present in the range by Little (1971).
DAC: small juniper woodlands are visible from US 93, southeast aspect only (1988–1994).

Snake Range (1)
elevation record: 6280–7200 ft
15 records in 1 county (White Pine) and reports in another (Lincoln)

Towns/Range/Section	Location—Habitat	County	Elevation	Collector	Herbarium/a
T11N R70E	Lexington Creek	WP	—	*Tausch 080193-1-10*	IRSR
T12N R70E	Snake Creek	WP	6280 ft	*Fred Landau 26*	UNLV
T13N R70E	S of Baker on SR 448, jct Ananda Way and Coffman, just outside Great Basin National Park	WP	—	*Joyce P Hardy 2535*	BRY
T16N R68E S2 SE/4	5.25 mi W of Mount Moriah, S side Negro Creek	WP	7200 ft	*JC Lyman 662 & MC Lyman*	UCR
—	Snake Range, 2 mi above fish hatchery, canyon side	WP	7000 ft	*I Ting 19387*	RENO
—	US National Park [Lehman Caves] near Ely	WP	—	*M Milner 9360*	UT

[DAC: (WP) T19N R68E R69E, T18N R68E R69E R70E, T17N R68E R70E, T16N R70E, T15N R68E R69E R70E, T14N R67E R68E R69E, T13N R68E R69E, T12N R68E R69E, T11N R68E R69E, T10N R68E R70E R71E; (LI) T9N R69E (Red Ledges)]

Reported as present in the range by Little (1971).
DAC: abundant in the pinyon-juniper woodlands at low to middle elevations; rare at best in adjacent Spring Valley to the west, which is occupied by extensive stands of *J. scopulorum* (1994).

Sonoma Range (1)
elevation record: 5000–8000 ft
3 records in 2 counties (Humboldt and Pershing)

Towns/Range/Section	Location—Habitat	County	Elevation	Collector	Herbarium/a
T35N R38E S13	Water Canyon drainage, on ridge	HU	8000 ft	*N Holmgren 1372 & J Reveal*	UTC
T34N R38E S15 NW/4	Sonoma Creek	PE	5000 ft	*Charlet 1875*	UCR, IRSR, RENO
T34N R38E S13	slopes above and E of Sonoma Creek	PE	6780 ft	*Charlet 1878*	UCR, IRSR, RENO

[DAC: (HU) T35N R39E S3 (Juniper Ridge)]

Reported as present in the range by Little (1971).
DAC: the label of *Holmgren 1372 & Reveal* records the collection site as R28E, but T35N R28E is not in the Sonoma Range, and there is no Water Canyon there. The Water Canyon in the Sonoma Range is at T35N R38E. Juniper woodlands are locally abundant in Water Canyon, in the Sonoma Creek drainage, and a small woodland occurs on Juniper Ridge. *Juniperus osteosperma* is the only conifer I have seen on the mountain (1994).

South Virgin Mountains (3)
Clark County
[Little: T19S R69E]

Reported as present in the range by Little (1971).

Spring Mountains (1)
elevation record: 5575–8860 ft
7 records in 1 county (Clark)

Towns/Range/Section	Location—Habitat	County	Elevation	Collector	Herbarium/a
T18S R56E S18	road crossing at Cold Creek	CL	5860 ft	*Robertson & Tiernan Ch3-74*	RENO
T18S R56E	Lee Canyon at jct w/Kyle Canyon Rd	CL	7800 ft	*Vasek 630716-9*	UCR
T19S R57E	Spring Mts, Kyle Canyon–Deer Creek	CL	8530 ft	*IW Clokey 5386*	UC, UTC
T19S R57E	Kyle Canyon hillside, juniper belt	CL	5575 ft	*IW Clokey 7824*	UTC, UT, BRY, RENO
T19S R57E	ridge S of Deer Creek; top of ridge	CL	8860 ft	*IW Clokey 7823*	BRY, UTC, UT, OGDF, RENO
T20S R57E	Harris Springs	CL	5770 ft	*IW Clokey 5389*	UTC, RENO
T24S R58E	Good Springs	CL	—	*ME Jones in 1905*	RSA

[DAC: T19S R56E (Lee Canyon)]
[Little: T18S R55E, T20S R55E R56E, T21S R56E R57E]
[Robin Tausch: T18S R55E R56E, T19S R55E]

Reported as present in the range by Little (1971).
DAC: the common juniper, and abundant, in pinyon-juniper woodlands at middle elevations; replaced by *J. scopulorum* in pinyon-juniper-mountain-mahogany woodlands at higher elevations (1982 & 1993).

Stillwater Range (1)
elevation record: 5135–7365 ft
14 records in 2 counties (Churchill and Pershing)

Towns/Range/Section	Location—Habitat	County	Elevation	Collector	Herbarium/a
T18N R32E	Mountain Well	CH	—	*Ann Pinzl 1015*	NSM
T23N R34E S1 NE/4	Copper Kettle Wash	CH	5380 ft	*Charlet 717 & Rust*	UCR
T23N R34E S14 NE/4	ridge above Copper Kettle Wash	CH	7365 ft	*Charlet 719 & Rust*	UCR, IRSR, RENO
T24N R34E S35	Copper Kettle Wash	CH	5135 ft	*Charlet 716 & Rust*	UCR
T27N R37E S11 SE/4	McKinney Pass, N end of range	PE	5770 ft	*Charlet 1445–1454 & Tausch*	IRSR

[DAC: (PE) T26N R37E, T25N R35E R36E R37E; (CH) T24N R35E R36E, T23N R35E, T22N R33E R34E, T21N R33E, T20N R32E R33E, T19N R33E]

Not reported as present in the range by Little (1971).
DAC: widespread and common in large pinyon-juniper woodlands in the central portion of the range, and in the juniper woodlands in the north (1990–1994).

Stonewall Mountain (1)
elevation record: 7400 ft
1 record in 1 county (Nye)

Towns/Range/Section	Location—Habitat	County	Elevation	Collector	Herbarium/a
T5S R43E	US 95, 3 mi S of jct w/SR 3, N of Scotty's Junction; w/ pinyon	NY	7400 ft	*Robert Rodin 9499*	UC
[Little: T5S R44E]					

Reported as present in the range by Little (1971) and Beatley (1976).

Sulphur Springs Range (1)
elevation record: 5500–6755 ft
12 records in 1 county (Eureka) and reports in another (Elko)

Towns/Range/Section	Location—Habitat	County	Elevation	Collector	Herbarium/a
T23N R52E	Garden Pass	EU	6755 ft	*Charlet 1374–1383 & Tausch*	IRSR
T26N R53E S5	——	EU	6500 ft	*Jon Carpenter 10-22-91*	UCR
T27N R52E S14	——	EU	5500 ft	*Jon Carpenter 10-22-91*	UCR
[DAC: (EL) T30N R53E, T29N R53E, T28N R53E, T27N R53E; (EU) T26N R52E R53E, T25N R52E R53E, T24N R52E, T22N R52E]					

Reported as present in the range by Little (1971).
DAC: in extensive pinyon-juniper woodlands (1993–1994).

Sweetwater Mountains (2)
Lyon County
[Little: T8N R24E, T7N R25E]

Reported as present in the range by Little (1971) and Lavin (1981).
DAC: common at lower elevations in pinyon woodlands (1985–1995).

****Sylvania Mountains (2)**
Esmeralda County
[RWR: T6S R39E]

Not reported as present in the range by Little (1971).
New report for the species in the range.

Timber Mountain (3)
Nye County
[McLane: T10S R49E]

Not reported as present in the range by Little (1971).
Reported as present in the range, with pinyon, by McLane (1978).

Timpahute Range (2)
Lincoln County
[RWR: T3S R57E R58E]

Reported as present in the range by Little (1971).

Toana John Mountains (1)
elevation record: 5380 ft
1 record in 1 county (Elko)

Towns/Range/Section	Location—Habitat	County	Elevation	Collector	Herbarium/a
T46N R65E S34	above Willow Creek	EL	5380 ft	*Charlet 1605*	UCR, IRSR, RENO
[Little: T46N R67E, T45N R66E R67E, T44N R66E]					

Reported as present in the range by Little (1971).
DAC: dominant in extensive juniper woodlands (1993).

Toana Range (1)
2 records in 1 county (Elko)

Towns/Range/Section	Location—Habitat	County	Elevation	Collector	Herbarium/a
T36N R68E	14 mi E of Oasis	EL	—	*Billings 1150*	RENO
T37N R67E S10	———	EL	—	*Ann Pinzl 8723*	UNLV, NSM
[DAC: T38N R68E, T37N R68E, T36N R67E, T35N R68E]					

Reported as present in the range by Little (1971).

****Tobin Range (1)**
elevation record: 5040–6280 ft
4 records in 1 county (Pershing)

Towns/Range/Section	Location—Habitat	County	Elevation	Collector	Herbarium/a
T28N R39E S4 S/4	Golconda Canyon	PE	5400 ft	*Charlet 1881*	UCR, IRSR, RENO, UC
T28N R39E S14	S of Golconda Pass	PE	5900 ft	*Charlet 1882*	UCR, IRSR, RENO
T28N R40E S31 NW/4	low end juniper woodlands	PE	5040 ft	*Charlet 1884*	UCR, IRSR, RENO
T32N R40E S34 W/4	China Creek	PE	6280 ft	*Charlet 1880*	UCR, IRSR, RENO
[DAC: T31N R39E S26, T31N R40E, T29N R39E, T28N R39E R40E, T27N R39E]					

Not reported as present in the range by Little (1971).
DAC: Utah juniper woodlands are small and isolated in the northern portion of the range but common in the south near Mount Tobin and the Golconda Pass area. *Juniperus occidentalis* is also present in the higher elevations (1994).
New record for the species in the range.

Toiyabe Range (1)
elevation record: 6500–8000 ft
17 records in 2 counties (Lander and Nye)

Towns/Range/Section	Location—Habitat	County	Elevation	Collector	Herbarium/a
T8N R42E	Peavine campground	NY	6500 ft	*Tiehm 2684 & M Williams*	RENO
T9N R42E	above Peavine Creek	NY	6560 ft	*Charlet 1342–1351 & Tausch*	IRSR
T12N R41E S6 SE/4	Stewart Creek	NY	7380 ft	*Tausch 4/22/94*	IRSR, UCR
T12N R42E	South Twin River [DAC: approaches *J. occidentalis*]	NY	7000 ft	*JM Linsdale & MAR Linsdale 14*	CAS
T12N R42E	mouth of N Fork Twin River	NY	—	*WH Henning 148*	UC, RSA, RENO
T13N R41E	Mohawk Ranger Station [DAC: approaches *J. occidentalis*]	NY	8000 ft	*JM Linsdale & MAR Linsdale 245*	CAS
T14N R42E	San Juan Creek, 1.7 mi above national forest boundary; streamside	NY	6960 ft	*W Padgett 2029 & M Manning*	OGDF
T18N R44E	17 mi from Austin, Birch Creek Canyon, common	LA	—	*WH Henning 35*	UTC, RENO

[DAC: (LA) T26N R46E R47E, T25N R46E R47E, T24N R45E R46E, T23N R45E R46E, T22N R44E R45E R46E, T21N R44E R45E, T20N R44E, T19E R44E R45E, T18N R43E R44E, T17N R43E R44E, T16N R43E R44E; (NY) T15N R42E R43E, T14N R42E, T13N R42E, T12N R40E, T11N R40E R41E R42E, T10N R40E R41E R42E, T9N R40E R41E]

Reported as present in the range by Little (1971) and Goodrich (1981).
DAC: in the northern part of the range *J. osteosperma* is only a minor component of pinyon-juniper woodland; in the southern part it is well mixed with *Pinus monophylla; J. occidentalis* (uncommon) and *J. scopulorum* (rare) are also present but seem to be restricted to canyons (1991–1994).

Toquima Range (1)
elevation record: 7000–9085 ft
16 records in 1 county (Nye) and reports in another (Lander)

Towns/Range/Section	Location—Habitat	County	Elevation	Collector	Herbarium/a
T8N R44E	Manhattan	NY	7000 ft	*Percy Train 3429*	RENO
T8N R44E	2–10 mi N of Manhattan	NY	—	*Goodner 1199 & Henning*	RENO
T10N R45E	Shoshone Mountain	NY	9000 ft	*Charlet 785*	UCR, IRSR
T10N R45E	Antone Creek	NY	8365 ft	*Charlet 800*	UCR, IRSR
T10N R45E	Antone Creek	NY	8000 ft	*Charlet 881 & J Carpenter*	IRSR
T11N R45E	N slope Mount Jefferson	NY	9085 ft	*Charlet 1235 & T Charlet*	UCR, IRSR, RENO
T12N R45E	Moores Summit	NY	8350 ft	*Charlet 905 & S Vander Wall*	IRSR
T12N R45E	Moores Summit	NY	7940 ft	*Charlet 1354 & Tausch*	IRSR
T12N R45E	Moores Summit	NY	7940 ft	*Charlet 1356–1363 & Tausch*	IRSR

[DAC: (LA) T16N R46E; (NY) T15N R46E, T14N R45E R46E, T13N R45E R46E, T11N R44E, T10N R44E, T9N R44E, T8N R43E, T7N R43E R44E]

Reported as present in the range by Little (1971) and Goodrich (1981).
DAC: *Juniperus osteosperma* is the common juniper in this range; abundant and with varying glandular expression. It grows at very high elevations, and some ridgetop individuals attain a krummholz form. Grows with pinyon and, at higher elevations, mountain-mahogany and limber pine. Generally restricted to elevations above 7000 ft (1991–1993).

****Trinity Range (1)**
elevation record: 5780–6160 ft
2 records in 1 county (Pershing)

Towns/Range/Section	Location—Habitat	County	Elevation	Collector	Herbarium/a
T29N R30E S34	low end juniper woods	PE	5780 ft	*Charlet 2067*	UCR, IRSR, RENO
T29N R30E S26	Poker Brown Spring	PE	6160 ft	*Charlet 2068*	UCR, IRSR, RENO, UC

Not reported as present in the range by Little (1971).
DAC: juniper woodlands occur around the head of Poker Brown Canyon (1993–1994).
New record for the species in the range.

Virginia Mountains (1)
elevation record: 4680–5100 ft
3 records in 1 county (Washoe)

Towns/Range/Section	Location—Habitat	County	Elevation	Collector	Herbarium/a
T24N R19E S11 W/2	juniper woodland just S of Dry Valley	WA	5060 ft	*Charlet 1821*	UCR, IRSR, RENO
T24N R20E S29	along Winnemucca Ranch Rd	WA	4680 ft	*Charlet 1791 & T Charlet*	UCR, IRSR, RENO, UC, BRY
T25N R20E	5.5 mi ESE of Pyramid [Post Office]	WA	5100 ft	*A Simontacchi 560*	UC
[DAC: T24N R20E R21E, T25N R19E, T26N R19E R20E]					

Reported as present in the range by Little (1971).
DAC: occasional at lower elevations, locally common in relict stands at high elevations (1973–1994).

Virginia Range (1)
elevation record: 4480–6890 ft
12 records in 3 counties (Carson City, Storey, and Washoe) and reports in another (Lyon)

Towns/Range/Section	Location—Habitat	County	Elevation	Collector	Herbarium/a
T15N R20E	5 mi E of Carson City	CC	—	*WP Cottam 9907*	UT
T16N R21E	2.1 mi N of US 50 on SR 17 near Virginia City	ST	4800 ft	*Vasek 620905-1*	UCR
T17N R21E	3 mi E of Washoe County, N side of hwy [SR 343?]	ST	—	*TC Frantz in 1951*	RENO
T18N R21E S9 N/8	high plateau W of Washington	ST	5580 ft	*Charlet 1802*	UCR, IRSR, RENO, UC
T19N R21E S27 S/4	tributary of Long Valley Creek	ST	4480 ft	*Charlet 1777*	UCR, IRSR, RENO
T17N R21E S8	upper Geiger Grade	WA	6890 ft	*Charlet 1728*	UCR, IRSR, RENO, UC
T17N R21E	top of Geiger Grade	WA	6500 ft	*R Sumner 111*	RENO
T18N R20E S35 NW/4	upper Geiger Grade	WA	6890 ft	*Charlet 1729*	UCR, UC, RENO
T19N R20E S14 S/2	Virginia Range	WA	4900 ft	*A Tiehm 4808 & M Williams*	CAS, RSA, UTC, RENO
T19N R20E S23 S/4	N of Hidden Valley County Park, just above mouth of canyon	WA	4700 ft	*Charlet 1801 & T Charlet*	UCR, IRSR, RENO
—	on Geiger Grade	ST	—	*AS Rhoads sn*	UTC
—	road between Reno & Virginia City [DAC: label says ST but must be WA at this elevation on this route]	WA	4700 ft	*WA Archer 5072*	RENO
[DAC: (LY) T19N R24E; (ST) T18N R22E, T19N R22E R23E; (WA) T17N R20E]					

Reported as present in the range by Little (1971).
DAC: abundant; the common juniper on the mountain, in pinyon-juniper woodland. Dominant at lower elevations, falling out before *P. monophylla* at the higher elevations (1973–1995).

Virgin Mountains (1)
elevation record: 4000–7500 ft
5 records in 1 county (Clark)

Towns/Range/Section	Location—Habitat	County	Elevation	Collector	Herbarium/a
T14S R71E	Cabin Canyon	CL	4000 ft	*JD Armstrong 830*	UNLV, NSM
T14S R71E	Cabin Creek	CL	—	*Richard W. Rust 54*	UCR
T15S R70E	Key West Mine; top of ridge	CL	6000 ft	*WP Cottam 8413*	UT
T15S R71E	———	CL	7500 ft	*Peter Herlan & Ann Pinzl in 1973*	NSM
T16S R71E	E Fork Yant Pit Canyon	CL	5500 ft	*PA Munz 16730*	RSA

Not reported as present in the range by Little (1971).

Wassuk Range (1)
elevation record: 6460–9200 ft
6 records in 1 county (Mineral)

Towns/Range/Section	Location—Habitat	County	Elevation	Collector	Herbarium/a
T6N R28E S1	W slope, W of Corey Peak, lower end of pinyon-juniper	MI	6980 ft	*Charlet 1957*	UCR, IRSR, RENO
T6N R29E S13	N face Buller Mountain (determined Vasek)	MI	9200 ft	*Charlet 704 & Rust*	UCR
T6N R29E S24	canyon below Powell Summit, w/*Cercocarpus* and pinyon	MI	7800 ft	*Charlet 1974*	UCR, IRSR, RENO
T6N R30E S6 NE/4	Alum Creek; uncommon	MI	6460 ft	*Charlet 1971*	UCR, IRSR, RENO
T8N R28E S28 W/4	N of Rattlesnake Canyon	MI	6820 ft	*Charlet 1958*	UCR, IRSR, RENO
T9N R28E	Cottonwood Canyon	MI	7000–9000 ft	*WA Archer 7160*	RENO

[DAC: T13N R27E, T12N R27E R28E, T11N R28E, T10N R28E (Cottonwood Creek), T9N R29E (Dutch Creek), T8N R29E, T7N R28E R29E R30E, T6N R30E, T5N R29E]

Reported as present in the range by Little (1971), but only south of Lucky Boy Pass and at the extreme northern end of the range. DAC: occasional throughout the range, generally a minor component with *Pinus monophylla* in pinyon-juniper woodland; *Charlet 704*, the high-elevation record for the species in Nevada, is a shrubby form and is well above the woodland limit, near *P. flexilis, P. contorta*, and *P. jeffreyi* (1990–1995).

Wellington Hills (2)
Douglas County
[Little, DAC: T10N R23E, T9N R23E]

Reported as present in the range by Little (1971) and Lavin (1981).
DAC: common in extensive pinyon-juniper woodlands throughout the range (1988–1994).

****West Humboldt Range (1)**
elevation record: 4800–4860 ft
2 records in 1 county (Pershing)

Towns/Range/Section	Location—Habitat	County	Elevation	Collector	Herbarium/a
T27N R32E S13 SW/4	2 mi S of Coal Canyon, low end juniper woodland	PE	4860 ft	*Charlet 1867 & Rust & Sanderson*	UCR, IRSR, RENO, UC
T27N R33E S16 SW/4	Coal Canyon	PE	4800 ft	*Charlet 1868 & Rust & Sanderson*	UCR, IRSR, RENO

Not reported as present in the range by Little (1971).
DAC: *Juniperus osteosperma* is the only conifer species on the mountain. Open juniper woodlands are present, primarily restricted to the higher elevations on north-facing slopes (1993–1994).

West Range (3)
Lincoln County
[Little: T2N R65E]

Reported as present in the range by Little (1971).

****Whistler Range (1)**
elevation record: 6160 ft
1 record in 1 county (Eureka)

Towns/Range/Section	Location—Habitat	County	Elevation	Collector	Herbarium/a
T20N R52E S23 S/8	Devils Gate, above Slough Creek	EU	6160 ft	*Charlet 2041*	UCR, IRSR
[DAC: T22N R52E, T21N R52E]					

Not reported as present in the range by Little (1971).
DAC: dominant; most of the range is covered with juniper woodlands (1994).
New record for the species in the range.

White Mountains (1)
elevation record: 6960–7800 ft
2 records in 1 county (Mineral) and reports in another (Esmeralda)

Towns/Range/Section	Location—Habitat	County	Elevation	Collector	Herbarium/a
T1N R32E	2 mi SE of Mount Montgomery Pass PO, on road to Tip Top Mine	MI	7800 ft	*Percy Train 3995*	UC, RENO
T2N R33E	E of Montgomery Pass	MI	6960 ft	*Charlet 1737*	UCR, IRSR, RENO
[DAC: (MI) T1N R33E; (ES) T1N R33E R34E, T1S R32E R33E R34E, T2S R34E]					

Reported as present in the range by Little (1971).
DAC: common throughout the range at middle elevations, in pinyon-juniper woodland (1993).

White Pine Range (1)
elevation record: 6640–7640 ft
6 records in 2 counties (White Pine and Nye)

Towns/Range/Section	Location—Habitat	County	Elevation	Collector	Herbarium/a
T11N R59E S5	in pinyon-juniper w/*Pinus ponderosa*	NY	6640 ft	*Charlet 1987*	UCR, IRSR, RENO
T16N R57E S9	Seligman Canyon	WP	7545 ft	*Richard W. Rust 55*	UCR
T17N R58E S21 S/8	open pinyon-juniper woods	WP	7640 ft	*Charlet 2030*	UCR, IRSR, RENO
T17N R59E S9	Illipah, next to Moorman Ranch mailbox; low end pinyon-juniper woods	WP	6820 ft	*Charlet 2035*	UCR, IRSR, RENO
T18N R58E S30	1.7 road miles W of Antelope Summit	WP	6880 ft	*Charlet 2037*	UCR, IRSR
T18N R58E S28 SW/4	Antelope Summit	WP	7433 ft	*Charlet 2036*	UCR, IRSR, RENO
[DAC: (WP) T19N R57E R58E, T18N R57E R59E, T17N R59E, T16N R58E R59E, T15N R57E R58E R59E, T14N R57E R58E R59E, T13N R58E R59E R60E, T12N R59E; (NY) T12N R58E, T11N R58E]					

Reported as present in the range by Little (1971).
DAC: codominant with pinyon in extensive pinyon-juniper woodlands throughout the lower and middle elevations (1994).

White River Range (1)
elevation record: 5960 ft
1 record in 1 county (White Pine)

Towns/Range/Section	Location—Habitat	County	Elevation	Collector	Herbarium/a
T12N R60E S13 NW/4	above White River; open juniper woodland	WP	5960 ft	*Charlet 2002*	UCR, IRSR, RENO
[RWR: T11N R61E]					

Reported as present in the range by Little (1971).
DAC: dominant in juniper woodlands throughout the range (1994).

White Rock Mountains (1)
elevation record: 6820 ft
1 record in 1 county (Lincoln)

Towns/Range/Section	Location—Habitat	County	Elevation	Collector	Herbarium/a
T5N R69E S36 NE/4	pinyon-juniper woodlands, near ponderosa pine stand	LI	6820 ft	*Charlet 2099*	UCR, IRSR, RENO
[Little: T5N R69E R70E, T4N R70E R71E, T3N R70E R71E, T2N R70E R71E, T1N R70E R71E]					

Reported as present in the range by Little (1971).
DAC: codominant with pinyon in extensive pinyon-juniper woodlands (1994).

Wilson Creek Range (1)
elevation record: 5080–7980 ft
2 records in 1 county (Lincoln)

Towns/Range/Section	Location—Habitat	County	Elevation	Collector	Herbarium/a
T2N R69E	Echo Valley, near reservoir; 19 mi NE of Panaca, S of Rose Valley	LI	5080 ft	*Leila Shultz 6232 & J Shultz*	UTC
T5N R68E S22 W/4	pinyon-juniper-mountain-mahogany woodlands	LI	7980 ft	*Charlet 2092*	UCR, IRSR, RENO
[DAC: T7N R68E, T6N R67E R68E R69E, T5N R67E R69E, T4N R67E R68E, T3N R68E R69E]					

Reported as present in the range by Little (1971).
DAC: codominant with pinyon in extensive pinyon-juniper woodlands throughout the lower and middle elevations (1994).

Windermere Hills (1)
elevation record: 6050 ft
2 records in 1 county (Elko)

Towns/Range/Section	Location—Habitat	County	Elevation	Collector	Herbarium/a
T38N R63E	9.9 mi N of Wells on US 93	EL	—	*Vasek 640810-1*	UCR
T38N R63E	N of Wells on US 93	EL	6050 ft	*Charlet 784*	UCR
[DAC: T40N R64E R65E, T39N R64E R65E]					

Reported as present in the range by Little (1971).
DAC: dominant throughout most of the range in juniper woodlands (1988–1994).

Wood Hills (2)
Elko County
[DAC: T36N R62E R63E, T37N R62E R63E]

Reported as present in the range by Little (1971).
DAC: codominant with pinyon in pinyon-juniper woodlands throughout most of the range (1992–1993).

Worthington Mountains (2)
Lincoln County
[RWR: T1N R57E, T1S R57E]

Reported as present in the range by Little (1971).

OTHER FEATURES IN NEVADA, 4500–7000 FT, RECORDS (34) AND REPORTS

Clark County
[Robin Tausch: T17S R56E]
[DAC: T19S R58E]

Douglas County
elevation record: 5000–6060 ft
7 records

Towns/Range/Section	Location—Habitat	Elevation	Collector	Herbarium/a
T11N R21E S19	pinyon-juniper woods	5360 ft	*Charlet 1908*	UCR, IRSR, RENO
T11N R21E S15	Double Springs Flat	5960 ft	*Charlet 1904*	UCR, IRSR, RENO
T11N R21E S29	1 mi S of Doud Springs	5880 ft	*Charlet 1905*	UCR, IRSR, RENO, UC
T11N R21E S32 S/2	pinyon-juniper woods near *Pinus jeffreyi*	6060 ft	*Charlet 1907*	UCR, IRSR, RENO
T11N R20E S24	E Fork Carson River	5340 ft	*Charlet 1911*	UCR, IRSR, RENO
T11N R20E S12	E Fork Carson River	5740 ft	*Charlet 1912*	UCR, IRSR, RENO
T12N R20E	E Fork Carson River, 6 mi S of Gardnerville on US 395	5000 ft	*Wm Priest in 1956*	RENO

[DAC: T14N R20E]

Elko County
elevation record: 5300–5500 ft
9 records

Towns/Range/Section	Location—Habitat	Elevation	Collector	Herbarium/a
T47N R69E S32	2 mi W of Trout Creek Ranch	5300 ft	*Jon Carpenter 10-16-91*	UCR
T47N R69E S32	2 mi W of Trout Creek Ranch	5300 ft	*Jon Carpenter 10-16-91*	UCR
T47N R69E S32	2 mi W of Trout Creek Ranch	5500 ft	*Jon Carpenter 10-16-91*	UCR
T46N R68E S2	Rancho Grande	5500 ft	*Jon Carpenter*	UCR
T46N R68E S15	Horse Creek	5500 ft	*Jon Carpenter 10-16-91*	UCR
T46N R68E S20	Horse Creek	5495 ft	*Charlet 1606*	UCR, IRSR, RENO
T46N R68E S20	Horse Creek	5495 ft	*Charlet 1607*	UCR, IRSR, RENO
T46N R68E S21	Horse Creek	5500 ft	*Jon Carpenter 10-16-91*	UCR
T47N R69E	5 mi S of Idaho, 7 mi W of Utah	—	*Verne Woodbury in 1951*	RENO

[Little: T28N R68E, T27N R59E R69E, T26N R60E]
[DAC: T32N R56E]

Eureka County
elevation record: 6300 ft
1 record

Towns/Range/Section	Location—Habitat	Elevation	Collector	Herbarium/a
T21N R50E S3	Kobeh Valley	6300 ft	*Jon Carpenter*	UCR

[DAC: T21N R51E (Kobeh Valley)]
[RWR: T20N R51E (Lone Mountain; very old trees occur here)]

Humboldt County
elevation record: 4700–5000 ft
2 records

Towns/Range/Section	Location—Habitat	Elevation	Collector	Herbarium/a
T36N R35E S27	Blue Mountain, SE side	4700 ft	*N Holmgren 3968 & P Holmgren*	UTC
T36N R35E S29	Blue Mountain, S side, low end juniper woodland	5000 ft	*Charlet 1192*	UCR, IRSR, RENO, UC

Not reported as present in the county by Little (1971).

Lincoln County
elevation record: 4500–6500 ft
1 record

Towns/Range/Section	Location—Habitat	Elevation	Collector	Herbarium/a
—	Panaca Valley & vicinity	4500–6500 ft	*Marie Gentry 109*	UC, CAS

[Little: T3S R71E, T4S R71E, T5S R69E R70E R71E]
[DAC: T7N R65E, T6N R65E (Dutch John Mountain), T7N R67E, T5N R64E, T4N R64E R69E R70E, T3N R64E R69E R70E, T2N R70E, T1N R70E]

Mineral County
elevation record: 7000 ft
10 records

Towns/Range/Section	Location—Habitat	Elevation	Collector	Herbarium/a
T4N R29E S31	Mono Basin; disjunct, ancient stand in sand, < 1 km from CA	7000 ft	*Charlet 1296–1305 & Tausch*	IRSR

Nye County
1 record

Towns/Range/Section	Location—Habitat	Elevation	Collector	Herbarium/a
—	Ione Valley Willow Creek	—	*Timmy Cass 74-667-3*	CAS

[DAC: T3N R58E (Garden Valley)]

Washoe County
elevation record: 5780–7000 ft
2 records

Towns/Range/Section	Location—Habitat	Elevation	Collector	Herbarium/a
T46N R22E S14 SW/4	Catnip Mountain; scattered in sage-brush, only Utah juniper seen in area	5780 ft	*Charlet 1846*	UCR, IRSR, RENO
—	Sheldon NWR	6000–7000 ft	*GH Greenway Nev 6*	RENO

DAC: virtually all the junipers in northern Washoe County are *J. occidentalis.* These collections are essentially eglandular and are the northernmost occurrences in Washoe County.

White Pine County
elevation record: 5800 ft
1 record

Towns/Range/Section	Location—Habitat	Elevation	Collector	Herbarium/a
T11N R67E	Spring Valley, 0.5 mi SSE of Shoshone	5800 ft	*KE Bradshaw in 1943*	UTC

[DAC: T16N R56E, T15N R56E, T13N R60E R64E]

Juniperus scopulorum Sargent

common name: Rocky Mountain juniper, Rocky Mountain red cedar
elevation: 5200–9200 ft (1585–2805 m)
zones: Canadian, Arid Transition, Upper Sonoran
102 Records for 6 Nevada counties (CL, EL, EU, LI, NY, WP)
Reports for 39 mountain ranges, Records for 29 of these
New Reports for 11 mountain ranges, New Records for 7 of these

Rocky Mountain juniper is well distributed throughout the Rocky Mountains of North America, from British Columbia south to Texas. Along the Pacific Mountain System of the United States, the species occurs only in the North Cascades and the Olympic Mountains of Washington. Rocky Mountain juniper occurs in all physiographic divisions (*sensu* Hunt 1967) of the Intermontane Plateaus region, from the Columbia Plateau of Washington and Oregon, south through the central and eastern Great Basin, to the Four Corners area of the Colorado Plateau. *Juniperus scopulorum* is closely related to *J. virginiana* L. and was once considered a subspecies of that taxon.

In Nevada, Rocky Mountain juniper is distributed throughout 6 central and eastern counties (Clark, Elko, Eureka, Lincoln, Nye, and White Pine) and 39 mountain ranges. I include new reports for 11 mountain ranges, 7 of which have records. Rocky Mountain juniper is generally restricted to deep canyons with perennial streams in east-central Nevada but is a major component of pinyon-juniper woodlands on hillsides and ridges in the Spring Mountains (Clark County). It also occurs on rocky sites in the mountains of northern Elko County, where it forms small juniper woodland stands in the Bull Run Mountains. Rocky Mountain juniper is commonly associated with curlleaf mountain-mahogany (*Cercocarpus ledifolius*) on rocky sites, and it grows on the xeric periphery of aspen stands in the Canadian Zone in Nevada. A unique population of Rocky Mountain juniper that occurs at relatively low elevations (5500–6000 ft) with *Sarcobatus* in Spring Valley (White Pine County) in eastern Nevada is probably an ecotype. Including these unique stands, the elevation range of the species in Nevada is 4000 ft (1220 m).

Antelope Range (4)
White Pine and Elko Counties

Not reported as present in the range by Little (1971).
DAC: not yet found, but likely to occur in the range along streams (1994).

Bruneau Range (1)
elevation record: 6500 ft
2 records in 1 county (Elko)

Towns/Range/Section	Location—Habitat	County	Elevation	Collector	Herbarium/a
T46N R56E S13	Burnt Timber Creek, W of Bruneau River	EL	6500 ft	*Peter Herlan & Ann Pinzl in 1976*	NSM
T46N R56E	41°51′19″N, 115°42′23″W, Meadow Creek, 2.2 mi S of Taylor Creek Rd	EL	—	*W Padgett & M Manning 2545*	RENO, UTC
[DAC: T46N R53E R54E]					

Reported as present in the range by Little (1971).
DAC: small, open stands are scattered throughout the range; common along the Bruneau River (1989–1994).

Buck Creek Mountains (2)
Elko County
[DAC: T46N 56E R57E R58E]

Reported as present in the range by Little (1971).
DAC: small, open woodlands of *J. scopulorum* are present; common along streams (1993–1994).

****Bull Run Mountains (1)**
elevation record: 6315 ft
1 record in 1 county (Elko)

Towns/Range/Section	Location—Habitat	County	Elevation	Collector	Herbarium/a
T43N R51E S12	3 km N of Wilson Peak	EL	6315 ft	*Charlet 1600*	UCR

Not reported as present in the range by Little (1971).
DAC: small stands are present in rocks on the western and eastern aspects north of Wilson Peak, south of Bull Run Creek (1991–1994).
New record for the species in the range.

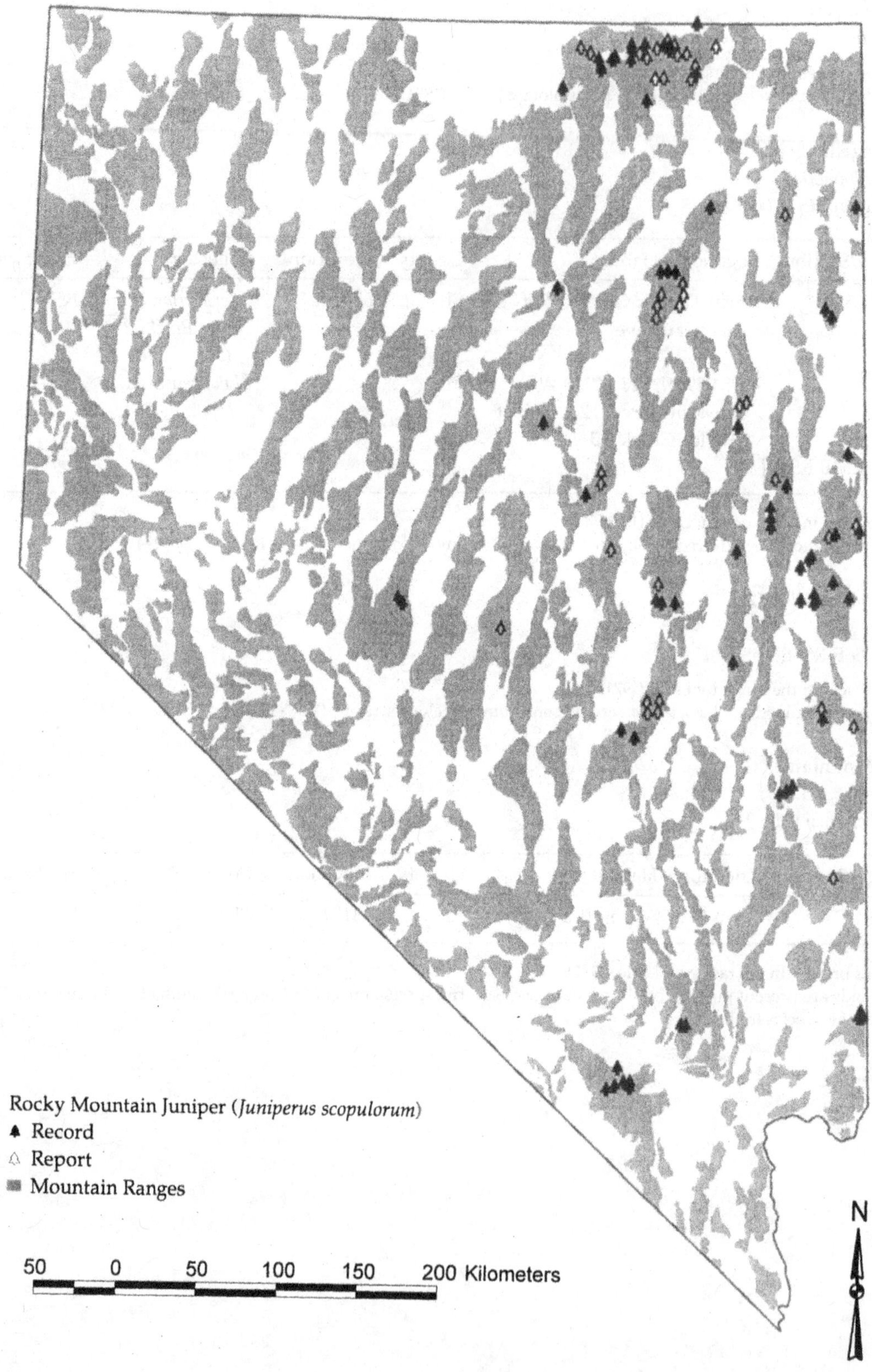

Rocky Mountain Juniper (*Juniperus scopulorum*)
Record
Report
Mountain Ranges
50 0 50 100 150 200 Kilometers
N

Cherry Creek Range (1)
elevation record: 6700 ft
1 record in 1 county (White Pine)

Towns/Range/Section	Location—Habitat	County	Elevation	Collector	Herbarium/a
T24N R62E S36	Cherry Creek, rare	WP	6700 ft	*Williams 84-44-2 & Tiehm*	CAS, RENO, RSA, NSM, UTC, BRY
[DAC: T25N R62E R63E (Paris Creek)]					

Reported as present in the range by Little (1971).
DAC: common in Paris Creek from the mouth of the canyon to 7600 ft; probably occurs in other drainages in the northern and eastern parts of the range (1994).

Clover Mountains (3)
Lincoln County
[Little: T6S R69E]

Reported as present in the range by Little (1971).

Copper Mountains (2)
Elko County
[Little, DAC: T44N R57E]

Reported as present in the range by Little (1971).
DAC: occasional at rocky sites and along streams (1988–1994).

Cortez Mountains (1)
1 record in 1 county (Eureka)

Towns/Range/Section	Location—Habitat	County	Elevation	Collector	Herbarium/a
T31N R51E	JD Ranch, Palisade (determined Billings)	EU	—	*Francis Case before 1951*	RENO

Not reported as present in the range by Little (1971).
DAC: not observed at the south end during my one brief visit (1993).

Diamond Mountains (3)
Eureka and White Pine Counties
[Little: T21N R54E, T22N R54E]

Reported as present in the range by Little (1971).

****Duck Creek Range (1)**
elevation record: 7700 ft
1 record in 1 county (White Pine)

Towns/Range/Section	Location—Habitat	County	Elevation	Collector	Herbarium/a
T17N R65E S31	Duck Creek; lone tree w/*Salix;* near aspen	WP	7700 ft	*Charlet 2022*	RENO

Not reported as present in the range by Little (1971).
DAC: uncommon along streams; generally below the fir forest (1994).
New record for the species in the range.

East Humboldt Range (1)
1 record in 1 county (Elko)

Towns/Range/Section	Location—Habitat	County	Elevation	Collector	Herbarium/a
T36N R61E	East Humboldt	EL	—	*AL Menzies in 1962*	CAS

Reported as present in the range by Little (1971).

Egan Range (1)
elevation record: 6000–8170 ft
2 records in 2 counties (White Pine and Nye)

Towns/Range/Section	Location—Habitat	County	Elevation	Collector	Herbarium/a
T8N R62E S12	Shingle Wash, near Granite Spring	NY	6000 ft	*Riordan in 1938*	RENO
T15N R63E	Sawmill Canyon	WP	8170 ft	*Richard W. Rust 5*	UCR

Reported as present in the range by Little (1971).
DAC: occasional along streams at middle elevations; the distribution is certainly much wider than represented by collections (1994).

Elk Mountains (3)
Elko County
[Little: T47N R61E]

Reported as present in the range by Little (1971).

****Fish Creek Range (1)**
elevation record: 7160 ft
1 record in 1 county (Eureka)

Towns/Range/Section	Location—Habitat	County	Elevation	Collector	Herbarium/a
T18N R53E S3 N/8	Prospect Mountain, New York Canyon	EU	7160 ft	*Charlet 2040*	RENO

Not reported as present in the range by Little (1971).
DAC: common on slopes with mountain-mahogany, bristlecone, and limber pine. *Charlet 2040* represents the lowest individual I observed on the mountain range (1994).
New record for the species in the range.

Fox Creek Range (1)
elevation record: 6820 ft
1 record in 1 county (Elko)

Towns/Range/Section	Location—Habitat	County	Elevation	Collector	Herbarium/a
T46N R58E S28 W/4 [DAC: T44N R58E]	above W Fork Jarbidge River w/mountain-mahogany	EL	6820 ft	*Charlet 2147 & T Charlet*	UC, RENO

Reported as present in the range by Little (1971).
DAC: occasional on rocky sites and along streams (1988–1994).

Goshute Mountains (1)
elevation record: 6900 ft
2 records in 1 county (Elko)

Towns/Range/Section	Location—Habitat	County	Elevation	Collector	Herbarium/a
T30N R68E S32	———	EL	6900 ft	*Cooper 95 & Holmgren & Lund*	UTC, BRY, RENO
T30N R68E S36	Hawk Ridge	EL	—	*Richard W. Rust 41*	UCR

Not reported as present in the range by Little (1971).
RWR: commonly scattered from middle to high elevations throughout the range, generally above *J. osteosperma.*

Grant Range (2)
Nye County
[DAC: T6N R57E R58E, T5N R57E R58E]

Reported as present in the range by Little (1971).
DAC: occasional along streams; observed generally above *J. osteosperma.*

Highland Range (1)
elevation record: 7200–9000 ft
3 records in 1 county (Lincoln)

Towns/Range/Section	Location—Habitat	County	Elevation	Collector	Herbarium/a
T1N R66E S35	Pine Spring	LI	7200 ft	*Ann Pinzl 3366*	NSM
T0 R66E	Highland Peak	LI	8000–9000 ft	*CA Purpus in 1898*	UC
T1S R66E	Comet Peak, Pioche	LI	8000 ft	*ME Jones in 1912*	CAS, RSA

Reported as present in the range by Little (1971).
DAC: occasional along streams and, at higher elevations, common on slopes with *Abies concolor.* Generally found above *J. osteosperma* (1994).

Ichabod Range (1)
elevation record: 5660–6740 ft
5 records in 1 county (Elko)

Towns/Range/Section	Location—Habitat	County	Elevation	Collector	Herbarium/a
T43N R57E S31 NE/4	between Mount Ichabod & Mason Mountain	EL	6740 ft	*Charlet 1602*	UCR
T45N R56E S18	Meadow Creek	EL	—	*Steve Anderson 1-4/1994*	UCR
T45N R56E S18	Meadow Creek	EL	—	*Steve Anderson 2-4/1994*	UCR
T46N R56E	Meadow Creek, 5 mi S of confluence w/Bruneau River	EL	—	*Ann Pinzl 10177*	NSM
T46N R56E S30	E side, Meadow Creek, ca. 0.8 km SSE of confluence w/Telephone Creek	EL	5660 ft	*MD Windham 91-335*	UT

Not reported as present in the range by Little (1971).
DAC: common along creeks in the north end of the range, particularly along Meadow Creek (1988–1994).

****Independence Mountains (1)**
elevation record: 5960 ft
1 record in 1 county (Elko)

Towns/Range/Section	Location—Habitat	County	Elevation	Collector	Herbarium/a
T44N R54E S5	E Fork Owyhee River	EL	5960 ft	*Charlet 1599*	UCR, RENO

Not reported as present in the range by Little (1971).
DAC: although it is common along East Fork Owyhee River and in nearby ranges in northern Elko County, I have not seen the species in any other drainage or feature on the mountain (1991–1994).
New record for the species in the range.

Jarbidge Mountains (1)
elevation record: 6200–8000 ft
4 records in 1 county (Elko)

Towns/Range/Section	Location—Habitat	County	Elevation	Collector	Herbarium/a
T46N R58E S16	W Fork Jarbidge River, 1 mi N of Jarbidge	EL	6200 ft	*Charlet 782*	UCR
T46N R58E	W Fork Jarbidge River, 2 mi S of Jarbidge	EL	6400 ft	*Percy Train 823*	RENO
T46N R58E	Snowslide Gulch	EL	8000 ft	*Guyer 5222B*	RENO
—	Jarbidge Mts	EL	—	*AL Menzies in 1962*	CAS

[DAC: T45N R59E, T46N R59E, T47N R58E]

Reported as present in the range by Little (1971).
DAC: locally common at lower elevations and along streams; uncommon on the northeast side of the range; not seen in the extreme south; most common along the forks of the Jarbidge River and at the north end of the range. Occurs in Cougar Creek (ca. 9000 ft) at rocky sites (1988–1994).

****Kern Mountains (1)**
elevation record: 7920 ft
1 record in 1 county (White Pine)

Towns/Range/Section	Location—Habitat	County	Elevation	Collector	Herbarium/a
T21N R69E S22	upper Rock Springs Canyon	WP	7920 ft	*Charlet 2107 & T Charlet*	UC, RENO

Not reported as present in the range by Little (1971).
DAC: present on high slopes in the range above pinyon-juniper woodlands (1994).
New record for the species in the range.

Marys River Range (1)
elevation record: 6560 ft
1 record in 1 county (Elko)

Towns/Range/Section	Location—Habitat	County	Elevation	Collector	Herbarium/a
T44N R60E S7 SE/4	Camp Creek	EL	6560 ft	*Charlet 2131 & T Charlet*	UCR, UC, RENO

[DAC: T44N R59E (Camp Creek), T45N R60E (Cottonwood Creek), T46N R59E (Slide Creek)]

Reported as present in the range by Little (1971).
DAC: scattered throughout the lower elevations, at rocky sites near streams and south-facing hillsides (1988–1994).

Monitor Range (3)
Nye County
[Kerry Heise: T11N R48E (Clear Creek Canyon, 7500–8000 ft)]

Not reported as present in the range by Little (1971) or Goodrich (1981).
Reported as present by Grayson (1993) and (apparently) Wells (1983).
DAC: I have seen no material, nor have I seen the species on the mountain, but it was reported to me by David Weixelman (USDA FS, Toiyabe NF, pers. comm. 1993) and by Kerry Heise (USDA, Toiyabe NF). Weixelman and Heise observed the species in Clear Creek Canyon above Little Fish Lake Valley, on the east side of the range, from 7500 to 8000 ft elevation (1993–1994).

Painted Point Range (7)
Washoe County

DAC: several individuals planted at the Bald Mountain headquarters at Sheldon National Wildlife Refuge, probably during the 1930s when the station was built, are now producing berries; likely to escape and naturalize from this location (1993–1994).

****Pancake Range (2)**
White Pine County
[RWR: T16N R55E (SE of Black Point)]

Not reported as present in the range by Little (1971).
New report for the species in the range.

****Pequop Mountains (2)**
Elko County
[RWR: T36N R65E (Sixmile Creek)]

Not reported as present in the range by Little (1971).
New report for the species in the range.

Pilot Range (1)
elevation record: 6000–7000 ft
2 records in 1 county (Elko)

Towns/Range/Section	Location—Habitat	County	Elevation	Collector	Herbarium/a
T36N R70E	Pilot Peak Range, 2 mi E of McCuiston Ranch, MacDonald Spring	EL	6000 ft	*Percy Train 3778*	UC, RENO, UTC
—	Pilot Mountain, near stream	EL	7000 ft	*WP Cottam 12694*	UT

Not reported as present in the range by Little (1971).

Quinn Canyon Range (1)
elevation record: 7100 ft
2 records in 1 county (Nye)

Towns/Range/Section	Location—Habitat	County	Elevation	Collector	Herbarium/a
T4N R56E	Willow Creek, frequent	NY	—	*B Maguire 25296 & A Holmgren*	UC, UTC
T4N R56E S36	Little Cherry Creek, frequent	NY	7100 ft	*Charlet 2009*	RENO

Reported as present in the range by Little (1971).
DAC: Rocky Mountain juniper probably occurs in most of the major drainages of the mountain. Frequent at higher elevations, at least at Willow and Little Cherry Creeks (1994).

****Roberts Mountains (1)**
elevation record: 7700 ft
1 record in 1 county (Eureka)

Towns/Range/Section	Location—Habitat	County	Elevation	Collector	Herbarium/a
T23N R50E S24 E/8	alone along Roberts Creek, above Summer Camp	EU	7700 ft	*Charlet 2091*	UC, RENO, BRY

Not reported as present in the range by Little (1971).
DAC: this was the only individual of the species seen on the mountain, although more are likely to occur along streams above the pinyon-juniper woodlands; *J. communis*, *J. occidentalis*, and *J. osteosperma* also occur in the range (1993–1994).
New record for the species in the range.

Ruby Mountains (1)
elevation record: 6700–7800 ft
5 records in 1 county (Elko)

Towns/Range/Section	Location—Habitat	County	Elevation	Collector	Herbarium/a
T32N R58E	middle Lamoille Canyon	EL	7800 ft	*ME Lewis 2102*	UTC, BRY
T32N R58E	Lamoille Canyon	EL	6700 ft	*C McMillan 244*	UT
T32N R58E	falls in Lamoille Canyon, south rocky slope	EL	7500 ft	*H Price 193*	UT
T32N R58E	Camp Lamoille, Lamoille Canyon	EL	7200 ft	*Lloyd Loope 756*	DUKE
—	Ruby Mountains	EL	—	*M Williams in 1961*	CAS

[Little: T31N R58E, T30N R57E R58E, T29N R57E R58E, T28N R57E]

Reported as present in the range by Little (1971).
DAC: occasional along most of the major streams (1988–1994).

Schell Creek Range (1)
elevation record: 7000–9185 ft
4 records in 1 county (White Pine)

Towns/Range/Section	Location—Habitat	County	Elevation	Collector	Herbarium/a
T17N R65E	N Fork Berry Creek, 2.5–4 mi above Berry Creek Ranger Station in scattered stands	WP	7000–8000 ft	*Percy Train 1038*	UC, RENO
T17N R65E	S Schell Creek Peak	WP	8650 ft	*Guyer 5099E & Nigro*	RENO
T17N R65E	———	WP	9185 ft	*Richard W. Rust 37*	UCR
T19N R66E	Meadow Canyon	WP	6000 ft	*Irwin Ting in 1959*	RENO

[DAC: T20N R65E (Kalamazoo Creek)]

Reported as present in the range by Little (1971).
DAC: occasional in most of the major drainages of the range (1994).

Seaman Range (4)
Lincoln and Nye Counties

Not reported as present in the range by Little (1971).
DAC: may yet be found in the range.

Sheep Range (1)
elevation record: 7500–9200 ft
6 records in 1 county (Clark)

Towns/Range/Section	Location—Habitat	County	Elevation	Collector	Herbarium/a
T15S R60E	Hidden Forest	CL	9200 ft	*PA Munz 16867*	CAS
T15S R60E	Hidden Canyon	CL	8000 ft	*LaRivers 614 & Hancock*	RENO
T15S R60E	S of Timber Canyon w/*Pinus longaeva*	CL	9000 ft	*T Ackerman 3511*	UNLV
T15S R60E	Wire Grass Spring	CL	8040 ft	*BH Banta in 1950*	RENO
T15S R60E S13	Wire Grass Spring	CL	8040 ft	*Richard W. Rust 52*	UCR
–	Sheep Range	CL	7500 ft	*OV Deming sn*	UTC

Reported as present in the range by Little (1971).
DAC: *Munz 16867* is the high-elevation record for the species in Nevada.

Snake Range (1)
elevation record: 6000–8500 ft
11 records in 1 county (White Pine)

Towns/Range/Section	Location—Habitat	County	Elevation	Collector	Herbarium/a
T12N R67E S1	Worthington spas	WP	—	*Kenneth Bradshaw KB325 & G Swallow*	OGDF
T12N R68E	Pole Canyon	WP	8500 ft	*I Ting 19401*	RENO
T12N R70E	Snake Creek	WP	6000 ft	*Ted Frantz in 1953*	RENO
T13N R69E	Lehman Creek	WP	—	*AL Menzies in 1962*	CAS
T13N R69E	Lehman Creek	WP	7800 ft	*Mozingo in 1962*	RENO
T13N R69E	Lehman Creek, 3 mi above Lehman Caves	WP	—	*Billings 1895*	RENO
T13N R69E	Lehman Caves	WP	8500 ft	*PA Lehenbauer in 1926*	RENO
T13N R69E	E slope below Wheeler Peak	WP	7840 ft	*Leila Shultz 6292 & J Shultz*	UTC, RENO
T16N R69E	4 mi W of Mount Moriah	WP	—	*JC Lyman 633 & Phillips Lyman*	UCR
T16N R70E S10	Mount Moriah, Hampton Creek	WP	7260 ft	*Charlet 2113 & T Charlet*	UC, RENO
—	Snake Range	WP	—	*AL Menzies in 1971*	CAS

[Lyman & Sanders: T16N R68E (Salt Marsh Canyon), T17N R70E (Smith Canyon)]

Reported as present in the range by Little (1971), in both the north and south.
DAC: common along most of the major streams (1994).

Spring Mountains (1)
elevation record: 7545–9085 ft
10 records in 1 county (Clark)

Towns/Range/Section	Location—Habitat	County	Elevation	Collector	Herbarium/a
T18S R56E	Lee Canyon, on ridge above Lee Canyon Guard Station	CL	8000 ft	*Vasek 630716-6*	UCR
T19S R55E	Upper Clark Canyon, NW of peak, SW face	CL	9000 ft	*J Beatley 13033 & Ackerman & Bamberg*	RENO
T19S R56E	ridge E of Deer Creek, Lee Canyon area, E face Spring Mts	CL	8000 ft	*J Beatley 9529 & Dainty & Jefferies*	RENO
T19S R57E	Deer Creek, 6 mi NE of Charleston Peak	CL	8500 ft	*LaRivers 473 & Hancock*	RENO
T19S R57E	ridge S of Deer Creek	CL	9085 ft	*IW Clokey 5387*	UNLV
T19S R57E	ridge S of Deer Creek	CL	8760 ft	*IW Clokey 5390*	UTC
T19S R57E	Kyle Canyon; flat hilltop, pinyon belt	CL	7545 ft	*IW Clokey 8260*	UT, UTC, BRY, RENO, DUKE
T19S R57E	ridge S of Deer Creek; hilltop	CL	8940 ft	*IW Clokey 7821*	UT, UTC, BRY, RENO, DUKE
T19S R57E S6 SE/4	———	CL	8200 ft	*AC Sanders 6880 et al.*	RSA
–	Water Canyon, Lee Canyon near spring	CL	8000 ft	*Percy Train 2100*	RENO

Reported as present in the range by Little (1971) and Beatley (1976).
DAC: common above 8000 ft with pinyon, white fir, and ponderosa pine (1982 & 1993).

Toiyabe Range (1)
elevation record: 7000–7200 ft
3 records in 1 county (Nye)

Towns/Range/Section	Location—Habitat	County	Elevation	Collector	Herbarium/a
T12N R42E	South Twin River	NY	7000 ft	*Linsdale & Linsdale 16*	CAS, UTC
T12N R42E	South Twin River	NY	7000 ft	*Linsdale & Linsdale 897*	CAS
T12N R42E S28	S Twin River, 2 mi above mouth of canyon along creek, 50–60 ft tall	NY	7200 ft	*T Frantz in 1954*	DUKE

DAC: I have not seen Rocky Mountain juniper on the mountain. I have not included the report by Linsdale et al. (1952) of its presence on Kingston Creek (*Linsdale 147*). I visited the collection locale in 1992 and 1993 and could not find the species, nor could I find the collection at CAS, UC, or AUSTIN, where it was reported to have been deposited. The species is definitely known only from the South Twin River, an occurrence verified by Kerry Heise (Toiyabe National Forest, pers. comm. 1995) (1991–1994).

Virgin Mountains (1)
elevation record: 5200–7300 ft
6 records in 1 county (Clark)

Towns/Range/Section	Location—Habitat	County	Elevation	Collector	Herbarium/a
T14S R71E S23	Cabin Canyon, seepage area, S end of creek	CL	6000 ft	*Mozingo 74-6*	RENO
T15S R71E S20	Big Springs	CL	7200 ft	*JD Armstrong 97*	UNLV
T15S R71E S7	Virgin Mts	CL	5200 ft	*J Armstrong 94*	UTC
T15S R71E	E flank of Virgin Peak in forest	CL	7300 ft	*Gary Baird 2861*	BRY
T15S R71E S13	along Nickel Creek	CL	5250 ft	*Gary Baird 3000*	BRY
T15S R70E S24	———	CL	7000 ft	*J Knight & S Cichowlaz in 1993*	UCR

Not reported as present in the range by Little (1971).
DAC: *Armstrong 94* is the low-elevation record for the species in the state.

White Pine Range (1)
elevation record: 7000–7800 ft
3 records in 1 county (Nye)

Towns/Range/Section	Location—Habitat	County	Elevation	Collector	Herbarium/a
T12N R58E S21 NE/4	N of White Pine Peak	NY	7800 ft	*Mozingo 76-54-11*	RENO, BRY
T12N R58E	Little Currant Creek, 9.8 mi E on Broom Creek Rd, N of White Pine Peak	NY	7300 ft	*Ann Pinzl 3474*	NSM
T12N R59E S20	streamside in pinyon-juniper w/*Abies concolor* & *Salix*	NY	7000 ft	*Charlet 1988*	RENO, UC
[Little: T14N R57E]					

Reported as present in the range by Little (1971).

****White Rock Mountains (2)**
Lincoln County
[RWR: T5N R70E]

Not reported as present in the range by Little (1971).
New report for the species in the range.

Wild Horse Range (1)
elevation record: 5900–6500 ft
7 records in 1 county (Elko)

Towns/Range/Section	Location—Habitat	County	Elevation	Collector	Herbarium/a
T45N R55E	Alleghany Creek (8000 ft) [DAC: Alleghany Creek does not ascend to 8000 ft; must be lower]	EL	—	*Aven Nelson & JF Macbride*	CAS
T45N R54E S13	3.5 mi W of Point of Rocks	EL	—	*Jon Carpenter 6-5-92*	UCR
T45N R54E S32	N aspect near *Abies lasiocarpa* & aspen	EL	6040 ft	*Charlet 2153*	UC, RENO, BRY
T45N R54E	1 mi SE of Wild Horse Crossing	EL	5900 ft	*Ann Pinzl 8781*	UNLV, NSM
T45N R54E	1.3 mi S of Wild Horse Campground	EL	—	*Vasek 640811-4*	UCR
T45N R55E	Humboldt National Forest, Bieroth Creek, E aspect, common	EL	6500 ft	*B Leonard 77BL*	OGDF
T45N R55E S7	2.5 mi W of Point of Rocks	EL	—	*Jon Carpenter 6-5-92*	UCR

Reported as present in the range by Little (1971).
DAC: common along East Fork Owyhee River; uncommon and scattered elsewhere (1991–1994).

****Wilson Creek Range (1)**
elevation record: 8140 ft
1 record in 1 county (Lincoln)

Towns/Range/Section	Location—Habitat	County	Elevation	Collector	Herbarium/a
T5N R68E S27 NW/4 [RWR: T6N R68E]	w/*Abies concolor*	LI	8140 ft	*Charlet 2095*	UC, RENO

Not reported as present in the range by Little (1971).
DAC: common in pinyon-juniper-mountain-mahogany woodlands bordering *Abies concolor* stands at high elevations (1994).
New record for the species in the range.

OTHER FEATURES IN NEVADA, RECORDS (13)

Elko County
elevation record: 6260–6320 ft
2 records

Towns/Range/Section	Location—Habitat	Elevation	Collector	Herbarium/a
T47N R60E S6 NE/4	Snake River Plains; above Sheep Creek	6260 ft	*Charlet 2128*	UC, RENO
T47N R60E S6 SW/4	Snake River Plains; west canyon rim above Pole Creek	6320 ft	*Charlet 2129*	UC, RENO

Nye County

DAC: Reese River valley; planted at USFS Reese River Guard Station, near Yomba Indian Reservation. Could escape into the adjacent Shoshone Mountains and/or the Toiyabe Range (1993–1994).

White Pine County
elevation record: 5500–6000 ft
11 records

Towns/Range/Section	Location—Habitat	Elevation	Collector	Herbarium/a
T18N R65E S30 NW/4	Duck Creek Valley, Berry Creek Crossing, SR 486	—	*Ann Pinzl 8840*	UNLV, RSA, NSM
—	Spring Valley, SE of Ely	—	*A Holmgren & D Holmgren 7827*	CAS
—	Spring Valley	6000 ft	*A Holmgren & D Holmgren 7826*	UCR
—	Spring Valley, W of Mount Wheeler, alkali valley bottom	5500 ft	*C McMillan 1350*	UT
T12N R67E	S Spring Valley, near Artesian Wells	5800 ft	*Chambers & Cain in 1970*	UTC
T12N R67E S24	Saline area, 0.75 mi E of Baking Powder Flat	6000 ft	*M Williams 82-51 & Rodd Hardy*	BRY, RENO
T14N R67E	Spring Valley, 38 mi E of Ely 39°6'30"N, 114°28'W	—	*TF Wieholdt 1253 & A Holmgren & J Major*	UTC
T15N R67E S34	Spring Valley	5800 ft	*RC Holmgren 542*	UTC, BRY
T15N R67E S26 NE/4	5 mi N of Crystal Queen Mine on big bend of Rt 6/50 in Spring Valley	5700 ft	*Mozingo 74-47*	RENO
—	Spring Valley	—	*Billings 1897*	RENO
—	Spring Valley	—	*T Frantz in 1953*	DUKE

DAC: the last 10 collections reported here are all from Spring Valley, west of the Snake Range and east of the Schell Creek Range. The species is widespread in the valley and well separated from the populations that occur along streams and mountain slopes in both the Schell Creek and Snake Ranges. The populations in Spring Valley have been called "swamp cedars" and are considered by Billings (1954) to be a possible ecotype of Rocky Mountain juniper (1994).

Pinaceae: Pine family

The pine family is represented by 17 native species in 5 genera (*Abies, Picea, Pinus, Pseudotsuga,* and *Tsuga*) in Nevada. Of the 3 Nevada species of *Abies,* only *A. concolor* is widespread, and it is restricted to the eastern and western borders of the state. The only spruce native to Nevada, *Picea engelmannii,* forms forest stands with limber pine in a few eastern mountain ranges and occurs in the Snake Range (White Pine County) as krummholz at timberline. Mountain hemlock (*Tsuga mertensiana*) is restricted to subalpine woodlands and timberline in far western Nevada. Rocky Mountain Douglas-fir (*Pseudotsuga menziesii* var. *glauca*) is present at a few isolated locations in eastern Nevada, and may have become locally extinct in Elko County during the last 50 years. A single, spectacular Pacific Douglas-fir (*Pseudotsuga menziesii* var. *menziesii*) on the east shore of Lake Tahoe (Douglas County) may be a relict of a considerably more widespread population in the Lake Tahoe basin that was destroyed by logging in the 1860s to provide timbers for the Comstock mines.

The 11 species of *Pinus* in Nevada (the most species of any conifer genus) have varied distribution patterns. Singleleaf pinyon (*Pinus monophylla*), present in 128 Nevada mountain ranges, is by far the most widespread of the pine species in the state, although *P. flexilis* is known from 50 mountain ranges.

Abies concolor (Gordon & Glendinning) Hildebrand

varieties in Nevada:
A. concolor var. *concolor* [Rocky Mountain white fir]
A. concolor var. *lowiana* (Gordon) Lemmon [California white fir]
common name: white fir
elevation: 6000–9500 ft (1830–2895 m)
zones: Hudsonian, Canadian, Arid Transition
118 Records for 10 counties (CC, CL, DO, EL, LI, LY, NY, ST, WA, WP) (*A. concolor* var. *lowiana* in CC, DO, LY, ST, WA; *A. concolor* var. *concolor* in CL, EL, LI, NY, WP)
Reports for 31 mountain ranges (*A. concolor* var. *lowiana* in 6, *A. concolor* var. *concolor* in 24), Records for 26 of these
New Reports for 7 mountain ranges and 1 other feature in Nevada, New Records from 4 of these

White fir is found in all the major mountain systems of 8 western states and northern Mexico. The typical, or Rocky Mountain, variety (*A. concolor* var. *concolor*) occurs in Idaho, Utah, Colorado, Arizona, New Mexico, Nevada, and southeastern California. California white fir (*A. concolor* var. *lowiana*) is found only in Nevada, Oregon, and California. The Rocky Mountain variety is much more widespread in Nevada, occurring from close to the southern border with California (Spring Mountains, Clark County) to near the northern border (Pilot Range, Elko County), while the California variety occurs only along the northern half of the western boundary with California. Richard S. Hunt (in Morin 1993) resurrected the old name *Abies lowiana* (Gordon) A. Murray in *Flora of North America* to elevate the California white fir to species rank. The varieties are well separated in Nevada because the western limit of the Rocky Mountain variety is at about longitude 116°W in the Spring Mountains of Clark County, and the eastern limit of the California variety is slightly west of longitude 119°W in the Sweetwater Mountains of Lyon County.

Within its range, white fir is least common in Idaho and Nevada, even though it occurs in 31 Nevada mountain ranges in 10 counties (Carson City, Clark, Douglas, Elko, Lincoln, Lyon, Nye, Storey, Washoe, and White Pine). The California variety is most common in the Carson Range, where it is codominant with Jeffrey pine in the Arid Transition Zone forests. The Rocky Mountain variety dominates Canadian Zone forests with limber pine in the high mountains of White Pine County, and occasionally reaches the Hudsonian Zone. Elsewhere in Nevada, white fir is generally restricted to canyons with perennial streams at low elevations and hollows at high elevations, giving the species an overall elevation range in Nevada of 3500 ft (1065 m). New reports of the species are made for 7 Nevada mountain ranges in Elko, White Pine, and Washoe Counties; new records are presented for 4 of these mountain ranges.

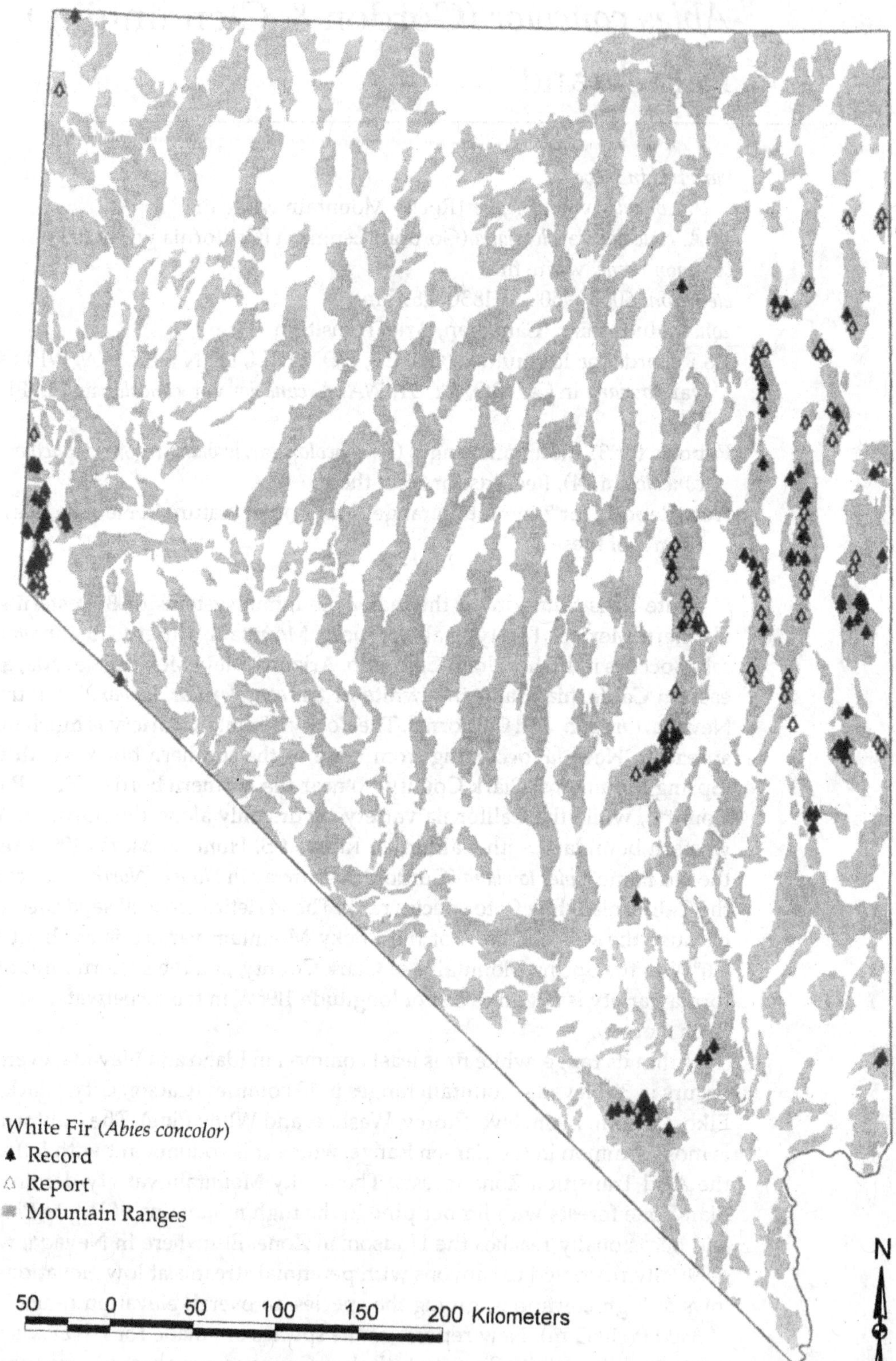
White Fir (*Abies concolor*)
Record
Report
Mountain Ranges
N
50 0 50 100 150 200 Kilometers

****Antelope Range (1)**
elevation record: 8740 ft
1 record in 1 county (White Pine)

Towns/Range/Section	Location—Habitat	County	Elevation	Collector	Herbarium/a
T24N R67E S4	nivation hollow, NW aspect w/*P. flexilis*	WP	8740 ft	*Charlet 2028*	DUKE, RENO
[DAC: T24N R67E S6 S7 (N of Rock Springs Pass), T24N R67E S21, T23N R67E]					

Not reported as present in the range by Little (1971).
DAC: var. *concolor* is common at high elevations; at the head of Tunnel Canyon and north of Rock Springs Pass (1994).
New record for the species in the range.

Carson Range (1)
elevation record: 6100–7800 ft
18 records in 3 counties (Carson City, Douglas, and Washoe)

Towns/Range/Section	Location—Habitat	County	Elevation	Collector	Herbarium/a
T13N R18E S13 N/8	N of Castle Rock	DO	7700 ft	*Charlet 2171 & Charlet*	DUKE, RENO, UC
T14N R18E S10	Glenbrook	DO	—	*PA Lehenbauer in 1929*	RENO
T14N R18E S22	1 air mile N of Cave Rock; w/*Pseudotsuga*	DO	6240 ft	*Charlet 2058 & T Charlet*	DUKE, RENO, UC
T14N R19E S4	Clear Creek campground, on US 50 6 mi W of jct w/US 395	DO	6500 ft	*W Martin 20132*	RSA
T14N R19E S6	Spooner Summit	DO	7700 ft	*RJ Holland in 1949*	RENO
T15N R19E S28	upper Kings Canyon	CC	6760 ft	*Charlet 1945 & Phoenix King*	DUKE, RENO, UC
T15N R18E S11	near Marlette reservoir	WA	6700 ft	*Charlet 1920 & T Charlet*	DUKE, RENO, UC, BRY, UNLV
T16N R18E S18	3 mi E of Brockway, N end Lake Tahoe	WA	6300 ft	*Nichols 680 & Lund*	RENO
T16N R18E	Incline, growing at meadow edge, N of hwy	WA	—	*TC Frantz in 1951*	RENO
T16N R19E	mountains W of Franktown	WA	6100 ft	*AA Heller in 1912*	CAS
T16N R19E	E of Little Valley	WA	—	*R Everett in 1970*	RENO
T17N R19E S9	Galena Creek, 7 mi W of Reno Hot Springs	WA	6200 ft	*Billings 1173*	RENO
T17N R19E S27	above Bower's Mansion	WA	—	*PA Lehenbauer 625*	RENO
T17N R19E S32	Prices Lake	WA	7200 ft	*TL Breene 68 & H Sampson*	RENO
T18N R18E S11	Hunter Creek Canyon	WA	7800 ft	*AA Heller 10373*	RENO
T18N R19E	Dutch Flat, 10 mi SW of Reno	WA	6800 ft	*Nichols & Lund 633*	CAS, RENO
T19N R18E	6 mi W of Reno on road to Hunter Creek	WA	6200 ft	*WA Archer 5135*	RENO
—	canyon near Mount Rose	WA	—	*NF Petersen 161*	RENO

[DAC: (WA) T19N R19E (Alum Creek), T17N R18E, T15N R19E; (CC) T15N R18E; (DO) T14N R19E, T13N R19E, T12N R19E]

Reported as present in the range by Little (1971).
DAC: var. *lowiana* is abundant in the middle elevational range of Jeffrey pine forest. The species suffered high mortality from bark beetles and drought between 1990 and 1994 (1971–1995).

Cherry Creek Range (1)
elevation record: 6800–8800 ft
2 records in 1 county (White Pine) and reports in another (Elko)

Towns/Range/Section	Location—Habitat	County	Elevation	Collector	Herbarium/a
T25N R62E S25	Paris Creek; lowest one observed	WP	6800 ft	*Charlet 2033*	DUKE, RENO, UC
T25N R63E S4	between Currie Summit & Goshute Basin	WP	8800 ft	*Charlet 2115 & T Charlet*	DUKE, RENO

[Loope: (EL) T27N R62E, T28N R62E R63E]
[DAC: (EL) T27N R63E S19 S20 S29, T26N R63E; (WP) T26N R63E, T25N R62E S1 S12 S23, T25N R63E S6]

Not reported as present in the range by Little (1971).
Reported as present in the range by Loope (1969).
DAC: var. *concolor* is abundant in the higher elevations, forming forest islands with limber pine (1994).

****Dolly Varden Mountains (2)**
Elko County
[RWR: T28N R66E S5 S8 S9 S16]

Not reported as present in the range by Little (1971).
RWR: common below Bald Peak, on north and east aspects.
New report for the species in the range.

****Duck Creek Range (1)**
elevation record: 8700–9260 ft
2 records in 1 county (White Pine)

Towns/Range/Section	Location—Habitat	County	Elevation	Collector	Herbarium/a
T16N R64E S24 N/4	N of Camel Peak, E slope in forest island	WP	8700 ft	*Charlet 2021*	DUKE, RENO
T16N R64E S13 E/4	ridge N of Camel Peak, subalpine woods	WP	9260 ft	*Charlet 2018*	DUKE, RENO

Not reported as present in the range by Little (1971).
DAC: var. *concolor* is abundant, forming large forest islands and occupying protected sites near the main ridge (1994).
New record for the species in the range.

Egan Range (1)
elevation record: 7500–9000 ft
5 records in 1 county (White Pine) and reports in another (Lincoln)

Towns/Range/Section	Location—Habitat	County	Elevation	Collector	Herbarium/a
T12N R63E S28 N/4	top of Sawmill Canyon at pass	WP	8240 ft	*Charlet 2001 & Walter Cripps*	DUKE, RENO
T15N R62E	NW slope of Ward Mountain	WP	9000 ft	*W Sandow in 1953*	RENO
T15N R62E S36	Lower terrace, W of Ward Mountain	WP	8170 ft	*Richard W. Rust 6*	UCR
T16N R62E	Murry Summit, w/pinyon-juniper	WP	7900 ft	*Guyer 5065 & Nigro*	RENO
T22N R62E S36	Telegraph Canyon Rd	WP	7500 ft	*Ann Pinzl 3473*	NSM
[DAC: (WP) T12N R63E S21 S33, T15N R63E, T14N R64E, T13N R63E, T10N R64E; (LI) T9N R64E (Mount Grafton)]					

Not reported as present in the range by Little (1971).
DAC: var. *concolor* is abundant in the northern portion of the range at higher elevations, where it forms small forest stands with limber pine (1994).

Goshute Mountains (1)
elevation record: 6800 ft
1 record in 1 county (Elko)

Towns/Range/Section	Location—Habitat	County	Elevation	Collector	Herbarium/a
T30N R68E S6	"Toana Mtn" [Goshute Mts]	EL	6800 ft	*Lund 87 & Holmgren & Cooper*	UTC, BRY
[Loope: T30N R68E (reported as "Toana Range" in text)]					

Reported as present in the range by Loope (1969) and Little (1971).
RWR: var. *concolor* was observed at T30N R68E S35.

Grant Range (1)
elevation record: 7200–9000 ft
5 records in 1 county (Nye)

Towns/Range/Section	Location—Habitat	County	Elevation	Collector	Herbarium/a
T6N R57E	Troy Canyon, abundant on S side of canyon	NY	—	*Rogers McVaugh 6053*	UC
T5N R57E	Troy Peak	NY	8850 ft	*Guyer 5039 & Nigro*	RENO
T5N R58E S15	Scofield Canyon	NY	7440 ft	*Charlet 2014*	DUKE, RENO
T5N R58E S18	E face Troy Peak, Scofield Canyon	NY	9000 ft	*NK Johnson in 1953*	DUKE
T4N R57E S19 E/4	between Burnt Canyon & Cherry Creek Summit	NY	7200 ft	*Charlet 2003*	DUKE, RENO

[DAC: T5N R57E S7 S17, T8N R58E S19 (Blue Eagle Mountain)]

Reported as present in the range by Little (1971).
DAC: var. *concolor* is abundant at higher elevations (1994).

Groom Range (1)
elevation record: 8000–8500 ft
1 record in 1 county (Lincoln)

Towns/Range/Section	Location—Habitat	County	Elevation	Collector	Herbarium/a
T6S R56E S6	Bald Mountain, E Groom-W Desert Valley drainage N slope; AEC Test Site, w/pinyon at 8000 ft	LI	8500 ft	*J Beatley & R Rasp 5292*	CAS, RENO

Reported as present in the range by Little (1971) and Beatley (1976).
DAC: var. *concolor;* although the label says "Timpahute Range," there is no Bald Mountain there, and 8500 ft exceeds the highest point in the range. Further, Beatley (1976) reports the species as present at Bald Mountain, Groom Range. This must be the site of the collection.

Hays Canyon Range (2)
Washoe County
[Critchfield & Allenbaugh: T42N R18E S12 (head, Fortynine Canyon, 6000 ft)]

Reported as present in the range by Critchfield & Allenbaugh (1966) and Little (1971).
DAC: I viewed this stand of some 4 large reproductives (probably var. *lowiana*) and at least 10 saplings from the pass between Vya and Cedarville. The stand is growing with aspen on a north aspect on White Pine Corporation property (1994).

Highland Range (1)
elevation record: 7200–9000 ft
3 records in 1 county (Lincoln)

Towns/Range/Section	Location—Habitat	County	Elevation	Collector	Herbarium/a
T1N R66E S35	Pine Spring	LI	7200 ft	*Ann Pinzl 3367*	NSM
T1N R66E S33	dominant on N slope Highland Peak	LI	< 9000 ft	*V Bostick 5869*	UNLV
T1S R66E S26	Comet Peak (determined DAC)	LI	—	*ME Jones in 1912*	RSA

Not reported as present in the range by Little (1971).
DAC: var. *concolor* was observed at Highland Peak along with a few *Abies lasiocarpa* (1994).

****Kern Mountains (1)**
elevation record: 8740 ft
1 record in 1 county (White Pine)

Towns/Range/Section	Location—Habitat	County	Elevation	Collector	Herbarium/a
T21N R69E S22	ridge above Tungstonia & Rock Springs	WP	8740 ft	*Charlet 2109*	DUKE, RENO, UC
[DAC: T21N R70E]					

Not reported as present in the range by Little (1971).
DAC: var. *concolor* is extensive with limber pine in many sections within T21N R70E in the southeast portion of the range.
New record for the species in the range.

Mosquito Mountains (1)
elevation record: 6000–6840 ft
3 records in 1 county (Washoe)

Towns/Range/Section	Location—Habitat	County	Elevation	Collector	Herbarium/a
T47N R19E S27 E/2	41°58′N, 119°50′W, canyon running NE toward Coleman Ranch; oldest trees > 500 yr, w/most < 100 yr; 10–15 acres, 50–60% canopy cover	WA	6000 ft	*Critchfield in 1965*	IFGP
T47N R19E S27 S/4	head of canyon 4 air miles NNW of Little Coleman Canyon	WA	6840 ft	*Charlet 1864 & John Conlan*	DUKE, UC, RENO, BRY, UNLV
T47N R19E S27 S/4	head of canyon 4 air miles NNW of Little Coleman Canyon	WA	6840 ft	*Charlet 1865 & John Conlan*	DUKE, UC, RENO, BRY, UNLV

Reported as present in the range by Critchfield & Allenbaugh (1969) and Little (1971).
DAC: var. *lowiana* forms this large, impressive stand, which had dead and dying individuals in May 1994, apparently from bark beetles. *Charlet 1864* and *Charlet 1865* were taken from the upper limit of the forest island. *Critchfield in 1965* is the low-elevation record for the species in the state (1994).

Peavine Mountain (1)
elevation record: 6300–7800 ft
5 records in 1 county (Washoe)

Towns/Range/Section	Location—Habitat	County	Elevation	Collector	Herbarium/a
T20N R18E	1 mi W of Peavine Peak	WA	–	*Billings 1447*	RENO
T20N R18E	ridge W of summit	WA	7800 ft	*Tiehm & McPherson in 1974*	RENO
T20N R18E	N slope Peavine	WA	6300 ft	*RJ Bandoni 4*	RENO
–	NE slope Peavine	WA	6500 ft	*N Wood in 1953*	RENO
–	Peavine Mountain	WA	7000 ft	*Ann Pinzl in 1974*	NSM

Not reported as present in the range by Little (1971).
DAC: var. *lowiana* is frequent along the northeast drainage below Peavine Peak (1971–1995).

Pequop Mountains (1)
elevation record: 8200–8800 ft
3 records in 1 county (Elko)

Towns/Range/Section	Location—Habitat	County	Elevation	Collector	Herbarium/a
T31N R63E	Spruce Mountain	EL	–	*Eugene Zavarian 728-737*	IFGP
T31N R63E S23	Spruce Mountain	EL	8200 ft	*Ann Pinzl 2073*	NSM
T31N R63E	Spruce Mountain, E slope, lower forest border	EL	8800 ft	*Lloyd Loope 1348*	DUKE
[Loope: T32N R65E, T30N R63E]					

Reported as present in the range by Little (1971).
DAC: var. *concolor* is the taxon here.

****Petersen Mountain (3)**
Washoe County
[Michael Phillips: T22N R18E S9 (S end of Long Valley)]

Not reported as present in the range by Little (1971).
DAC: var. *lowiana* was reported by Michael Phillips (USDI, BLM, pers. comm. 1991), but I did not verify. A large fire burned Long Valley in June 1994, and these trees may have been extirpated (1990–1994).
New report for the species in the range.

Pilot Range (1)
elevation record: 7000–8500 ft
3 records in 1 county (Elko)

Towns/Range/Section	Location—Habitat	County	Elevation	Collector	Herbarium/a
T36N R70E	upper pinyon-juniper limit, W slope of range	EL	—	*Lloyd Loope 1373*	DUKE
T37N R70E	W aspect, bristlecone woodland	EL	8500 ft	*Guyer 5194 & Nigro*	RENO
—	Pilot Mountain, near stream	EL	7000 ft	*WP Cottam 12693*	UT

Reported as present in the range by Loope (1969) and Little (1971).
DAC: Loope (1969) reported var. *concolor* here.

****Quinn Canyon Range (1)**
elevation record: 6700–7320 ft
2 records in 1 county (Nye)

Towns/Range/Section	Location—Habitat	County	Elevation	Collector	Herbarium/a
T3N R57E S6	Little Cherry Creek	NY	6700 ft	*Charlet 2006*	DUKE, RENO, UC
T3N R56E S10 E/4 [DAC: T3N R55E]	Sawmill Canyon	NY	7320 ft	*Charlet 2010*	DUKE, RENO

Not reported as present in the range by Little (1971).
DAC: var. *concolor* is common at higher elevations and along streams (1994).

Ruby Mountains (1)
elevation record: 7200–8500 ft
1 record in 1 county (Elko)

Towns/Range/Section	Location—Habitat	County	Elevation	Collector	Herbarium/a
T32N R58E S18	Seitz Canyon, only known locality for the species in this range	EL	7200 ft	*Lloyd Loope 676*	DUKE
[Critchfield & Allenbaugh: Seitz Canyon 8500 ft, 40°39′N, 115°20′W]					

Reported as present in the range by Critchfield & Allenbaugh (1969), Loope (1969), and Little (1971).
DAC: Loope (1969) reports the taxon to be var. *concolor.*

Schell Creek Range (1)
elevation record: 6500–9500 ft
9 records in 1 county (White Pine) and reports in another (Lincoln)

Towns/Range/Section	Location—Habitat	County	Elevation	Collector	Herbarium/a
T16N R65E S23	Kolchek Basin, limestone cliffs	WP.	7380 ft	*Charlet 1998*	DUKE, RENO
T16N R65E S29 E/4	Steptoe Creek, grove at base of limestone cliffs, w/*Cercocarpus*	WP	7800 ft	*Charlet 2017*	DUKE
T17N R65E S10	N Fork, Berry Creek, 2.5–4 mi above Berry Creek Ranger Station	WP	8000 ft	*Percy Train 1041*	UTC, RENO, DUKE
T17N R65E S10	Berry Creek; N aspect, open forest	WP	8650 ft	*Guyer 5099C & Nigro*	RENO
T17N R66E S14 SW/4	S Fork Taft Creek in pinyon-juniper	WP	6500 ft	*Charlet 1996*	DUKE, RENO, UC
T19N R65E S2 N/8	Kalamazoo Summit	WP	8940 ft	*Charlet 2103 & T Charlet*	DUKE, RENO, UC
T20N R65E S25 S/4	Kalamazoo Creek	WP	6740 ft	*Charlet 2025*	DUKE
T21N R65E	Aurum	WP	—	*ME Jones in 1891*	RSA
T21N R64E	Indian Creek, 16 mi N of McGill	WP	8000–9500 ft	*Percy Train 981*	UC, RENO

[DAC: (WP) T16N R60E, T17N R66E, T18N R65E R66E, T19N R65E; (LI) T9N R65E S27 (limestone terraces around Quaker Spring)]
[RWR: (WP) T24N R65E S33 (head Dry Canyon), T15N R65E S14 (Cooper Summit), T10N R64E S36; (LI) T9N R64E (Mount Grafton)]

Reported as present in the range by Loope (1969) and Little (1971).
DAC: var. *concolor* is abundant throughout much of the north-central portion of the range and is common at high elevations in the south (1994).

Seaman Range (4)
Lincoln and Nye Counties

Not reported as present in the range by Little (1971).
DAC: var. *concolor* may yet be found here.

Sheep Range (1)
elevation record: 7545–8040 ft
3 records in 1 county (Clark)

Towns/Range/Section	Location—Habitat	County	Elevation	Collector	Herbarium/a
T14S R60E S25	Basin Spring	CL	7545 ft	*T Ackerman 6701*	UNLV
T15S R60E S13	Hidden forest, Wire Grass Spring	CL	8040 ft	*BH Banta in 1950*	RENO
T15S R60E	"Hidden Forest Canyon" [Deadman Canyon]	CL	—	*OV Deming in 1947*	UNLV

Not reported as present in the range by Little (1971).
DAC: the taxon here is var. *concolor.*

Snake Range (1)
elevation record: 6800–9500 ft
14 records in 1 county (White Pine)

Towns/Range/Section	Location—Habitat	County	Elevation	Collector	Herbarium/a
T11N R69E S3	S of Wheeler Peak, along road to Lexington Arch; w/aspen	WP	8000 ft	*Wilbert Blackburn 65*	BRY
T12N R68E S15	Pole Canyon, W side of Mount Washington	WP	9000 ft	*N Holmgren 3766 & A Holmgren*	UTC
T12N R69E S7	Snake Creek, 2 mi below Johnson Lake	WP	9500 ft	*I Ting 19394*	RENO
T13N R68E S6	ridge top in *Cercocarpus*, on road to Mount Wheeler	WP	—	*S Welsh 2241 & G Moore*	BRY
T13N R68E	Wheeler Peak, E slope w/pinyon-juniper and mountain-mahogany	WP	6800 ft	*E McClintock in 1969*	CAS
T13N R69E S7	NE slope Wheeler Peak, creek bottom, Baker Creek headwaters	WP	8000–8200 ft	*Percy Train 1147*	RENO, RSA
T13N R69E S21	Baker Creek, N aspect	WP	8500 ft	*Frank Gray 190*	OGDF
T13N R69E S5	1 mi N of Lehman Creek campground, 32°1.5′N, 114°15.5′W	WP	8200 ft	*Critchfield & Steinhoff (1-4)-9/8/64*	IFGP
T13N R69E S9	below Wheeler Peak, 6 mi W of Baker	WP	7840 ft	*Leila Shultz 6293 & J Shultz*	UTC
T13N R69E S16	Lehman Caves	WP	8500 ft	*PA Lehenbauer in 1926*	RENO
T16N R70E S16	Hampton Creek; common along creek, at quarries	WP	7300 ft	*Charlet 2112 & T Charlet*	DUKE, RENO

[Lyman & Sanders (1993): T16N R69E (Cricket Canyon), T17N R69E (head of Smith Canyon)]

Reported as present in the range by Little (1971).
DAC: var. *concolor* is abundant in forests with *Picea engelmannii* and *Pinus flexilis;* abundant also along streams with *Pinus ponderosa* var. *scopulorum*. *I Ting 19394* is the high-elevation record for the species in the state (1994).

Spring Mountains (1)
elevation record: 6400–8690 ft
19 records in 1 county (Clark)

Towns/Range/Section	Location—Habitat	County	Elevation	Collector	Herbarium/a
T18S R56E S25	Lee Canyon	CL	8100 ft	*Critchfield & Steinhoff (1-6)-9/21/64*	IFGP
T18S R57E S19	Lee Canyon	CL	7000 ft	*Vasek 770304-5*	UCR
T19S R55E S8 NE/4	Clark Canyon, 0.5 mi E of jct w/Wheeler Pass Rd	CL	6400 ft	*Vasek 790918-1*	UCR
T19S R55E	Clark Canyon	CL	7000 ft	*Vasek 770304-1*	RSA
T19S R55E	Clark Canyon	CL	7000 ft	*Vasek 790918-5*	RSA
T19S R56E S7	Upper Clark Canyon	CL	7600 ft	*J Beatley 10716 & Reveal*	RENO
T19S R56E S35	Kyle Canyon (little falls)	CL	7955 ft	*IW Clokey A134*	OGDF
T19S R57E S17	Kyle Canyon–Deer Creek, bottom of box canyon	CL	8200 ft	*IW Clokey 8261*	UC, RSA, UT, UTC, RENO
T19S R57E S5	Deer Creek, 6 mi NE of Charleston Peak	CL	8500 ft	*LaRivers 481 & Hancock*	RENO
T19S R57E S6	Deer Creek trail to Lee Canyon	CL	8400 ft	*EK Balls 19289 & R Straw*	RSA
T19S R57E S10	ridge S of Deer Creek	CL	8690 ft	*IW Clokey 5383*	UTC
T19S R57E S28	Kyle Canyon	CL	7000 ft	*Vasek 770304-6*	UCR
T19S R57E S28	Kyle Canyon	CL	7000 ft	*Vasek 770304-10*	RSA
T20S R58E S15	N of La Madre Mountain	CL	7200 ft	*J Butler 4071 & PJ Leary 4071*	UNLV

[DAC: T19S R56E S10 (end of Lee Canyon Road)]

Reported as present in the range by Little (1971) and Beatley (1976).
DAC: var. *concolor* is abundant in forests with *Pinus ponderosa* and *P. flexilis* (1982 & 1983).

Sweetwater Mountains (1)
elevation record: 7500 ft
1 record in 1 county (Lyon)

Towns/Range/Section	Location—Habitat	County	Elevation	Collector	Herbarium/a
T7N R24E S12	Sweetwater Creek, w/*Pinus jeffreyi* and *P. monophylla*	LY	7500 ft	*Billings 1846*	RENO

Reported as present in the Nevada portion of the range by Little (1971).
DAC: var. *lowiana* is common in aspen and Jeffrey pine woods near Cottonwood and Quaking Aspen Creeks in the southeast part of the range (1985–1995).

Toana Range (3)
Elko County
[Little: T36N R68E]

Reported as present in the range by Little (1971).

Virginia Range (1)
elevation record: 5500–6800 ft
5 records in 2 counties (Storey and Washoe)

Towns/Range/Section	Location—Habitat	County	Elevation	Collector	Herbarium/a
T17N R21E S20	steep NE slope in canyon, 1 mi NW of Virginia City	ST	6800 ft	*Billings 1772*	RENO
T17N R21E S20 NW/4	N slope of Cedar Hill Canyon, NW of Virginia City	ST	6780 ft	*Jan Nachlinger 1493 & Billings & Billings*	DUKE
T17N R21E S20	0.5 mi NW of Fivemile House, NW of Virginia City	WA	6500 ft	*KE Bradshaw 185*	UC
T17N R21E S20	N slope, on altered rock	WA	6400 ft	*Billings 2020*	RENO
T17N R21E S6	altered andesite	WA	5500 ft	*Billings 1769*	RENO

Not reported as present in the range by Little (1971).
DAC: var. *lowiana* is rare in the Virginia Range, restricted to only 2 known locations and 4 individuals (1973–1995).

Virgin Mountains (1)
elevation record: 7000–7800 ft
3 records in 1 county (Clark)

Towns/Range/Section	Location—Habitat	County	Elevation	Collector	Herbarium/a
T15S R71E	North Virgin Peak	CL	7800 ft	*Ann Pinzl in 1974*	NSM
T15S R70E S24	just NE of Virgin Peak	CL	7000 ft	*Mozingo 71-5*	RENO
T15S R70E S24	small stand near peak	CL	7000 ft	*J Knight & SD Cichowlaz in 1993*	DUKE

Not reported as present in the range by Little (1971).
DAC: var. *concolor* is the taxon here.

White Pine Range (1)
elevation record: 7300–9500 ft
4 records in 2 counties (White Pine and Nye)

Towns/Range/Section	Location—Habitat	County	Elevation	Collector	Herbarium/a
T12N R58E S21 NE/4	end of Broom Creek Rd, N of White Pine Peak	NY	7800 ft	*Mozingo 76-54-7*	RENO
T12N R58E	Little Currant Creek	WP	7300 ft	*Ann Pinzl 3473*	NSM
T12N R59E S20	along stream w/*Salix*, near *J. scopulorum*	WP	7000 ft	*Charlet 1989*	DUKE, RENO, UC
T16N R57E	pure groves, ridge SW of Hamilton [Pogonip Ridge]	WP	9500 ft	*Billings 2194*	RENO

[DAC: (WP) T12N R59E (S of Red Mountain), T12N R59E S21 S35 (Red Mountain), T12N R58E S11 (Currant Mountain), T14N R58E S3, T15N R57E S2 S11 (Mount Hamilton), T15N R58E S34, T16N R57E S3 S10]

Not reported as present in the range by Little (1971).
DAC: var. *concolor* is abundant and sometimes dominant across much of the higher elevations. In July 1994, a 40-acre stand on Mount Hamilton, with nesting flammulated owls (Susie Dunham, pers. comm. 1994), was being cut to make way for a new mine. *Billings 2194* is the high-elevation record for the species in the state (1994).

****White Rock Mountains (2)**
White Pine County
[RWR: T5N R70E S34, T4N R70E S3]

Not reported as present in the range by Little (1971).
RWR: var. *concolor* occurs in large stands on the western slope of White Rock Peak.
New report for the species in the range.

Wilson Creek Range (1)
elevation record: 7800–9500 ft
4 records in 1 county (Lincoln)

Towns/Range/Section	Location—Habitat	County	Elevation	Collector	Herbarium/a
T5N R68E S27 NW/4	about 100–150 reproductives in small grove	LI	8140 ft	*Charlet 2094*	DUKE, RENO, UC
T5N R68E S31	Wilson Mountain, W slope, dominant	LI	8400–9500 ft	*Percy Train 2563*	RENO
T6N R68E S2	E of Pony Springs, Rosencrans Mts	LI	8500 ft	*G Christensen in 1952*	DUKE
—	Wilson Creek Rd	LI	7800 ft	*Ann Pinzl 3409*	NSM

[RWR: T5N R68E S32, T4N R68E S5 S16 S21 S27]

Reported as present in the range by Little (1971).
DAC: abundant, and sometimes dominant, in forest stands throughout the higher elevations, usually at north aspect sites (1994).

OTHER FEATURE IN NEVADA, REPORT

Lincoln County

[DAC: T7N R65E S27 (Dutch John Mountain)]

Not reported as present on the mountain by Little (1971).
DAC: small stands (probably var. *concolor*) line limestone terraces on the east face of Dutch John Mountain. This site burned in the summer of 1994, but many individuals remain (1994).
New report for the species on the mountain.

Abies lasiocarpa (Hooker) Nuttall

common name: subalpine fir, alpine fir; known as balsam fir in Elko County
elevation: 6040–10,000 ft (1840–3050 m) (observed at 10,500 ft)
zones: Arctic-Alpine, Hudsonian, Canadian
25 Records for 4 Nevada counties (EL, WP, NY, LI)
Reports for 15 mountain ranges, Records for 13 of these
New Record for 1 mountain range

Subalpine fir occurs throughout the Rocky Mountains, from latitude 63°N in Alaska, through the mountains of western Canada, and south through the Rocky Mountain System to near the Mexico border in Arizona and New Mexico. The species is also found in the Pacific Coastal System in the United States from Washington to the Oregon-California border. Only the typical variety is found in Nevada. Richard S. Hunt (in Morin 1993) separated the typical variety into 2 species, with *Abies lasiocarpa* in the Cascades and northern California, and *Abies bifolia* A. Murray occurring throughout Nevada and the Rocky Mountains.

Subalpine fir is absent from the Sierra Nevada, but it occurs in 15 mountain ranges in Nevada, mostly in the Jarbidge Mountains complex in northern Elko County, and in small populations near timberline in the tall mountains of White Pine, Nye, and Lincoln Counties. In Elko County, subalpine fir is often dominant and sometimes codominant with whitebark pine in the subalpine forests and woodlands of the Hudsonian Zone. It forms nearly pure forest stands in the upper Canadian Zone of the Jarbidge Mountain complex (Elko County). The species adopts a mat form above timberline, and so extends into the Arctic-Alpine Zone, giving it an overall elevation range in Nevada of 3960 ft (1210 m). A new record for the species is presented for the Wild Horse Range in Elko County.

Little (1971) indicates that subalpine fir is found only in Elko County in Nevada, but herbaria searches revealed collections (*Arthur Cronquist 11112* at BRY, *Walter P. Cottam 3320 & Biddulph* at BRY, and *Vernon Bostick 5870* at UNLV) from 3 eastern Nevada counties (Lincoln, Nye, and White Pine) at timberline. Look for subalpine fir at timberline in the other high ranges of these counties.

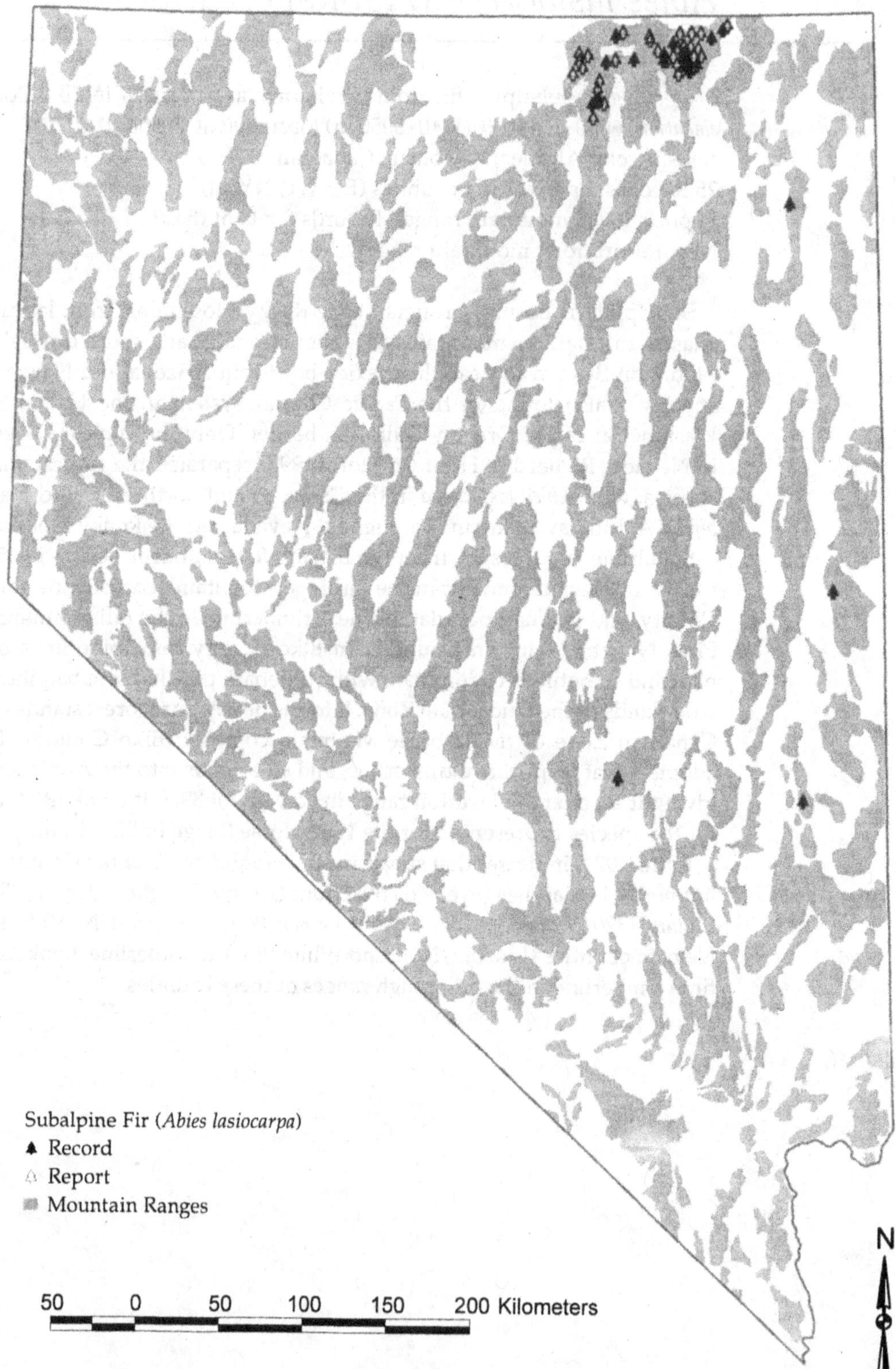
Subalpine Fir (*Abies lasiocarpa*)
Record
Report
Mountain Ranges
50 0 50 100 150 200 Kilometers
N

Bruneau Range (1)
elevation record: 7180 ft
1 record in 1 county (Elko)

Towns/Range/Section	Location—Habitat	County	Elevation	Collector	Herbarium/a
T46N R54E S12	NW slope of Merritt Mountain	EL	7180 ft	*Charlet 2152 & T Charlet*	DUKE, RENO
[Loope: T47N R55E]					

Not reported as present in the range by Little (1971).
Reported as present by Loope (1969).
DAC: a few small stands and several scattered individuals occur on Merritt Mountain, mostly on the northern aspect (1988–1994).

Buck Creek Mountains (2)
Elko County
[Loope: T46N R57E, T47N R57E]

Reported as present in the range by Little (1971).
DAC: isolated pockets of the species occur at high elevations (1993–1994).

Bull Run Mountains (2)
Elko County
[Loope: T46N R52E, T45N R52E R53E, T44N R52E]

Reported as present in the range by Critchfield & Allenbaugh (1969), Loope (1969), and Little (1971).
DAC: large forest stands grow at high elevations (1991–1994).

Copper Mountains (1)
elevation record: 8200–8840 ft
2 records in 1 county (Elko)

Towns/Range/Section	Location—Habitat	County	Elevation	Collector	Herbarium/a
T45N R57E	head of Copper Basin	EL	8200–8800 ft	*Percy Train 653*	RENO
T45N R57E S14	W of Coon Creek Summit	EL	8840 ft	*Charlet 2149 & T Charlet*	DUKE, RENO, UC

Not reported as present in the range by Little (1971).
DAC: extensive forest stands are mixed here with whitebark pine (1988–1994).

Elk Mountains (1)
elevation record: 8240 ft
1 record in 1 county (Elko)

Towns/Range/Section	Location—Habitat	County	Elevation	Collector	Herbarium/a
T47N R61E S32	*Abies* forest island with aspen	EL	8240 ft	*Charlet 2126 & T Charlet*	DUKE, RENO, UC

Not reported as present in the range by Little (1971).
Reported as present in the range by Loope (1969).
DAC: isolated pockets of the species occur at the middle and high elevations of the range (1988–1994).

Fox Creek Range (1)
elevation record: 7500 ft
1 record in 1 county (Elko)

Towns/Range/Section	Location—Habitat	County	Elevation	Collector	Herbarium/a
T46N R58E S29	below & N of Bear Creek Summit	EL	7500 ft	*Charlet 2148 & T Charlet*	DUKE, RENO, UC
[DAC: T45N R58E 57E, T44N R58E, T43N R58E]					

Reported as present in the range by Loope (1969) and Little (1971).
DAC: extensive forest stands are mixed with whitebark pine (1988–1994).

Highland Range (1)
elevation record: 9000 ft
1 record in 1 county (Lincoln)

Towns/Range/Section	Location—Habitat	County	Elevation	Collector	Herbarium/a
T0 R66E	Highland Peak	LI	> 9000 ft	*V Bostick 5870*	UNLV

Not reported as present in the range by Little (1971).
DAC: rare; observed along Highland Ridge with *Pinus longaeva* and *Abies concolor* (1994).

Ichabod Range (1)
elevation record: 6500 ft
1 record in 1 county (Elko)

Towns/Range/Section	Location—Habitat	County	Elevation	Collector	Herbarium/a
T45N R55E S25	Waterlog Summit, lone tree in aspen	EL	6500 ft	*Peter Herlan in 1976*	NSM

Not reported as present in the range by Little (1971).
DAC: absent throughout most of the range; uncommon in the Meadow Creek area; abundant on Pine Mountain (1988–1994).

Independence Mountains (1)
elevation record: 6900–9700 ft
4 records in 1 county (Elko)

Towns/Range/Section	Location—Habitat	County	Elevation	Collector	Herbarium/a
T42N R53E S2 NE/4	N slope above N Fork Humboldt River, between Cole Canyon & Jack Creek Summit	EL	7400 ft	*P Packard 80-149 & J Grimes & L Smithman*	BRY
T42N R53E S10	summit to Jack Creek	EL	6900 ft	*Ann Pinzl 4486*	NSM
T42N R53E S11	N Fork–Jack Creek divide, SE of pass, 10 mi WNW of North Fork (town)	EL	8600 ft	*N Holmgren 6299 & P Holmgren*	UTC, BRY
T42N R53E S16	1 mi S of Jack Creek Summit, open stand	EL	9700 ft	*Lloyd Loope 677*	DUKE
[Loope, DAC: T44N R53E, T43N R53E]					
[Loope: T41N R53E]					

Reported as present in the range by Critchfield & Allenbaugh (1969), Loope (1969), and Little (1971).
DAC: locally forms small forest islands in drainages in the northern part of the range, especially on the forks of Mill Creek; common at higher elevations with whitebark and limber pines (1991–1994).

Jarbidge Mountains (1)
elevation record: 8855–8940 ft
3 records in 1 county (Elko)

Towns/Range/Section	Location—Habitat	County	Elevation	Collector	Herbarium/a
T45N R58E S13 W/4	W Fork Jarbidge River	EL	8900 ft	*Charlet 2144 & T Charlet*	DUKE, RENO
T45N R59E S19 W/4	Emerald Basin	EL	8940 ft	*Charlet 2139 & Agee Smith*	DUKE, RENO
T45N R59E S7 SW/4	Cougar Creek	EL	8855 ft	*Charlet 2142 & Agee Smith*	DUKE, RENO, UC, BRY
[DAC: T47N R58E R59E, T46N R58E R59E]					

Reported as present in the range by Critchfield & Allenbaugh (1969), Loope (1969), and Little (1971).
DAC: widespread throughout the range; often dominant in extensive forests, usually with whitebark pine. Occurs in subalpine woodlands at higher elevations and extends to form mats at and above treeline. At lower elevations, occurs as island forest patches in north aspect nivation hollows (1988–1994).

Marys River Range (1)
elevation record: 6960–9070 ft
7 records in 1 county (Elko)

Towns/Range/Section	Location—Habitat	County	Elevation	Collector	Herbarium/a
T44N R59E	T Creek, 50 mi N of Deeth (determined Billings)	EL	—	*Bill Miller in 1951*	RENO
T44N R59E S2	N Fork Camp Creek	EL	6960 ft	*Charlet 2133 & Agee Smith*	DUKE, RENO
T44N R59E S7 NE/4	Marys River Basin	EL	8500 ft	*Charlet 2137 & T Charlet*	DUKE, RENO
T45N R58E S25	Marys River Peak	EL	9070 ft	*Charlet 334*	UCR
T45N R58E S25	Marys River Peak	EL	9070 ft	*Charlet 335*	UCR
T45N R59E S32	Camp Creek	EL	8260 ft	*Charlet 2136*	DUKE, RENO, UC
T46N R60E S15	near Canyon Creek Pockets	EL	7880 ft	*Charlet 2124*	DUKE, RENO, UC
[DAC: T45N R59E R60E, T46N R60E, T47N R59E]					

Reported as present in the range by Little (1971).
DAC: locally dominant at higher elevations; often mixed with whitebark pine (1988–1994).

Mormon Mountains (5)
Lincoln County

Not reported as present in the range by Little (1971).
Reported as present by Kartesz (1988), but he was unable to provide the source for the report, which may be based on the Lincoln County record for the Highland Range (*Bostick 5870* at UNLV).

Pequop Mountains (1)
elevation record: 8000 ft
1 record in 1 county (Elko)

Towns/Range/Section	Location—Habitat	County	Elevation	Collector	Herbarium/a
T36N R65E	N slopes, Pequop Mts	EL	8000 ft	*WP Cottam 3135 & Biddulph*	BRY

Not reported as present in the range by Little (1971).

Quinn Canyon Range (1)
elevation record: 7545 ft
1 record in 1 county (Nye)

Towns/Range/Section	Location—Habitat	County	Elevation	Collector	Herbarium/a
T2N R55E or R56E	small NE aspect ravine NW of Cottonwood Summit, ca. 165 km SW of Ely [Cronquist label: "Small trees to nearly 15m tall, the bark becoming rough but not very thick. Lvs conspicuously green above, conspicuously pale in 2 bands beneath. Ovulate cones ca. 10+ cm long. Possibly contaminated w/*Abies concolor.* Note resin canals."]	NY	7545 ft	*Arthur Cronquist 11112*	BRY

Not reported as present in the range by Little (1971).
DAC: *Cottam 3320* is the high-elevation record for the species in the state.

Snake Range (1)
elevation record: 10,000 ft
1 record in 1 county (White Pine)

Towns/Range/Section	Location—Habitat	County	Elevation	Collector	Herbarium/a
T13N R68E	timberline [Wheeler Peak]	WP	10,000 ft	*Cottam 3320 & Biddulph*	BRY

Not reported as present in the range by Little (1971).

****Wild Horse Range (1)**
elevation record: 6040 ft
1 record in 1 county (Elko)

Towns/Range/Section	Location—Habitat	County	Elevation	Collector	Herbarium/a
T45N R54E S32	N end of range, N aspect with aspen	EL	6040 ft	*Charlet 2154*	DUKE, RENO
[DAC: T45N R54E S24 (Rough Mountain, N slope)]					

Not reported for the range by Critchfield & Allenbaugh (1969), Loope (1969), or Little (1971).
DAC: *Charlet 2154* (the low-elevation record for the species in Nevada) can be observed from SR 225, east of the Owyhee River. Two reproductives and several saplings occur with aspen at a mesic site; also associated with *Juniperus scopulorum* (1991–1994).
New record for the species in the range.

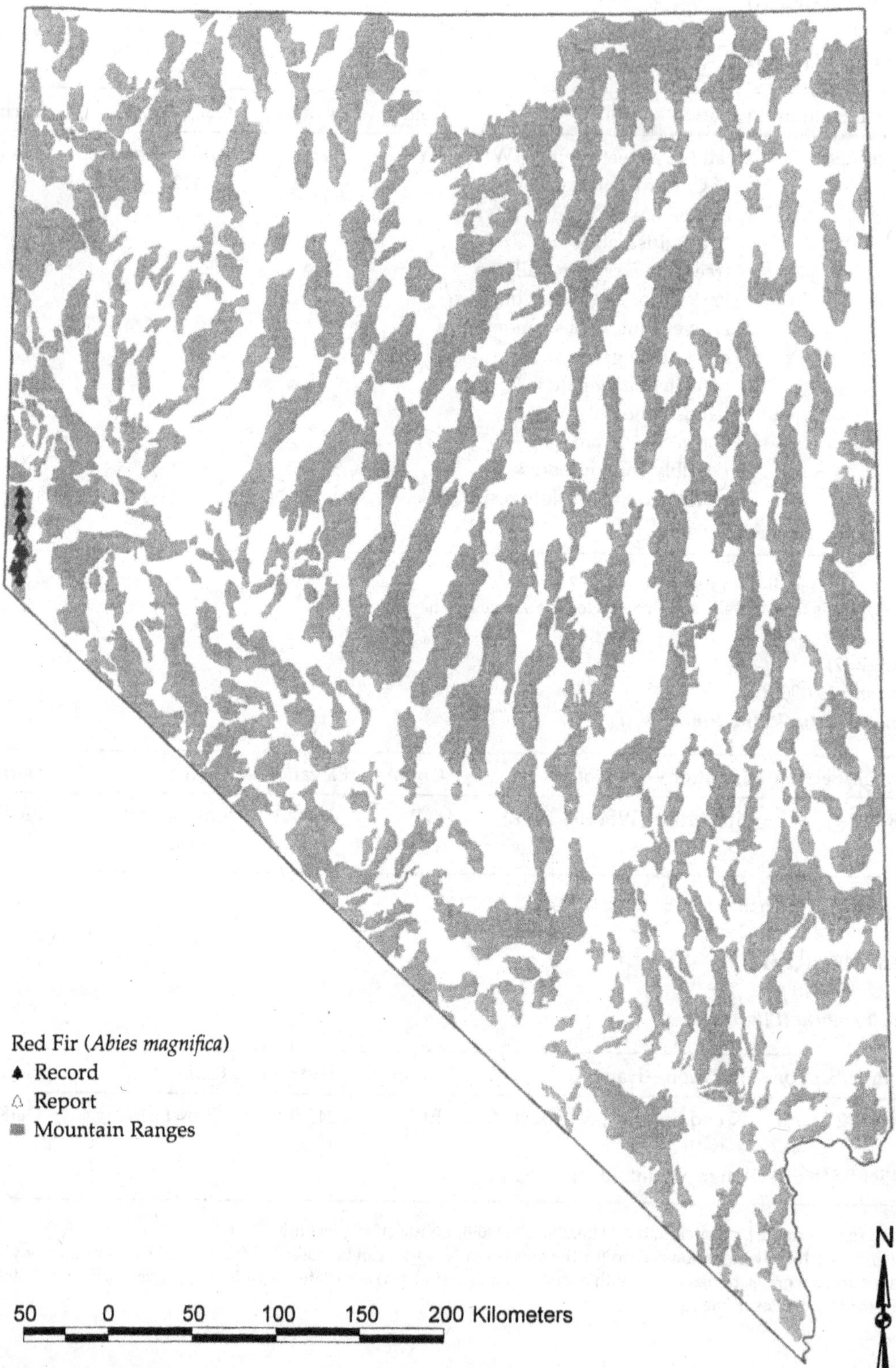
Red Fir (*Abies magnifica*)
Record
Report
Mountain Ranges
50 0 50 100 150 200 Kilometers
N

Abies magnifica A. Murray

common name: red fir, California red fir
elevation: 6500–8880 ft (1980–2705 m)
zones: Hudsonian, Canadian
12 Records for 3 Nevada counties (CC, DO, WA) and 1 mountain range

Red fir is principally a tree of the Sierra Nevada; it occurs in the Cascades only in northern California and southern Oregon but is also present in mountains inland from the northern California coast. In Nevada, red fir is the most narrowly distributed of all the firs, being restricted to the Carson Range of the Sierra Nevada. It is the only Sierran conifer besides Pacific Douglas-fir that occurs in Nevada but not on Peavine Mountain. There are records for the species only from Carson City, Douglas, and Washoe Counties. The narrow geographic distribution is reflected in the species' relatively narrow elevation range of 2380 ft (725 m). Red fir can be dominant, however, in the upper Canadian Zone forests of the Sierra Nevada and is a principal component there of the subalpine forests and woodlands in the Hudsonian Zone. The species is not included in *Intermountain Flora* (Cronquist et al. 1972).

Carson Range (1)
elevation record: 6500–8880 ft
12 records in 3 counties (Carson City, Douglas, and Washoe)

Towns/Range/Section	Location—Habitat	County	Elevation	Collector	Herbarium/a
T13N R18E S13 N/8	N of Castle Rock, Tahoe basin	DO	7700 ft	*Charlet 2170 & T Charlet*	DUKE, RENO, UC
T14N R18E S10	Glenbrook	DO	—	*PA Lehenbauer in 1929*	RENO
T14N R19E S6	near summit on US 50, W side	DO	7300 ft	*Carl Wolf 3369*	RSA
T14N R19E S29	James Canyon, with *Pinus monticola*	DO	8000 ft	*Charlet 2175*	DUKE, RENO, UC
T15N R19E S18	1 mi N of Snow Valley Peak, dominant	CC	8000 ft	*Percy Train 3310*	RENO
T15N R18E S11 SW/4	E of Marlette reservoir	WA	6680 ft	*Charlet 1919 & T Charlet*	DUKE, RENO, UC, BRY, UNLV
T15N R18E S13	Marlette Lake	WA	6500 ft	*PB Kennedy in 1904*	RENO
T17N R18E S24	Mount Rose; headwaters of Galena Creek	WA	8880 ft	*Charlet 2157 & Rust*	DUKE
T17N R18E	SE slopes of Mount Rose	WA	8200 ft	*Billings 1437*	RENO
T17N R19E	Slide Mountain	WA	7500 ft	*AA Heller 10666*	RENO, RSA
T18N R18E	12 mi W of Reno Hot Springs	WA	7200 ft	*WA Archer 6646*	RENO
T19N R18E S36	Hunter Creek Canyon	WA	6500 ft	*AA Heller in 1911*	RENO
[DAC: (WA) T16N R18E (ridge W of Little Valley)]					

Reported as present in the range by Little (1971).
DAC: abundant at high elevations throughout the range, often a dominant forest tree; associated with *Pinus contorta, P. albicaulis*, and *P. monticola* (1971–1995).

Picea engelmannii Parry ex Engelmann

common name: Engelmann spruce
elevation: 8000–11,600 ft (2440–3535 m)
zones: Arctic-Alpine, Hudsonian
14 Records for 2 Nevada counties (WP, EL) and 4 mountain ranges

Engelmann spruce is common in the Rocky Mountains from northern British Columbia to New Mexico and southern Arizona (Chiricahua Mountains). In the Pacific Mountain System, Engelmann spruce is known only from the Cascades and the Klamath Mountains in northwestern California and southwestern Oregon. In the Intermontane Plateaus, it is known from the Wallowa Mountains, the Colorado Plateau, and as isolated stands in the central and eastern Great Basin.

In Nevada, Engelmann spruce is mostly restricted to near timberline in both the Hudsonian and Arctic-Alpine Zones in 4 mountain ranges in White Pine and Elko Counties. In the Snake Range (White Pine County), the species forms impressive forests associated with *Pinus flexilis* and *Pseudotsuga menziesii* var. *glauca,* and grows as krummholz above treeline. Given that the species' presence is very restricted in the Ruby Mountains, the Schell Creek Range, and the Pilot Range, it may yet be found locally in other rarely visited high mountain ranges in eastern Nevada. Engelmann spruce grows over an elevation range of 3600 ft (1095 m) in Nevada.

Mormon Mountains (5)
Clark County

DAC: *Picea engelmannii* is reported by Kartesz (1988) to occur here, but I have seen no material, and Kartesz was unable to provide a voucher. This is an improbable location, and I found no record for *Picea* in Clark County.

Painted Point Range (7)
Washoe County

DAC: a single individual was planted at Bald Mountain headquarters, Sheldon National Wildlife Refuge. There is little danger of this species escaping from this location. *Picea engelmannii* is not known to occur in Washoe County outside of cultivation (1994).

Pilot Range (1)
elevation record: 8500 ft
1 record in 1 county (Elko)

Towns/Range/Section	Location—Habitat	County	Elevation	Collector	Herbarium/a
T36N R70E	steep W slope of range, amid sandstone boulders	EL	8500 ft	*Lloyd Loope 1372*	DUKE
[Loope: T37N R70E]					

Not reported as present in the range by Little (1971).
Reported as present by Loope (1969).

Pioche Hills (7)
Lincoln County

Towns/Range/Section	Location—Habitat	County	Elevation	Collector	Herbarium/a
T1N R67E	Pioche, upper Sonoran Zone [cultivated]	LI	6000 ft	*K Brown in 1961*	UNLV

Not reported as present in the range by Little (1971).
DAC: a most unlikely location and habitat; this collection is undoubtedly from the cultivated tree I observed in a residential yard in the town. *Picea engelmannii* is not known to occur in Lincoln County outside of cultivation (1994).

Ruby Mountains (1)
elevation record: 9600 ft
1 record in 1 county (Elko)

Towns/Range/Section	Location—Habitat	County	Elevation	Collector	Herbarium/a
T32N R59E S8	Thorpe Canyon, only locality in the Ruby Mts	EL	9600 ft	*Lloyd Loope 1278*	DUKE

Reported as present in the range by Critchfield & Allenbaugh (1969), Loope (1969), and Little (1971).
DAC: the Thorpe Canyon stand is still the only known location of the species in the Ruby Mountains (1988–1994).

Engelmann Spruce (*Picea engelmannii*)
Record
Report
Mountain Ranges
N
50 0 50 100 150 200 Kilometers

Schell Creek Range (1)
elevation record: 8500–10,500 ft
4 records in 1 county (White Pine)

Towns/Range/Section	Location—Habitat	County	Elevation	Collector	Herbarium/a
T21N R65E S10	Aurum	WP	—	*ME Jones in 1891*	RSA
T17N R65E S10	N Fork Berry Creek, 2.5–4 mi above Berry Creek Ranger Station	WP	8500 ft	*Percy Train 1039*	RENO
T17N R65E S10	Berry Creek	WP	8650 ft	*Guyer & Nigro 5099B*	RENO
T17N R65E S14	upper Berry Creek	WP	10,500 ft	*Richard W. Rust 35*	UCR
[RWR: T15N R65E S1]					
[W Sandow: T17N R66E (Taft Creek)]					

Reported as present in the range by Little (1971).
DAC: reported on *W Sandow in 1953* at RENO (*Pinus ponderosa*) label; at 7500 ft.

Snake Range (1)
elevation record: 8000–11,600 ft
8 records in 1 county (White Pine)

Towns/Range/Section	Location—Habitat	County	Elevation	Collector	Herbarium/a
T13N R68E S36	Pyramid Peak, Johnson Lake	WP	10,600 ft	*Guyer & Nigro 5171*	RENO
T12N R68E S14	S slopes of Mount Washington, Liberty Mine	WP	10,400 ft	*N Holmgren 3764 & A Holmgren*	UTC
T13N R68E S11	Wheeler Peak campground, spruce-aspen forest	WP	9980 ft	*Charlet 2034 & Christine Strehl*	DUKE, RENO
T13N R68E S11	trail between Wheeler Peak campground & Lake Teresa, w/limber pine	WP	9975 ft	*M Williams 81-139-1 & Tiehm*	RENO
T13N R68E	Wheeler Peak	WP	11,000 ft	*Critchfield 6-9/12/61*	IFGP
T13N R68E S23	SE Wheeler Peak	WP	—	*JM Tucker 4427 & WR Hildreth*	UNLV
T13N R69E	Baker Creek headwaters, NE slope Wheeler Peak [Lehman Creek]	WP	8000 ft	*Percy Train 1144*	RENO
T16N R69E S3 NE/4	Mount Moriah	WP	11,600 ft	*JC Lyman 431B*	UCR
[DAC: T12N R69E S6 (SE of Pyramid Peak), T14N R68E, T13N R69E]					

Reported as present in the range by Little (1971).
DAC: abundant; dominant in large forests at higher elevations, to krummholz form at upper treeline (1994).

Spring Mountains (5)
Clark County

Reported as present in the range by Billings (1954); this report cited in turn by both Little (1956) and Kartesz (1988). Billings's initial report was in error, however, and the species is almost certainly not in the range (WD Billings, in litt. 1994).
Not reported as present by Clokey (1951) or Little (1971).
DAC: I have been unable to verify the presence of the species here and have seen no material in herbaria. No naturally occurring *Picea engelmannii* are known to occur in Clark County (1982 & 1993).

Pinus albicaulis Engelmann

common name: whitebark pine
elevation: 6800–10,750 ft (2070–3275 m)
zones: Arctic-Alpine, Hudsonian, Canadian
43 Records for 7 Nevada counties (CC, DO, EL, HU, MI, ST, WA)
Reports for 17 mountain ranges, Records for 15 of these
New Record for 1 mountain range

Whitebark pine is widely distributed in western Canada and extends south into high elevations of the northern Rocky Mountains, the Cascades, and the Sierra Nevada, but the species is not known from Colorado or Utah. In the Intermontane Plateaus, whitebark pine is found at high elevations in the mountains of the Columbia Plateau, and also in the Warner Mountains of Oregon and California in the northwestern Great Basin.

In Nevada, whitebark pine is known from 17 mountain ranges in 7 counties (Carson City, Douglas, Elko, Humboldt, Mineral, Storey, and Washoe) in northern and western Nevada. Except for the Wassuk Range (Mineral County), where it is known from only 1 stand, whitebark pine is a very important component of the mountain ecosystem wherever it occurs. It so dominates the subalpine forests of the Pine Forest Range that limber pine was not known to occur there until now (probably because both species have similar life forms and vegetative characteristics), even though it is present at both the upper and lower limits of whitebark pine. When sympatric with subalpine fir and limber pine (in the mountains of northern Elko County), whitebark pine dominates the fir on drier sites, is often codominant with it in more mesic situations, and usually occurs above limber pine. A new record from the Elk Mountains of Elko County is presented here, and whitebark pine may yet be found in the Nevada portion of the White Mountains. Individuals or small stands may be present in the Santa Rosa Range (Humboldt County), although I was unable to find any there in 8 visits. The elevation range of the species in Nevada is 3950 ft (1205 m).

Buck Creek Mountains (2)
Elko County
[Loope: T46N R57E R58E]

Reported as present in the range by Loope (1969) and Little (1971).
DAC: uncommon; restricted to high elevations and nivation pockets (1988–1994).

Bull Run Mountains (1)
elevation record: 8600 ft
1 record in 1 county (Elko)

Towns/Range/Section	Location—Habitat	County	Elevation	Collector	Herbarium/a
T44N R52E S21	Near White Rock, 41°42'N, 116°7.8'W; w/*P. flexilis*	EL	8600 ft	*SA Scott in 1965*	IFGP
[Loope: T44N R53E, T45N R52E, T46N R52E]					

Reported as present in the range by Loope (1969) and Little (1971).
DAC: present at high elevations (1990–1994).

Carson Range (1)
elevation record: 8200–10,750 ft
11 records in 3 counties (Carson City, Douglas, and Washoe)

Towns/Range/Section	Location—Habitat	County	Elevation	Collector	Herbarium/a
T14N R19E S30	Genoa Peak	DO	8800 ft	*JT Howell 50856 & True & M Williams*	CAS
T15N R19E S19	Snow Valley Peak	CC	9000 ft	*Ann Pinzl 2352*	NSM
T17N R18E	Mount Rose	WA	9650 ft	*AA Heller 10656*	CAS, RSA, RENO
T17N R18E	Mount Rose, w/white pine & lodgepole pine	WA	9000 ft	*WD Billings 1263*	UC, RENO
T17N R18E	Mount Rose	WA	9000 ft	*PA Lehenbauer*	RENO
T17N R18E S12	Mount Rose, N slope, rocky slope above timberline	WA	10,750 ft	*DG Cooney 112*	RENO
T17N R18E S25	Tahoe Meadows, Mount Rose	WA	8500 ft	*WA Archer 6449*	RENO
T17N R19E S12	Snow Basin [E slope Mount Rose]	WA	9500–10,000 ft	*Percy Train 4478*	UTC, RENO
T17N R19E S19	Slide Mountain	WA	8500 ft	*EL Little 19142*	CAS
T17N R19E S19	Mount Rose road summit	WA	9200 ft	*WW Schwartzkopf in 1967*	RENO
—	Carson Range	WA	8200 ft	*JT Howell 53539*	CAS
[DAC: (WA) T16N R19E S8 (Little Valley), T15N R19E, T16N R18E, T18N R18E, T17N R18E S32 (Rose Knob)]					

Not reported as present in the Nevada portion of the range by Little (1971).
DAC: abundant, sometimes dominant, at high elevations, with a suite of subalpine conifers; grows alone at the highest sites. *Cooney 112* is the high-elevation record for the species in the state (1971–1995).

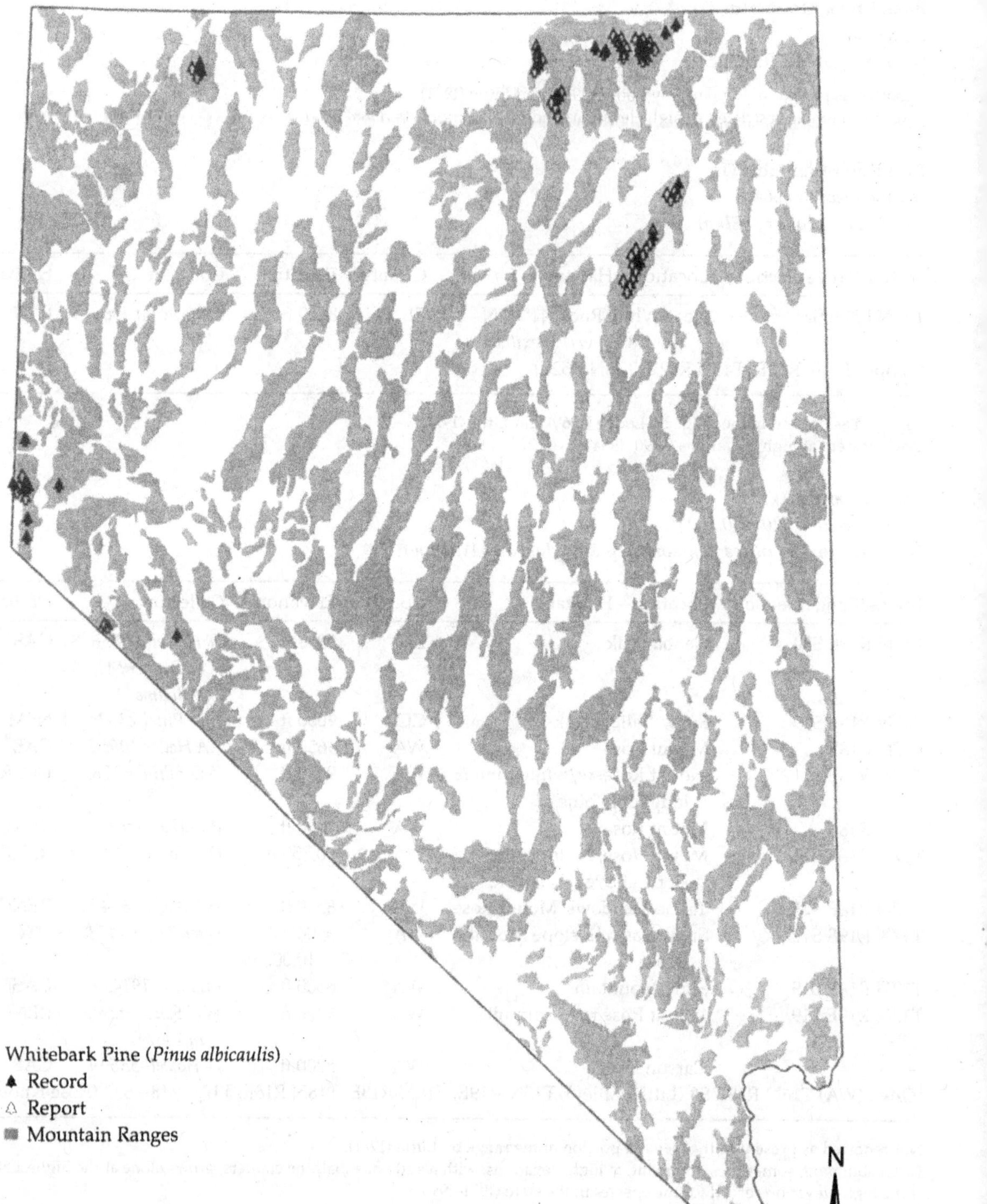
Whitebark Pine (*Pinus albicaulis*)
Record
Report
Mountain Ranges
50 0 50 100 150 200 Kilometers
N

Copper Mountains (1)
elevation record: 8840 ft
1 record in 1 county (Elko)

Towns/Range/Section	Location—Habitat	County	Elevation	Collector	Herbarium/a
T45N R57E S14	West of Coon Creek Summit, w/*Abies lasiocarpa*	EL	8840 ft	*Charlet 2150 & T Charlet*	DUKE, RENO

Not reported as present in the range by Little (1971).
Reported as present by Loope (1969).
DAC: abundant in subalpine fir forests and higher (1988–1994).

East Humboldt Range (1)
elevation record: 9000 ft
1 record in 1 county (Elko)

Towns/Range/Section	Location—Habitat	County	Elevation	Collector	Herbarium/a
T36N R61E S4 [Loope: T35N R60E R61E, T36N R61E]	Angel Lake	EL	9000 ft	*P Raven 20623*	CAS, RSA

Reported as present in the range by Little (1971).
DAC: abundant at high elevations (1988–1994).

****Elk Mountains (1)**
elevation record: 8260 ft
1 record in 1 county (Elko)

Towns/Range/Section	Location—Habitat	County	Elevation	Collector	Herbarium/a
T47N R61E S32	on periphery of *Abies lasiocarpa*–aspen forest island	EL	8260 ft	*Charlet 2127*	DUKE, RENO, UC

Not reported as present in the range by Loope (1969) or Little (1971).
DAC: uncommon; widely scattered on high rocky sites (1994).
New record for the species in the range.

Fox Creek Range (1)
elevation record: 7300 ft
1 record in 1 county (Elko)

Towns/Range/Section	Location—Habitat	County	Elevation	Collector	Herbarium/a
T45N R58E S4 [Loope: T44N R58E] [DAC: T46N R58E, T45N R58E]	Pine Creek Rd, about 0.5 mi S of Jarbidge, 41°48'N, 115°24'W	EL	7300 ft	*Critchfield 2-9/9/61*	IFGP

Reported as present in the range by Loope (1969) and Little (1971).
DAC: commonly associated with subalpine fir (1988–1994).

Ichabod Range (1)
elevation record: 8600 ft
2 records in 1 county (Elko)

Towns/Range/Section	Location—Habitat	County	Elevation	Collector	Herbarium/a
T45N R56E	Pine Mountain, Martin Creek, Humboldt Reserve (8800 ft)	EL	—	*PB Kennedy in 1913*	CAS, RENO
T45N R56E S23	Pine Mountain, pure stand at summit	EL	8600 ft	*Lloyd Loope 1371*	DUKE

Not reported as present in the range by Little (1971).
Reported as present by Loope (1969).
DAC: the maximum elevation of Pine Mountain (the highest feature in the range) is 8650 ft, so the 8800-ft elevation reported by Kennedy must be in error. Apparently the species is present only in the most northern part of this range (1988–1994).

Independence Mountains (1)
elevation record: 9200–9300 ft
2 records in 1 county (Elko)

Towns/Range/Section	Location—Habitat	County	Elevation	Collector	Herbarium/a
T42N R53E S15	2 mi S of Jack Creek Summit, W slope of range	EL	9200 ft	*Lloyd Loope 992*	DUKE
T42N R53E S23	E Fork Mill Creek, on glacial moraine, near head of cirque	EL	9300 ft	*Lloyd Loope 993*	DUKE
[Loope, DAC: T42N R53E R54E, T41N R53E, T40N R53E]					

Not reported as present in the range by Little (1971).
Reported as present by Loope (1969).
DAC: dominant and abundant throughout the higher, northern part of the range, to treeline (1991–1994).

Jarbidge Mountains (1)
elevation record: 8855–8940 ft
3 records in 1 county (Elko)

Towns/Range/Section	Location—Habitat	County	Elevation	Collector	Herbarium/a
T45N R58E S13 W/4	W Fork Jarbidge River	EL	8900 ft	*Charlet 2145 & T Charlet*	DUKE, RENO
T45N R59E S19 W/4	Emerald Basin	EL	8940 ft	*Charlet 2138 & Agee Smith*	DUKE, RENO
T45N R59E S7 SW/4	Cougar Creek	EL	8855 ft	*Charlet 2143 & Agee Smith*	DUKE, RENO, UC
[Loope, DAC: T46N R58E R59E, T45N R59E, T46N R59E]					

Reported as present in the range by Loope (1969) and Little (1971).
DAC: associated with *Abies lasiocarpa* in forests, extending to the highest elevations as krummholz. Dominant in subalpine woodlands at drier sites (1988–1994).

Marys River Range (1)
elevation record: 7880–10,500 ft
5 records in 1 county (Elko)

Towns/Range/Section	Location—Habitat	County	Elevation	Collector	Herbarium/a
T45N R58E S25 SE/4	Marys River headwaters, N of Marys River Peak	EL	8880 ft	*Charlet 2140 & T Charlet*	DUKE, RENO
T45N R59E S21	Gods Pocket Peak	EL	9600 ft	*Charlet 38*	RENO, UC
T45N R59E S30	just below Marys River Peak	EL	10,500 ft	*Charlet 327*	RENO
T45N R59E S32	Camp Creek, near Camp Creek Spring	EL	8240 ft	*Charlet 2135*	DUKE, RENO, UC
T46N R60E S15	near Canyon Creek Pockets, in nivation hollow	EL	7880 ft	*Charlet 2125*	DUKE, RENO
[Loope, DAC: T45N R60E]					
[DAC: T44N R59E, T45N R59E, T46N R59E]					

Not reported as present in the range by Little (1971).
Reported as present by Loope (1969).
DAC: abundant with and above *Abies lasiocarpa* in the higher portions of the range; in subalpine forests, woodlands, treeline situations, and above treeline as krummholz (1988–1994).

Peavine Mountain (1)
elevation record: 8200–8270 ft
4 records in 1 county (Washoe)

Towns/Range/Section	Location—Habitat	County	Elevation	Collector	Herbarium/a
T20N R18E S23	Peavine summit	WA	8270 ft	*Billings 1451*	RENO
T20N R18E S23	Peavine Mountain summit	WA	8200 ft	*M Williams 72-45-2*	RENO, CAS
T20N R18E	N slope of peak (5000 ft)	WA	—	*R Miller & W Priest in 1956*	RENO
T20N R18E	N slope of peak (5000 ft)	WA	—	*JW Vetter in 1956*	RENO

Not reported as present in the range by Little (1971).
DAC: the 5000 ft reported elevation must be in error. Williams et al. (1992) report a few individuals at the summit at 8200 ft. I have seen these (1971–1995).

Pine Forest Range (1)
elevation record: 8000 ft
3 records in 1 county (Humboldt)

Towns/Range/Section	Location—Habitat	County	Elevation	Collector	Herbarium/a
T43N R28E S1	[DAC: Blue Lake]	HU	8000 ft	*Virgil Hart in 1950*	RENO
T43N R29E S18	near Leonard Creek Lake	HU	8000 ft	*N Holmgren 3752 & A Holmgren*	UTC
T44N R28E S36	rocky slope along a small stream 1 mi N of Blue Lake	HU	8000 ft	*A Cronquist 8659*	UTC
[DAC: T43N R28E S16, T44N R28E S34]					

Reported as present in the range by Little (1971).
DAC: dominant throughout extensive subalpine forests in the Duffer Peak area. At lower elevations, mixed with aspen and occasionally with limber pine. At higher elevations forms large, pure stands, mostly absent from southern aspect. One stand east of Duffer Peak and Bare Pass is mixed with old limber pines (1990–1994).

Ruby Mountains (1)
elevation record: 9000–10,000 ft
3 records in 1 county (Elko)

Towns/Range/Section	Location—Habitat	County	Elevation	Collector	Herbarium/a
T31N R58E S1	Lamoille Canyon	EL	9000 ft	*Mozingo sn*	RENO
T32N R58E S35	Thomas Creek Canyon	EL	9500 ft	*N Holmgren 6271 & P Holmgren*	UTC
T33N R59E S14	N Fork Cold Creek, near timberline at head of cirque	EL	10,000 ft	*Lloyd Loope 888*	DUKE
[Loope: T33N R59E, T32N R58E R59E, T31N R58E, T30N R57E R58E, T29N R57E]					

Reported as present in the range by Loope (1969) and Little (1971).
DAC: the common conifer in upper Lamoille Canyon; abundant at high elevations in the northern portion of the range (1988–1994).

Sweetwater Mountains (2)
Lyon County
[Matt Lavin, DAC: T7N R24E, T8N R24E]

Not reported as present in the Nevada portion of the range by Little (1971).
Reported as present in the range by Lavin (1981).
DAC: abundant at higher elevations throughout the range; the treeline tree (1985–1995).

Virginia Range (1)
elevation record: 6800 ft
4 records in 1 county (Storey)

Towns/Range/Section	Location—Habitat	County	Elevation	Collector	Herbarium/a
T17N R21E S20	Christmas Tree Canyon, 1 mi N of Virginia City, N slope, altered soil	ST	6800 ft	*J Darling 1*	RENO
T17N R21E S20	in canyon 1 mi NW of Virginia City, altered andesite, steep NE slope	ST	6800 ft	*Billings 1770*	RENO
T17N R21E	2 mi NW of Fivemile House (WA 6000 ft)	ST	6800 ft	*TC Adams 207*	UC
T17N R21E	N-facing slope, Cedar Hill Canyon, altered andesite soil (6000 ft)	ST	6800 ft	*W Miller in 1952*	DUKE

Reported as present in the range by Little (1971).
DAC: only 1 tree has been found in this range, at 6800 ft in Storey County (WD Billings, in litt. 1994). These collections are the lowest elevation reported for the species in Nevada (1973–1995).

Wassuk Range (1)
elevation record: 9200 ft
1 record in 1 county (Mineral)

Towns/Range/Section	Location—Habitat	County	Elevation	Collector	Herbarium/a
T8N R29E S30	middle Cat Canyon, 1 mi E of Laphan Meadows, SW of Mount Grant; isolated stand	MI	9200 ft	*Percy Train 4321*	UC, RENO

Not reported as present in the range by Little (1971).
DAC: *Percy Train 4321* was annotated and its identity confirmed by Billings at RENO.

White Mountains (4)
Esmeralda and Mineral Counties

DAC: a collection of whitebark pine made by Dean Taylor from the ridge above Middle Creek is at RSA (Morefield et al. 1988). The collection location is in California, but very close to the border with Nevada. Whitebark pine should be looked for at high elevations in the Nevada portion of the range; it may have been overlooked because of its resemblance to limber pine.

Pinus contorta Douglas ex Loudon var. *murrayana* (Greville & Balfour) Engelmann in S. Watson

common name: lodgepole pine, tamarack pine, black pine
elevation: 5000–9200 ft (1525–2805 m)
zones: Hudsonian, Canadian, Arid Transition
33 Records in 6 Nevada counties (CC, DO, LY, MI, ST, WA)
Report for 1 additional county (ES)
Reports for 9 mountain ranges, Records for 8 of these
New Records in 3 mountain ranges and 1 county (ST)

Lodgepole pine is widely distributed in western North America, from the Arctic to Baja California. It is present in the northern Columbia Plateau but is mostly absent from the Great Basin and Colorado Plateaus. Three varieties are recognized: the shore pine (var. *contorta*), restricted to coastal areas of the Pacific Northwest; the Rocky Mountain lodgepole (var. *latifolia*); and the Sierran lodgepole (var. *murrayana*), which occurs in the Middle and South Cascades and the Sierra Nevada. Only var. *murrayana* occurs in Nevada, although var. *latifolia* is present in a large stand near the Jarbidge Mountains (Elko County) along Shoshone Creek in Idaho, just 19 km north of the Nevada border.

The species is restricted to the far western part of the state, but there it is rather widely distributed in 9 high mountain ranges in 7 counties (Carson City, Douglas, Esmeralda, Lyon, Mineral, Storey, and Washoe), all west of the Lahontan Basin. A new record for the species, and the high-elevation record for the species in the state, is reported for the Wassuk Range (Mineral County), where a stand of more than 500 reproductive individuals was discovered growing with Jeffrey and limber pines at 9200 ft (*Charlet 702 & Rust* at UCR). A new record for the Flowery Range and Storey County, as well as the low-elevation record for the species in the state, was acquired in Sixmile Canyon at 5000 ft (*Charlet 1816 & T Charlet* at DUKE, RENO, UC, BRY, UNLV).

Lodgepole pine (probably the Rocky Mountain variety, *P. contorta* var. *latifolia* Engelmann) was recently planted at the Pole Creek Ranger Station (Jarbidge Mountains) and at Wild Horse Creek Campground (Independence Mountains), both in the Humboldt National Forest in Elko County. Once these individuals reach reproductive age, the species is likely to naturalize at these locations. The species was also planted, probably in the 1930s, in the Santa Rosa Range at Hinkey Summit (Humboldt County, Humboldt National Forest) and at the old Bald Mountain headquarters at Sheldon National Wildlife Refuge in the Painted Point Range (Washoe County), and these individuals are now producing cones. Whereas it seems possible for the species to naturalize from the Santa Rosa Range location, it is unlikely that it will successfully reproduce in the Painted Point Range.

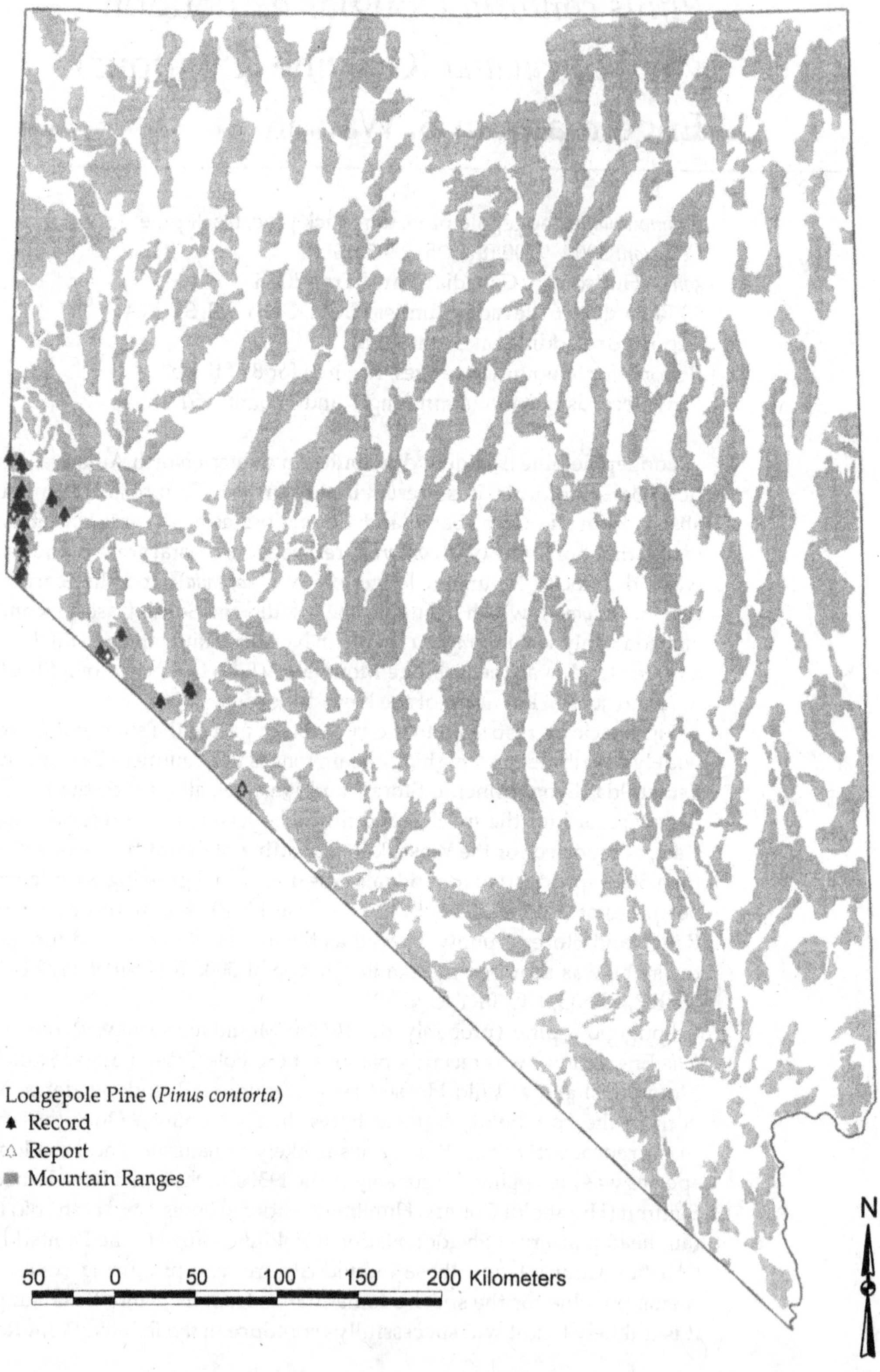
Lodgepole Pine (*Pinus contorta*)
Record
Report
Mountain Ranges
50 0 50 100 150 200 Kilometers
N

Bodie Mountains (1)
elevation record: 7500 ft
1 record in 1 county (Mineral)

Towns/Range/Section	Location—Habitat	County	Elevation	Collector	Herbarium/a
T5N R27E S11	Bodie Creek Canyon, 4 mi E of California	MI	7500 ft	*Billings 1676*	RENO

Reported as present in the range by Little (1971).
DAC: large trees are present, but uncommon, along Bodie Creek (1991–1995).

Carson Range (1)
elevation record: 6680–9000 ft
20 records in 3 counties (Carson City, Douglas, and Washoe)

Towns/Range/Section	Location—Habitat	County	Elevation	Collector	Herbarium/a
T13N R18E S24	1 mi W of Kingsbury Grade, W slope	DO	7000 ft	*Percy Train 3157*	RENO
T13N R19E S7	N of Kingsbury Grade	DO	8150 ft	*Ann Pinzl 9999*	RSA
T15N R19E S31	Clear Creek Canyon	CC	7600 ft	*RJ Holland in 1949*	RENO
T15N R19E S31	US 50 E of Spooner Summit	CC	—	*Ann Pinzl in 1973*	NSM
T15N R19E S18	divide 1 mi S of Marlette Lake, 1 mi N of Snow Valley Peak	CC	8000 ft	*Percy Train 3309*	UTC, RENO
T15N R18E S11 SW/4	E of Marlette reservoir	WA	6680 ft	*Charlet 1918 & T Charlet*	DUKE, RENO, UC
T16N R19E S8	Little Valley, W of Franktown	WA	—	*NF Petersen 168*	RENO
T16N R19E	Little Valley, W of Franktown	WA	—	*PB Kennedy in 1904*	RENO
T17N R18E	SE side of Mount Rose	WA	—	*Ann Pinzl in 1974*	NSM
T17N R18E S10	Bronco Creek, W of Mount Rose	WA	—	*NF Petersen 512*	RENO
T17N R18E S13	S slope Mount Rose	WA	9000 ft	*Billings 1262*	RENO
T17N R19E S17	Galena Creek headwaters	WA	8000 ft	*WA Archer 6679*	RENO
T17N R19E S17	upper Galena Creek, w/Washoe pine	WA	7240 ft	*Charlet 1916 & T Charlet*	DUKE, RENO, UC
T17N R19E	Mount Rose	WA	7500 ft	*P Herlan in 1971*	NSM
T17N R19E S19	1 mi below Mount Rose summit	WA	8500 ft	*WW Schwartzkopf in 1967*	RENO
T18N R18E S12	Hunter Creek Canyon	WA	7500 ft	*AA Heller 10350*	RENO
T18N R18E S23	mountain meadow near Hunters Lake	WA	8500 ft	*WA Archer 6370*	BRY, RENO
T18N R18E S23	Hunters Lake	WA	8500 ft	*WA Archer 6371*	RENO
T18N R18E	9 mi S, 15 mi W of Reno	WA	8900 ft	*J Henrichs 177*	RENO
—	canyon near Mount Rose	WA	—	*NF Petersen 229*	RENO

[DAC: (WA) T16N R18E; (DO) T14N R19E]

Reported as present in the range by Little (1971).
DAC: common, and sometimes dominant, at high elevations to near treeline (1971–1995).

****Flowery Range (1)**
elevation record: 5000 ft
1 record in 1 county (**Storey)

Towns/Range/Section	Location—Habitat	County	Elevation	Collector	Herbarium/a
T17N R21E S27 S/4	Sixmile Canyon Creek, below Sugarloaf, in willows along stream	ST	5000 ft	*Charlet 1816 & T Charlet*	DUKE, RENO, UC, BRY, UNLV

Not reported as present in the range by Little (1971).
DAC: only 1 individual was found in this mountain range. Surprisingly, this collection represents the low-elevation record for the species in the state (1973–1994).
New record for the species in the range. New record for the species in Storey County.

Independence Mountains (7)
Elko County

DAC: there are 40+ saplings at Wild Horse Crossing campground, apparently planted by the USFS. These individuals are not yet reproductive, but once they bear cones they are very likely to naturalize in this area (1991–1994).

Jarbidge Mountains (7)
Elko County

DAC: the many saplings and young trees present at Pole Creek Ranger Station were planted there. Once this population reaches reproductive age, naturalization is likely. The Pole Creek Ranger Station is technically in the Marys River Range (1988–1994).

Painted Point Range (7)
Washoe County

DAC: 6 individuals were planted at Bald Mountain headquarters at Sheldon National Wildlife Refuge. There is little danger of their escape from this location (1993–1994).

Peavine Mountain (1)
elevation record: 5700–8000 ft
4 records in 1 county (Washoe)

Towns/Range/Section	Location—Habitat	County	Elevation	Collector	Herbarium/a
T20N R18E	altered andesite, S side of peak	WA	5700 ft	*Billings 1425*	RENO
T20N R18E S23	just N of Peavine Peak	WA	8000 ft	*M Williams 72-89-8*	RENO
T20N R18E	SE of Peavine Peak	WA	6600 ft	*M Williams & LaRivers in 1974*	RENO
T20N R18E S36 SW/4 NW/4	altered andesite	WA	6200 ft	*Jan Nachlinger 1482 & Billings*	DUKE

Not reported as present in the range by Little (1971).
DAC: rare in the range; known only from the Peavine Peak area (1973–1995).

****Pine Grove Hills (1)**
elevation record: 7740 ft
1 record in 1 county (Lyon)

Towns/Range/Section	Location—Habitat	County	Elevation	Collector	Herbarium/a
T 9N R26E S17 NE/2	above Rockland Mines, near nevé cirque on steep northeast slope	LY	7740 ft	*Charlet 2208 & T Charlet*	RENO, UC, DUKE

Not reported as present in the range by Little (1971).
DAC: rare; a stand of about 200 individuals occurs with *Pinus flexilis* and *Pinus jeffreyi* above Rockland Mines, east of Pine Grove Summit (1989–1995).
New report for the species in the range.

Pine Nut Mountains (5)
Douglas County

Erroneously reported as present in the range by Little (1971).
DAC: I saw no material and was unable to verify the presence here of *Pinus contorta.* The report is most likely in error, perhaps resulting from misplacement of a dot intended to represent the Virginia Range, Washoe County, since that early report from Billings (1954) is not represented on Little's (1971) map (1989–1994).

Santa Rosa Range (7)
Humboldt County

DAC: planted at Hinkey Summit, probably during the 1930s. The trees are of reproductive age, but no seedlings seen (1991–1994).

Sweetwater Mountains (1)
elevation record: 7200 ft
1 record in 1 county (Lyon)

Towns/Range/Section	Location—Habitat	County	Elevation	Collector	Herbarium/a
T8N R24E S9	Desert Creek, S of Desert Creek Peak	LY	7200 ft	*Billings 1855*	RENO
[DAC: T8N R24E S33 S34]					

Not reported as present in the Nevada portion of the range by Little (1971).
DAC: common at higher elevations, forming small forest stands mostly in the California portion of the range. Occasional along streams, generally above *Pinus jeffreyi* (1985–1995).

Virginia Range (1)
elevation record: 5380–5700 ft
3 records in 1 county (Washoe)

Towns/Range/Section	Location—Habitat	County	Elevation	Collector	Herbarium/a
T18N R20E S26 S/4	——	WA	5380 ft	*Billings 1428 & Hudspeth*	RENO
T18N R20E S26 S/4	——	WA	5380 ft	*Billings 2059 & Darling*	RENO
T18N R20E S35 NE/4	——	WA	5700 ft	*Marshall Humphreys in 1951*	RENO

Not reported as present in the range by Little (1971).
DAC: one of the few individuals here (*Billings 1428*) was cut down for firewood in 1991. All collections are from altered andesite soils on the western slope (1973–1995).

****Wassuk Range (1)**
elevation record: 8280–9200 ft
2 records in 1 county (Mineral)

Towns/Range/Section	Location—Habitat	County	Elevation	Collector	Herbarium/a
T6N R29E S13	Buller Mountain	MI	9200 ft	*Charlet 702 & Rust*	UCR
T6N R29E S11	ridge N of Buller Mountain	MI	8280 ft	*Charlet 1977*	DUKE, RENO

Not reported as present in the range by Little (1971).
DAC: a stand of 500+ reproductive individuals is present with Jeffrey pine, limber pine, and mountain-mahogany in a nevé cirque, immediately north of Buller Mountain. Individuals occur sporadically along the ridge northward for about 1.5 mi, to a point about 3 mi southeast of Lucky Boy Pass (1990–1995).
New records for the species in the range.

White Mountains (3)
Esmeralda County
[Hall (1991): T1S R33E]

Reported to be present, and collected and verified in Middle Creek cirque by Elliot-Fisk and Peterson (in Hall 1991); however, I have seen no Nevada material in herbaria. James Morefield (botanist, Nevada Natural Heritage Program, pers. comm. 1995) told me it is likely to occur in Nevada along Middle Creek.
Not reported as present in the Nevada portion of the range by Little (1971).

Pinus edulis Engelmann var. *fallax* Little

common name: pinyon, nut pine, Colorado pinyon
zone: Upper Sonoran

Colorado pinyon, the principal pinyon of the Four Corners states, barely enters most of the border states (Wyoming, Oklahoma, Texas, and California) and Mexico. Three collections from 2 mountain ranges in Clark and Lincoln Counties were identified by D. K. Bailey as *Pinus edulis* var. *fallax.* However, these were all identified by their collectors (W. D. Billings, Ann Pinzl, and Hugh Mozingo) as *P. monophylla,* and my own assessment of the material leads me to conclude that they are indeed *P. monophylla.* There is a large incidence of double needles in *P. monophylla* in White Pine County, and a few of these individuals (with < 5% double needles) have been identified by other workers as *P. edulis.* I have annotated these collections and report them in this book as *P. monophylla.* I therefore do not include *Pinus edulis* as one of the native conifers of Nevada.

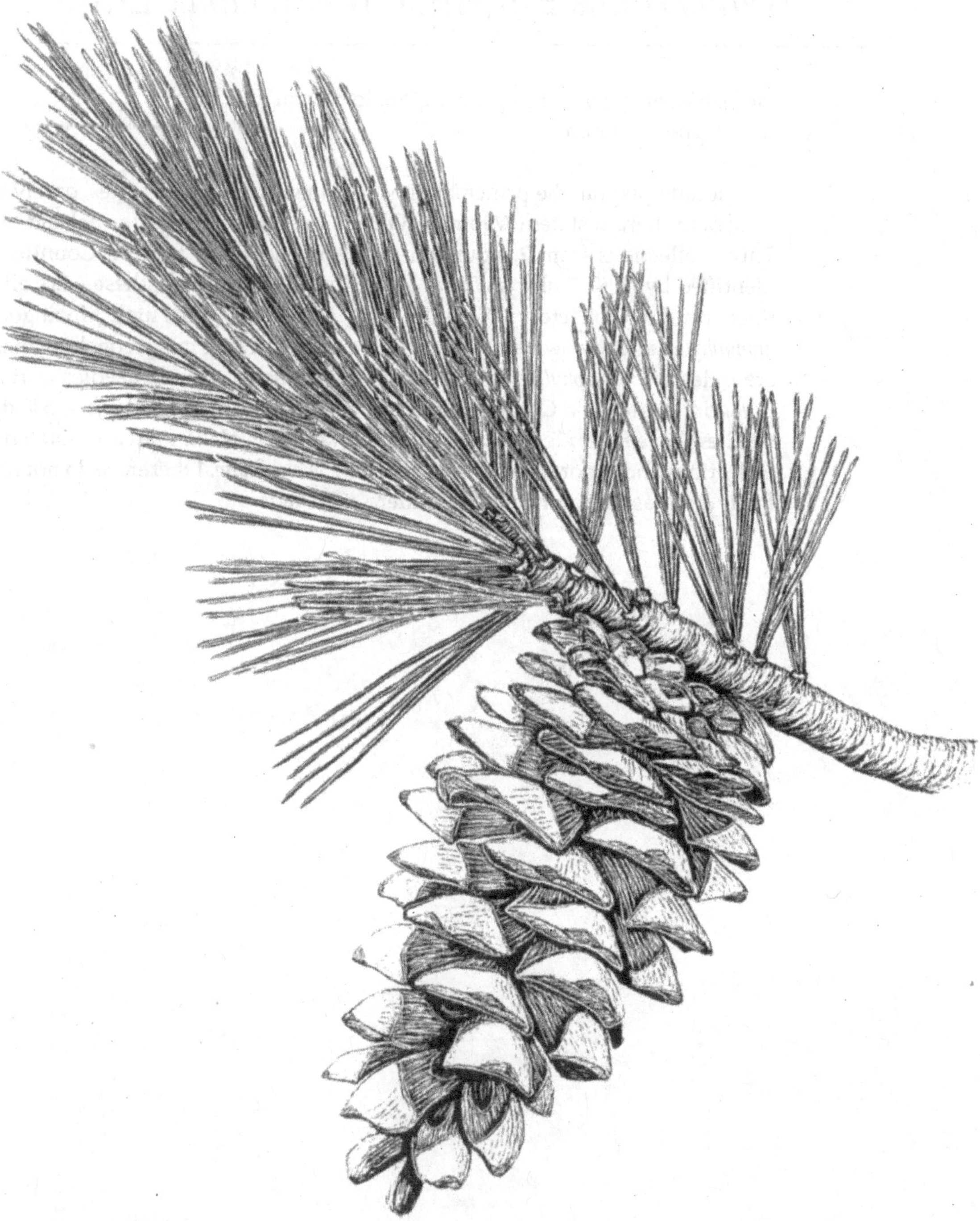

Pinus flexilis James

common name: limber pine, Rocky Mountain white pine
elevation: 6000–11,500 ft (1830–3505 m)
zones: Arctic-Alpine, Hudsonian, Canadian
124 Records for 11 Nevada counties (CL, EL, ES, EU, HU, LA, LI, LY, MI, NY, WP)
Reports for 51 mountain ranges, Records for 39 of these
New Records for 4 mountain ranges

Limber pine is principally a tree of the Rocky Mountains, where it is distributed from about latitude 52°N in Canada to latitude 35°N in central New Mexico. It enters the Pacific Coast Mountain System only in California, where it occurs in the southern Sierra Nevada, the Transverse Ranges, and the southern Coast Ranges. In the Intermontane Plateaus, limber pine occurs primarily in the Great Basin, with isolated occurrences in the Columbia and Colorado Plateaus.

Pinus flexilis is the most widely distributed high-elevation conifer in Nevada; it is known from 51 mountain ranges in 11 counties (Clark, Elko, Esmeralda, Eureka, Humboldt, Lander, Lincoln, Lyon, Mineral, Nye, and White Pine). Its presence is recorded here for the first time in the Antelope Range and Kern Mountains (both in White Pine County) and the Fish Creek Range (Eureka County). The most notable range extension is illustrated by the records for the Pine Forest Range (Humboldt County) (*Charlet 1036, Charlet 1256, Charlet 1533,* all at UCR), where the only conifer previously known was its look-alike, *Pinus albicaulis.* Limber pine grows over an impressive elevation range in Nevada of 5500 ft (1675 m).

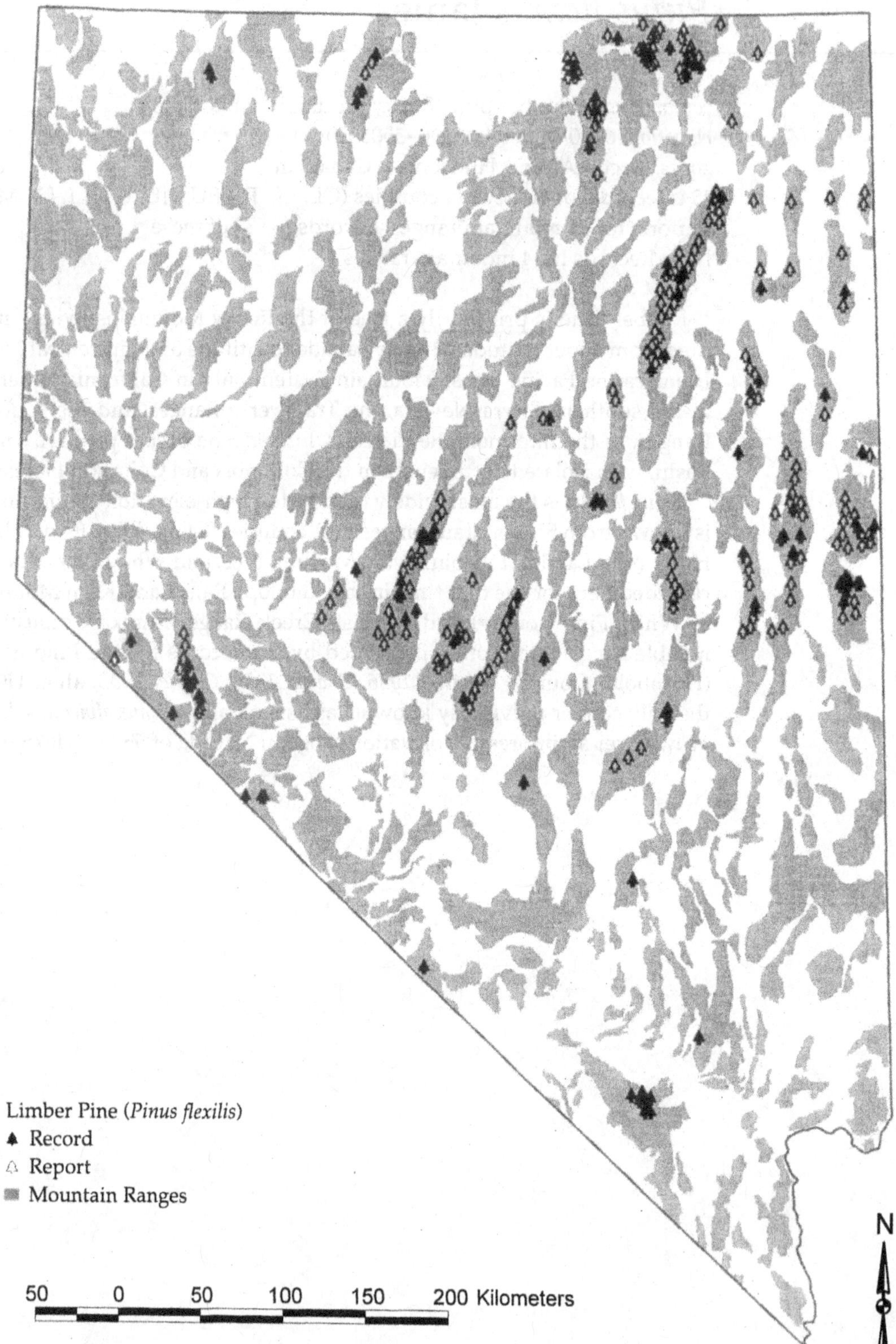
Limber Pine (*Pinus flexilis*)
Record
Report
Mountain Ranges
N
50 0 50 100 150 200 Kilometers

****Antelope Range (1)**
elevation record: 8740 ft
1 record in 1 county (White Pine)

Towns/Range/Section	Location—Habitat	County	Elevation	Collector	Herbarium/a
T24N R67E S4 [DAC: T23N R67E]	nivation hollow, NW aspect	WP	8740 ft	*Charlet 2027*	DUKE, RENO, UC

Not reported as present in the range by Critchfield & Allenbaugh (1969) or Little (1971).
DAC: restricted to the highest elevations, usually associated with *Abies concolor* or *Cercocarpus ledifolius* (1994).
New record for the species in the range.

Bodie Mountains (1)
elevation record: 9200 ft
1 record in 1 county (Mineral)

Towns/Range/Section	Location—Habitat	County	Elevation	Collector	Herbarium/a
T5N R27E S25	Brawley Peak	MI	9200 ft	*Marshall Humphreys 4*	RENO

Reported as present in the range by Little (1971).
DAC: a small stand of limber pines occur near Brawley Peak (1991–1995).

Bruneau Range (1)
1 record in 1 county (Elko)

Towns/Range/Section	Location—Habitat	County	Elevation	Collector	Herbarium/a
T46N R55E [Little: T46N R54E]	41°50′4″N, 115°46′32″W N slope Vincent Creek	EL	—	*B Leonard BL66*	UTC, OGDF

Reported as present in the range by Little (1971).
Not reported as present by Loope (1969).
DAC: I observed *Pinus flexilis* at high elevations in a few isolated subalpine fir stands (1988–1994).

Buck Creek Mountains (2)
Elko County
[Loope: T47N R57E, T46N R57E, T45N R57E]

Reported as present in the range by Loope (1969) and Little (1971).
DAC: I observed limber pine at high elevations in a few isolated subalpine fir stands (1988–1994).

Bull Run Mountains (1)
elevation record: 8600 ft
1 record in 1 county (Elko)

Towns/Range/Section	Location—Habitat	County	Elevation	Collector	Herbarium/a
T44N R52E S21	near White Rock, 41°42'N, 116°7.8'W, w/*Pinus albicaulis*	EL	8600 ft	*SA Scott in 1965*	IFGP
[Loope: T45N R52E R53E, T44N R53E]					

Reported as present in the range by Loope (1969) and Little (1971).

Cherry Creek Range (1)
elevation record: 6760–8800 ft
2 records in 1 county (White Pine) and reports in another (Elko)

Towns/Range/Section	Location—Habitat	County	Elevation	Collector	Herbarium/a
T25N R62E S25	Paris Creek; lowest one seen; in pinyon-juniper woodland	WP	6760 ft	*Charlet 2031*	DUKE, RENO
T25N R63E S4	between Currie Summit and Goshute Basin	WP	8800 ft	*Charlet 2116 & T Charlet*	DUKE, RENO
[Loope: (EL) T27N R63E, T26N R63E]					
[Little: (WP) T26N R63E]					
[RWR: (WP) T24N R62E R63E]					

Reported as present in the range by Loope (1969) and Little (1971).
DAC: abundant at the higher elevations in *Abies concolor* forests (1994).

Copper Mountains (1)
elevation record: 10,000 ft
1 record in 1 county (Elko)

Towns/Range/Section	Location—Habitat	County	Elevation	Collector	Herbarium/a
T45N R57E	2.5 mi S of Coon Creek Ranger Station	EL	10,000 ft	*Percy Train 697*	RENO, CAS

Not reported as present in the range by Little (1971).
Reported as present by Loope (1969).
DAC: uncommon; in rocky sites at high elevations (1988–1994).

Diamond Mountains (3)
Eureka and White Pine Counties
[Little: (EU) T25N R54E; (WP) T25N R55E]
[McLane: (EU, WP) T20N R55E (Diamond Peak)]

Reported as present in the range by Little (1971).
DAC: Alvin McLane told me (pers. comm. 1994) that a stand of very old trees occurs on Diamond Peak.

Duck Creek Range (1)
elevation record: 9260 ft
1 record in 1 county (White Pine)

Towns/Range/Section	Location—Habitat	County	Elevation	Collector	Herbarium/a
T16N R64E S24 N/4 [DAC: T17N R64E]	ridge N of Camel Peak	WP	9260 ft	*Charlet 2019*	DUKE, RENO

Reported as present in the range by Little (1971).
DAC: abundant at middle to high elevations, usually with *Abies concolor,* but also with *Pinus longaeva* on the ridge north of Camel Peak (1994).

East Humboldt Range (1)
elevation record: 8400–9700 ft
4 records in 1 county (Elko)

Towns/Range/Section	Location—Habitat	County	Elevation	Collector	Herbarium/a
T36N R61E S4	Angel Lake	EL	8400 ft	*P Raven & Otto Solbrig 13473*	UC
T36N R61E S4	Angel Lake	EL	8400 ft	*Critchfield & Allenbaugh 1-8/29/63*	IFGP
T36N R61E S4	Angel Lake	EL	—	*J Darling in 1954*	DUKE
—	near Deeth	EL	9700 ft	*AA Heller 10226*	RENO
[Loope: T37N R61E, T36N R60E R61E, T35N R60E R61E, T34N R61E]					

Reported as present in the range by Critchfield & Allenbaugh (1969), Loope (1969), and Little (1971).

Egan Range (1)
elevation record: 7000–8135 ft
4 records in 1 county (White Pine)

Towns/Range/Section	Location—Habitat	County	Elevation	Collector	Herbarium/a
T12N R63E S16	Lone Pine Spring, Sawmill Canyon	WP	7820 ft	*Charlet 2000 & Walter Cripps*	DUKE, RENO
T12N R63E S28	Sawmill Canyon	WP	8135 ft	*Richard W. Rust 7*	UCR
T15N R62E	NW slope of Ward Mountain, 10 mi S of Ely	WP	7000 ft	*W Sandow in 1953*	RENO
T21N R62E S18	N of Telegraph Peak	WP	—	*Ann Pinzl 10071 & Jim Morefield*	NSM
[Little: T15N R63E, T14N R63E, T13N R63E, T11N R63E, T10N R62E R63E]					

Reported as present in the range by Little (1971).
DAC: common at high elevations, usually with *Abies concolor.* The species probably occurs at more locations than reported, particularly north of Ely (1994).

Elk Mountains (3)
Elko County
[Little: T47N R61E]

Reported as present in the range by Little (1971).
Not reported as present by Critchfield & Allenbaugh (1969) or Loope (1969).

Fish Creek Range (1)
elevation record: 7200–8060 ft
3 records in 1 county (Eureka)

Towns/Range/Section	Location—Habitat	County	Elevation	Collector	Herbarium/a
T18N R53E S15	Prospect Mountain, SE of Prospect Peak, in *Cercocarpus* woodland	EU	8040 ft	*Charlet 2038*	DUKE, RENO
T18N R53E S11	Hamburg Ridge, emergent in mountain-mahogany	EU	8060 ft	*Charlet 2083*	DUKE, RENO
T18N R54E	S of Eureka	EU	7200 ft	*Margaret Williams 5-5-68*	RENO

Not reported as present in the range by Little (1971).
DAC: common on the ridges of Prospect Mountain, descending to *Cercocarpus* woodlands. *Charlet 2038* was the lowest individual I found (1994).

Fox Creek Range (1)
elevation record: 7000 ft
1 record in 1 county (Elko)

Towns/Range/Section	Location—Habitat	County	Elevation	Collector	Herbarium/a
T45N R58E S5	Pine Creek Rd, 0.5 mi N of Dry Gulch, S of Jarbidge, 41°49'N, 115°24.5'W	EL	7000 ft	*WB Critchfield 1-9/9/61*	IFGP
[Loope: T44N R58E, T46N R58E]					

Not reported as present in the range by Little (1971).
Reported as present by Loope (1969).

Goshute Mountains (1)
elevation record: 7500 ft
1 record in 1 county (Elko)

Towns/Range/Section	Location—Habitat	County	Elevation	Collector	Herbarium/a
T31N R68E S35	E aspect, not common	EL	7500 ft	*Cooper 143 & Lund & Holmgren*	UTC, RENO
[Little: T33N R68E]					
[RWR: T30N R68E S35 (Hawk Ridge)]					

Reported as present in the range by Critchfield & Allenbaugh (1969) (referred to as "Toano Range") and Little (1971).
Not reported as present by Loope (1969).

Grant Range (1)
elevation record: 8850–10,000 ft
4 records in 1 county (Nye)

Towns/Range/Section	Location—Habitat	County	Elevation	Collector	Herbarium/a
T5N R57E S1	Troy Peak, W slope	NY	8850 ft	*Guyer 5039 & Nigro*	RENO
T5N R58E S6	Troy Peak	NY	9400 ft	*Guyer 5041 & Nigro*	RENO
T5N R58E S18	E face Troy Peak, Scofield Canyon	NY	9000 ft	*ND Johnson in 1953*	DUKE
T6N R58E S30	near top of Timber Mountain	NY	10,000 ft	*M Williams 79-110-7 & Birdsey & Tiehm*	RENO

Reported as present in the range by Little (1971).
DAC: abundant at high elevations (1994).

Grapevine Mountains (1)
elevation record: 8200 ft
1 record in 1 county (Nye)

Towns/Range/Section	Location—Habitat	County	Elevation	Collector	Herbarium/a
T11S R43E S36	Wahguyhe Ridge	NY	8200 ft	*M Kurzius 1112 & K Kingsley*	UNLV

Reported as present in the range by Little (1971).

Groom Range (1)
elevation record: 7400–9200 ft
1 record in 1 county (Lincoln)

Towns/Range/Section	Location—Habitat	County	Elevation	Collector	Herbarium/a
T6S R56E S6	NE & E slopes of Bald Mountain	LI	7400–9200 ft	*JL Reveal 1482 & JC Beatley*	CAS, RENO

Reported as present in the range by Little (1971) and Beatley (1976) (N slope Bald Mountain).

Highland Range (4)
Lincoln County

Not reported as present in the range by Little (1971).
DAC: although my only visit to the mountain did not yield the species, and I have seen no material in herbaria, it seems likely that it is here. If not, the suitable sites are occupied instead by *Pinus longaeva* and *Abies concolor* (1994).

Hot Creek Range (1)
elevation record: 8800 ft
1 record in 1 county (Nye)

Towns/Range/Section	Location—Habitat	County	Elevation	Collector	Herbarium/a
T9N R51E S24	Morey Peak	NY	8800 ft	*V Bostick 6411*	UNLV

Reported as present in the range by Little (1971).

Ichabod Range (1)
elevation record: 8800 ft
1 record in 1 county (Elko)

Towns/Range/Section	Location—Habitat	County	Elevation	Collector	Herbarium/a
T45N R56E S26	Pine Mountain, Martin Creek, Humboldt Reserve	EL	8800 ft	*PB Kennedy in 1913*	CAS, RENO

Not reported as present in the range by Loope (1969) or Little (1971).
DAC: apparently present only at the north end of the range (1988–1994).

Independence Mountains (1)
elevation record: 7400–9000 ft
4 records in 1 county (Elko)

Towns/Range/Section	Location—Habitat	County	Elevation	Collector	Herbarium/a
T39N R53E S4	Wheeler Mountain, above head of Waterpipe Canyon, 17.5 mi SW of North Fork (town)	EL	8200 ft	*N Holmgren 6077*	RENO
T41N R53E	10 mi W of North Fork	EL	9000 ft	*Nichols 328 & Lund*	RENO
T42N R53E S2 NE/4	above Cole Canyon, below Jack Creek summit	EL	7400 ft	*PL Packard 80-150 & J Grimes*	BRY
T42N R53E S35	Humboldt National Forest, Pratt Creek	EL	8500 ft	*Dale Kinnaman DK-21*	OGDF
[Loope: T42N R53E R54E, T41N R53E, T40N R53E, T37N R53E]					

Reported as present in the range by Critchfield & Allenbaugh (1969), Loope (1969), and Little (1971).
DAC: uncommon; found with mountain brush on slopes above both forks of Mill Creek (1991–1994).

Jarbidge Mountains (1)
elevation record: 9000 ft
1 record in 1 county (Elko)

Towns/Range/Section	Location—Habitat	County	Elevation	Collector	Herbarium/a
T46N R58E S21	Jenny Creek headwaters	EL	9000 ft	*Ann Pinzl in 1976*	NSM
[Loope: T46N R58E R59E, T47N R58E]					

Reported as present in the range by Critchfield & Allenbaugh (1969), Loope (1969), and Little (1971).
DAC: uncommon; generally restricted to canyons at lower elevations (1988–1994).

Kawich Range (1)
elevation record: 8500 ft
1 record in 1 county (Nye)

Towns/Range/Section	Location—Habitat	County	Elevation	Collector	Herbarium/a
T1N R49E S2	SE of Kawich Peak, E face of N Kawich; small trees, in *Cercocarpus*	NY	8500 ft	*J Beatley 11525*	CAS, RENO

Not reported as present in the range by Little (1971).
Reported as present by Beatley (1976) (Kawich Peak area).

****Kern Mountains (1)**
elevation record: 8740 ft
1 record in 1 county (White Pine)

Towns/Range/Section	Location—Habitat	County	Elevation	Collector	Herbarium/a
T21N R69E S22	ridge above Tungstonia & Rock Springs	WP	8740 ft	*Charlet 2108*	DUKE, RENO
[DAC: T21N R70E]					

Not reported as present in the range by Little (1971).
DAC: uncommon on rocky ridges at high elevations, above *Abies concolor* forest (1994).
New record for the species in the range.

Marys River Range (1)
elevation record: 6580–8300 ft
4 records in 1 county (Elko)

Towns/Range/Section	Location—Habitat	County	Elevation	Collector	Herbarium/a
T44N R59E S2	North Fork Camp Creek	EL	6960 ft	*Charlet 2132 & Agee Smith*	DUKE, RENO
T44N R59E S29	ridge near T Creek, 50 mi N of Deeth (determined Billings)	EL	—	*Bill Miller in 1951*	RENO
T44N R60E S7 SE/4	Camp Creek Canyon	EL	6580 ft	*Charlet 2130 & T Charlet*	DUKE, RENO
T45N R59E S32 SE/4	above Camp Creek	EL	8300 ft	*Charlet 2134*	DUKE, RENO, UC

[Loope, DAC: T44N R59E S1 (Camp Creek)]
[DAC: T45N R60E S17 (Middle Fork Cottonwood Creek)]

Not reported as present in the range by Little (1971).
Reported as present by Loope (1969).
DAC: present in Cottonwood Creek and Camp Creek Canyons; common along the lower drainages. *Charlet 2134* was the highest individual I found in the range (1988–1994).

Medicine Range (4)
Elko County

Not reported as present in the range by Critchfield & Allenbaugh (1969), Loope (1969), or Little (1971).
DAC: I observed stands of pines south and southwest of High Bald Peaks (T27N R60E S1 S12) that probably were *P. flexilis* (1994).

Monitor Range (1)
elevation record: 8500–9115 ft
3 records in 1 county (Nye)

Towns/Range/Section	Location—Habitat	County	Elevation	Collector	Herbarium/a
T6N R45E S1	Top of Big Ten Peak, ca. 25 mi NE of Tonopah	NY	9115 ft	*E Neese 9795 & S White*	RENO, BRY
T12N R46E S12	Morgan Creek	NY	—	*sn, no year; prob. Steele in 1917*	AUSTIN
T13N R49E S19	Dobbin Basin, N end of range	NY	8500 ft	*GW Gullion 491*	UC

[Little: T13N R48E, T12N R48E, T11N R48E, T10N R48E, T9N R47E, T8N R47E, T7N R46E R47E, T6N R46E]

Reported as present in the range by Little (1971) and Goodrich (1981).
DAC: common at high elevations throughout the range (1994).

Paradise Range (3)
Nye County
[Little: T13N R37E, T12N R37E, T11N R37E]

Reported as present in the range by Little (1971).

Pequop Mountains (1)
elevation record: 8300–10,200 ft
2 records in 1 county (Elko)

Towns/Range/Section	Location—Habitat	County	Elevation	Collector	Herbarium/a
T31N R63E S23	top Spruce Mountain	EL	10,200 ft	*M Williams 81-84-2 & Tiehm*	CAS, RENO
T31N R63E S23 NW/4	NW slope Spruce Mountain	EL	8300 ft	*Ann Pinzl 10141*	NSM
[Little, Critchfield & Allenbaugh: T32N R65E, T34N R65E]					

Reported as present in the range by Critchfield & Allenbaugh (1969), Loope (1969), and Little (1971).

Pilot Range (3)
Elko County
[Loope: T37N R70E, T36N R70E]

Reported as present in the range by Critchfield & Allenbaugh (1969), Loope (1969), and Little (1971).

****Pine Forest Range (1)**
elevation record: 7150–9105 ft
3 records in 1 county (Humboldt)

Towns/Range/Section	Location—Habitat	County	Elevation	Collector	Herbarium/a
T43N R28E S3 NE/4	Onion Reservoir	HU	7150 ft	*Charlet 1036*	UCR
T43N R28E S13 W/4	E of Bare Pass	HU	9105 ft	*Charlet 1256*	UCR
T43N R28E S14 SE/4	E of Bare Pass	HU	9020 ft	*Charlet 1533*	UCR

Not reported as present in the range by Little (1971).
DAC: Cronquist et al. (1972), Wells (1983), Lanner (1984), and Grayson (1993) all say that limber pine does not occur here. Occasional to rare at lower and upper ends of whitebark pine distribution (1990–1994).
New record for the species in the range.

****Pine Grove Hills (1)**
elevation record: 7740 ft
1 record in 1 county (Lyon)

Towns/Range/Section	Location—Habitat	County	Elevation	Collector	Herbarium/a
T 9N R26E S17 N/2	above Rockland Mines, in nevé cirque on steep northeast slope	LY	7740 ft	*Charlet 2209 & T Charlet*	RENO, DUKE

Not reported as present in the range by Little (1971).
DAC: rare; a stand of about 50 individuals occurs with *Pinus contorta* and *Pinus jeffreyi* above Rockland Mines, east of Pine Grove Summit (1989–1995).
New record for the species in the range.

Quinn Canyon Range (3)
Nye County
[Little: T3N R55E R56E, T4N R56E]

Reported as present in the range by Little (1971) and Lewis (1975b).

Roberts Mountains (1)
elevation record: 8580–9780 ft
3 records in 1 county (Eureka)

Towns/Range/Section	Location—Habitat	County	Elevation	Collector	Herbarium/a
—	Roberts Mts	EU	—	*C Condit 10*	UC
T23N R50E S12 E/8	lower end of distribution, E of high peaks	EU	8580 ft	*Charlet 2090*	DUKE, RENO
T23N R50E S12	ridge E of high peaks	EU	9780 ft	*Charlet 2087*	DUKE, RENO
[DAC: T23N R50E S2 (Cooper Peak)]					

DAC: Possibly reported as present in the range by Little (1971), but the dot on Little's map is actually over the north end of the Simpson Park Range, where (at that location) *Pinus flexilis* probably does not occur. The species is common at upper elevations, particularly on the northern aspect of the high peaks, but also at lower elevations on both southern and northern aspects, along the Cooper Peak ridge (1993–1994).

Ruby Mountains (1)
elevation record: 6000–10,000 ft
8 records in 1 county (Elko)

Towns/Range/Section	Location—Habitat	County	Elevation	Collector	Herbarium/a
T26N R57E S18	just N of Sherman Mountain, near WP County	EL	9000 ft	*N Holmgren 3639 & P Kern*	UTC
T27N R57E S24	Cave Creek Canyon, 2 mi N of Cave Creek PO	EL	6000 ft	*HL Mason 4811*	UC
T31N R58E S35	40°34.5'N, 115°23.5'W	EL	10,000 ft	*Critchfield & Allenbaugh 2-8/28/63*	IFGP
T32N R58E S35	Thomas Creek Canyon	EL	9500 ft	*N Holmgren 6270 & P Holmgren*	UTC, BRY
T32N R58E S6	Lamoille Canyon	EL	6500 ft	*Nichols & Lund 589*	RENO
T32N R58E S6	Lamoille Canyon	EL	6500 ft	*C McMillan 237*	RSA, UT
T32N R59E S30	Lamoille Canyon, Terrace Ranger Station	EL	8400 ft	sn	UTC
—	Ruby Mts, Star Canyon [DAC: I cannot find a reference to this canyon]	EL	—	*AA Heller 10226*	CAS
[Loope: T33N R58E R59E, T32N R57E R58E R59E, T30N R57E R58E, T29N R57E, T28N R57E, T27N R57E, T26N R57E]					

Reported as present in the range by Critchfield & Allenbaugh (1969), Loope (1969), and Little (1971).
DAC: common in the middle and upper elevations in canyons and at high elevations throughout the range. *Mason 4811* is the low-elevation record for the species in the state (1988–1994).

Salmon River Range (3)
Elko County
[Critchfield & Allenbaugh: T45N R63E S16]

Reported as present in the range by Critchfield & Allenbaugh (1969) and Little (1971).
Not reported as present by Loope (1969).
DAC: the map location noted by Critchfield & Allenbaugh (1969) corresponds to Ellen D Mountain. With binoculars, I observed small, open stands of a conifer at this location (1988–1994).

Santa Rosa Range (1)
elevation record: 8925 ft
3 records in 1 county (Humboldt)

Towns/Range/Section	Location—Habitat	County	Elevation	Collector	Herbarium/a
T42N R38E S10	Dog Creek, ridge top in rocks	HU	—	*Bryant Martineau 41*	RENO
T43N R38E S24	above Eagle Creek	HU	8925 ft	*Charlet 777 & Rust*	UCR
T45N R39E S14	Buckskin Mountain	HU	—	*G Austin & Ann Pinzl in 1981*	NSM

[DAC: T44N R39 S20 (Granite Peak), T43N R38E S35 (Santa Rosa Peak), T45N R39E S35 (headwaters, N Fork Cabin Creek)]

Reported as present in range by Little (1971).
DAC: in woodland approaching subalpine forest only on Buckskin Mountain; small populations of old, mature trees grow in open parkland on Santa Rosa Peak (ca. 150) and Granite Peak (ca. 75 individuals).

Schell Creek Range (1)
elevation record: 8000–9500 ft
4 records in 1 county (White Pine)

Towns/Range/Section	Location—Habitat	County	Elevation	Collector	Herbarium/a
T15N R65E S14 SE/4	———	WP	—	*Ann Pinzl 10097*	NSM
T16N R65E S2	Kalamazoo Summit, in low sage	WP	8680 ft	*Charlet 2023*	DUKE, RENO
T16N R65E	Success Pass	WP	9035 ft	*Peter Herlan in 1974*	NSM
T16N R66E	Indian Creek, 16 mi N of McGill	WP	8000–9500 ft	*Percy Train 982*	RENO

[Little: T21N R65E, T20N R65E, T19N R66E, T18N R65E R66E, T17N R65E R66E, T16N R65E R66E, T15N R66E, T14N R65E, T13N R65E, T12N R65E, T11N R65E, T10N R64E]

Reported as present in the range by Little (1971).
DAC: common at high elevations in the central and northern portions of the range; in forests with *Abies concolor* and on exposed sites (1994).

Seaman Range (4)
Lincoln and Nye Counties

Not reported as present in the range by Little (1971).
DAC: although no collections or reports exist, the species is likely to occur here and may yet be found in the higher portions of the range.

Sheep Range (1)
elevation record: 9100 ft
1 record in 1 county (Clark)

Towns/Range/Section	Location—Habitat	County	Elevation	Collector	Herbarium/a
T15S R60E	Sheep Range, Hidden Forest, 36°39'N, 115°12'W	CL	9100 ft	*E Rockwell, B Henry & J Spaulding in 1977*	IFGP

Not reported as present in the range by Little (1971).

Shoshone Mountains (1)
elevation record: 10,000 ft
1 record in 1 county (Nye)

Towns/Range/Section	Location—Habitat	County	Elevation	Collector	Herbarium/a
T14N R39E	top of S Shoshone Peak	NY	10,000 ft	*V Woodbury in 1951*	RENO

Not reported as present in the range by Little (1971).

Simpson Park Range (3)
Lander County
[Little: T23N R48E S3 (vicinity of McClusky Peak)]

Reported as present in the range by Little (1971).

Snake Mountains (3)
Elko County
[Loope, Critchfield & Allenbaugh: T41N R62E S17]

Reported as present in the range by Critchfield & Allenbaugh (1969), Loope (1969), and Little (1971).

Snake Range (1)
elevation record: 6000–11,500 ft
16 records in 1 county (White Pine)

Towns/Range/Section	Location—Habitat	County	Elevation	Collector	Herbarium/a
T12N R68E S14	S slopes of Mount Washington	WP	10,400 ft	*N Holmgren 3763 & A Holmgren*	UTC
T13N R68E S10	39°N, 114°19'W, w/*Pinus longaeva* & *Picea engelmannii* [Mount Wheeler]	WP	11,000 ft	*DL Zohner in 1960*	IFGP
T13N R68E S10	NW ridge of Wheeler Peak, wet meadow	WP	11,500 ft	*E Neese 9108 & S White*	BRY
T13N R68E S10	saddle N of Wheeler Peak, w/mixed spruce & aspen	WP	11,500 ft	*Critchfield 1-9/12/61*	IFGP
T13N R68E S15	Wheeler Peak area	WP	10,500 ft	*Mozingo 74-41a*	RENO
T13N R68E S36	S hillside of Johnson Lake	WP	11,000 ft	*Irwin Ting 19393*	RENO
T13N R69E	Baker Canyon, 5 mi SW of Baker; canyon bottom	WP	—	*SJ Preece, Jr 1458*	UT
T13N R69E S7	Lehman Creek headwaters, NE slope Mount Wheeler	WP	8500 ft	*Percy Train 1155*	RENO
T13N R69E S14	S of Lehman Creek	WP	6000 ft	*T Frantz in 1953*	DUKE
T13N R69E S15	Lehman Caves	WP	8500 ft	*PA Lehenbauer 1269*	RENO
T13N R69E S15	Lehman Creek, above cave	WP	7800 ft	*O Reifschneider in 1950*	RENO
T14N R68E S36	on road to Mount Wheeler; ridge top in *Cercocarpus*	WP	—	*SL Welsh 2235 & G Moore*	BRY
T16N R69E S3 NE/4	slopes of Mount Moriah above Table Mountain	WP	—	*JC Lyman 430B*	UCR
T17N R69E S28 SE/4	Big Canyon, 2.5 mi NNW of Mount Moriah summit	WP	10,000 ft	*JC Lyman 707 et al.*	UCR
T17N R70E S34	Horse Canyon	WP	7000 ft	*T Frantz sn*	DUKE
—	Mount Wheeler slopes	WP	8000 ft	*WP Cottam 8128*	UT

[DAC: T16N R70E (Hampton Creek)]
[Little: T19N R68E, T18N R68E R69E, T17N R68E, T16N R68E, T15N R68E R69E, T12N R69E, T11N R68E R69E]

Reported as present in the range by Little (1971).
DAC: abundant in spruce forests and in *Abies concolor* forests (1994).

Spring Mountains (1)
elevation record: 7600–10,000 ft
9 records in 1 county (Clark)

Towns/Range/Section	Location–Habitat	County	Elevation	Collector	Herbarium/a
T18S R56E S35	Lee Canyon	CL	8200 ft	*IW Clokey 5380*	UTC
T19S R56E S6	Kyle Canyon Peak trail	CL	9185 ft	*IW Clokey A132*	OGDF
T19S R56E S15	115°42'W, 36°18'N, between ski area & saddle at head of Wallace Canyon	CL	9600 ft	*AC Sanders et al.*	UCR
T19S R56E S36	Kyle Canyon slopes	CL	8850 ft	*IW Clokey 8625*	RENO, RSA
T19S R57E S8	Deer Creek camp	CL	7600 ft	*L Benson 15291*	RSA
T19S R57E S18	above Deer Creek camp	CL	8500 ft	*L Benson 15072*	RSA
T20S R56E S1	Spring Mountains	CL	10,000 ft	*Ann Pinzl 4850*	NSM
T20S R57E S5	ridge S of Kyle Canyon	CL	9360 ft	*IW Clokey 7818*	RENO, UT
—	Spring Mountains	CL	9500 ft	*Edmund Jaegaer in 1926*	RSA

Reported as present in the range by Little (1971).
DAC: abundant at the higher elevations, often associated with *Abies concolor* (1982 & 1993).

Sweetwater Mountains (3)
Lyon County
[Lavin: T8N R24E]

Reported as present in the California portion of the range by Little (1971).
Reported as present in the Nevada portion of the range by Lavin (1981).

Toana Range (3)
Elko County
[Critchfield & Allenbaugh: T36N R68E]

Reported as present in the range by Critchfield & Allenbaugh (1969) and Little (1971).
Not reported as present by Loope (1969).

Toiyabe Range (1)
elevation record: 8000–10,260 ft
11 records in 2 counties (Lander and Nye)

Towns/Range/Section	Location—Habitat	County	Elevation	Collector	Herbarium/a
T11N R42E S8	S Fork S Twin River, open rocky hillside	NY	8800 ft	*Guyer 5151*	RENO
T12N R42E S25	N Fork S Twin River	NY	8200 ft	*V Bostick 6623*	UNLV
T13N R41E S1	Head of Ophir Creek, summit of Toiyabe Range	NY	10,000 ft	*Percy Train 3347*	UTC, RENO
T13N R41E S35	ridge between Crane Creek & Ophir Wash, W side Toiyabe Range	NY	9700 ft	*Billings 1734*	RENO
T13N R41E S36 NE/4	N of Ophir Pass	NY	10,260 ft	*Charlet 2048 & John Brack*	DUKE, RENO
T16N R43E	Bunker Hill Canyon	LA	8000 ft	*PB Kennedy 4094*	RENO
T16N R43E S8	Mahogany Gulch of Kingston Canyon	LA	—	*Goodrich 11574*	BRY
T16N R43E S14	Bunker Hill Peak	LA	9500 ft	*P Herlan in 1971*	NSM
T16N R43E S14	Bunker Hill, rocky hillside	LA	9250 ft	*Guyer 5140 & Nigro*	RENO
T16N R43E S16	0.5 mi E of Kingston Canyon, Forest Service guard station, common	LA	—	*Goodner 214 & Henning*	UTC, RENO
T17N R43E S36	3 mi NNE of Bunker Hill, NW side of ridge, 39°17'N, 117°6'W	LA	10,200 ft	*WB Critchfield 2-9/16/61*	IFGP

[DAC: T17N R43E (above Big Creek)]
[Little: (LA) T18N R44E, T17N R43E R44E (Toiyabe Peak), T15N R42E R43E; (NY) T15N R42E R43E, T14N R42E, T13N R42E, T12N R42E, T11N R41E, T10N R40E R41E R42E]

Reported as present in the range by Little (1971).
DAC: the only subalpine pine I found on the mountain range; forms subalpine woodland islands, usually on north or east aspects at high elevations (1991–1994).

Toquima Range (1)
elevation record: 10,000–10,170 ft
3 records in 1 county (Nye)

Towns/Range/Section	Location—Habitat	County	Elevation	Collector	Herbarium/a
T10N R45E	S slope Mount Jefferson	NY	10,000 ft	*M Williams 76-95-1*	RENO
T10N R45E S4 W1/4	below S summit Mount Jefferson	NY	10,170 ft	*Charlet 852*	UCR
T11N R45E S22	near head of Pine Creek	NY	10,000 ft	*Gary Pierce 1983*	UTC

[Little: T14N R46E]
[DAC: T11N R45E (N face, North Summit Mount Jefferson), T10N R44.5E (Shoshone Mountain), T9N R45E]

Reported as present in the range by Little (1971).
DAC: the only subalpine pine I found on the mountain range; forms subalpine forests and woodlands at high elevations, particularly on the north and east faces of Mount Jefferson (1991–1993).

Wassuk Range (1)
elevation record: 6600-10,520 ft
11 records in 1 county (Mineral)

Towns/Range/Section	Location—Habitat	County	Elevation	Collector	Herbarium/a
T6N R29E S11	ridge N of Buller Mountain	MI	8280 ft	*Charlet 1978*	DUKE, RENO, UC
T6N R29E S13	Buller Mountain	MI	9200 ft	*Charlet 703 & Rust*	UCR
T7N R28E S1	Big Indian Mountain	MI	10,080 ft	*Ann Pinzl in 1974*	NSM
T7N R29E S9	5 mi from mouth, up Cory Creek	MI	6600–7200 ft	*WA Archer 6919*	UTC, RENO
T7N R29E S9	2–4 mi below head of Cory Canyon, rare	MI	—	*WH Henning 312*	RENO
T7N R29E S19	0.5 mi SE of Cory Peak, 38°27'N, 118°46.5'W	MI	9500 ft	*Critchfield & Allenbaugh 1-7/5/63*	IFGP
T7N R29E S19	Cory Peak	MI	10,520 ft	*Dorothy Mason in 1975*	RENO
T7N R29E S19	S side Cory Peak, isolated here	MI	9600 ft	*Billings 1679*	RENO
T7N R29E S20	Microwave tower peak E of Cory Peak	MI	9400 ft	*Mozingo 77-64*	RENO
T8N R29E S30	Cat Creek	MI	9300 ft	*AM Alexander & L Kellogg 4396*	UC
T8N R29E S18	Mount Grant, near Cat Canyon	MI	—	*Ken Hawkings in 1981*	RENO

[Little: T11N R28E, T10N R28E, T9N R28E]
[DAC: T8N R28E S27 (Lapon Canyon), T6N R29E S36, T5N R29E S1 (Powell Mountain)]

Reported as present in the range by Little (1971).
DAC: scattered throughout the range, mostly in small woodland patches at high elevations. Small stands were observed on the north and east slopes of Powell Mountain, near the peak (1990–1995).

White Mountains (1)
elevation record: 9500–11,000 ft
4 records in 1 county (Esmeralda)

Towns/Range/Section	Location—Habitat	County	Elevation	Collector	Herbarium/a
T1S R32E	Boundary Peak	ES	11,000 ft	*VK Johnson 1*	RENO
T1S R32E	Boundary Peak, N slopes	ES	9500–11,000 ft	*Billings 1683*	RENO
T1S R33E S17	1.3 mi N 55° E of Boundary Peak	ES	10,200 ft	*JD Morefield 4400 et al.*	RSA
T1S R33E	NE side Boundary Pk	ES	9900 ft	*Ann Pinzl 2400 & Pinzl*	NSM
T1S R33E S17	N slope, Trail Canyon	ES	—	*VK Johnson 2*	RENO
—	White Mts	ES	10,500 ft	*Roxana Ferris 6763*	CAS

Reported as present in the range by Little (1971).

White Pine Range (1)
elevation record: 8500–9500 ft
1 record in 1 county (White Pine) and reports in another (Nye)

Towns/Range/Section	Location—Habitat	County	Elevation	Collector	Herbarium/a
T16N R57E S24	ridge SW of Hamilton [Babylon Ridge]	WP	8500–9500 ft	*Billings 2193*	RENO
[Little: (WP) T15N R58E, T14N R58E, T13N R58E, T12N R58E; (NY) T12N R58E, T11N R58E]					
[RWR: (WP) T16N R57E S21 S27 S35 (Pogonip Ridge), T13N R58E S11 (Currant Mountain)]					

Reported as present in the range by Little (1971).

Wood Hills (3)
Elko County
[Critchfield & Allenbaugh, Loope: T36N R63E]

Reported as present in the range by Critchfield & Allenbaugh (1969), Loope (1969), and Little (1971).

Pinus jeffreyi Greville & Balfour

common name: Jeffrey pine, western yellow pine
elevation: 4500–9100 ft (1370–2775 m)
zones: Canadian, Arid Transition, Upper Sonoran
62 Records in 6 Nevada counties (CC, DO, LY, MI, ST, WA)
Records in 13 mountain ranges
New Records in 3 mountain ranges

Jeffrey pine is a dominant tree throughout the upper Arid Transition Zone and lower Canadian Zone forests in the Sierra Nevada, and is otherwise rather widely distributed in the northwestern and southwestern California mountains. The species extends north into southwestern Oregon and south into the mountains of Baja California.

In Nevada, Jeffrey pine occurs in 13 mountain ranges in 6 far western counties (Carson City, Douglas, Lyon, Mineral, Storey, and Washoe); all except Storey border California. Jeffrey pine is the dominant tree in the Arid Transition Zone forests of the Carson Range. It is commonly associated with white fir, and occasionally with ponderosa pine and sugar pine at lower elevations; and with red fir, lodgepole pine, and Washoe pine at higher elevations. In Nevada's Great Basin ranges, Jeffrey pine is restricted to ranges within or west of the Lahontan Basin, where it occurs in canyons in the pinyon-juniper woodlands of the Upper Sonoran Zone, in hollows or ridges in the Canadian Zone, and in altered andesite at lower elevations within the Upper Sonoran Zone. New records are presented for the Fort Sage Mountains and Seven Lakes Mountain in far western Washoe County and for the Pine Grove Hills in Lyon County. Jeffrey pine grows over an elevation range of 4600 ft (1405 m) in Nevada.

Jeffrey Pine (*Pinus jeffreyi*)
Record
Report
Mountain Ranges
50 0 50 100 150 200 Kilometers
N

Bodie Mountains (1)
elevation record: 5500–6680 ft
2 records in 1 county (Mineral)

Towns/Range/Section	Location—Habitat	County	Elevation	Collector	Herbarium/a
T6N R26E S3 SW/4	above Red Wash in altered andesite w/*Pinus pinderosa*	MI	6680 ft	*Charlet 2199 & Ann Buell*	DUKE, RENO, UC
T6N R26E S6 NW/4	2 mi SW of Sonoma on road to Masonic	MI	5500 ft	*Billings 2191*	RENO

Reported as present in the range by Little (1971).
Reported as present with *Pinus ponderosa* on altered andesite above Sonoma Canyon on *Nachlinger 1497* DUKE label.
DAC: only 1 individual seen in Red Wash stand with over 50 reproductive *Pinus ponderosa.* More may occur at additional Red Wash Canyon altered andesite locations (1991–1995).

Carson Range (1)
elevation record: 5300–9100 ft
16 records in 3 counties (Carson City, Douglas, and Washoe)

Towns/Range/Section	Location—Habitat	County	Elevation	Collector	Herbarium/a
T13N R18E S24	1 mi W of Kingsbury Grade summit, W slope Sierra Nevada	DO	7000 ft	*Percy Train 3156*	RENO
T14N R18E	Lake Tahoe, lakeshore	DO	6240 ft	*RK Vickery, Jr 1034*	UT
T15N R19E S19	Snow Valley Peak, 39°9.2'N, 119°53.1'W, w/*Tsuga*	CC	9100 ft	*LC Johnson 76-192-3*	IFGP
T15N R19E S23	Kings Canyon (determined Greene)	CC	5575 ft	*CF Baker 326 in 1902*	RSA
T15N R19E S31	US 50 E of Spooner Summit	CC	7000 ft	*Ann Pinzl in 1973*	NSM
T15N R18E S11 SW/4	E of Marlette reservoir	WA	6680 ft	*Charlet 1921 & T Charlet*	DUKE, RENO, UC
T16N R19E S10	mountains W of Franktown	WA	5300 ft	*AA Heller 10641*	RENO
T17N R18E S24	headwaters of Galena Creek	WA	8920 ft	*Charlet 2156 & Rust*	DUKE, RENO, UC
T17N R19E	Mount Rose, 5–6 mi from summit	WA	7500 ft	*WW Schwartzkopf in 1967*	RENO
T17N R19E S4	Jones Creek, next to 40-year-old burn area	WA	6000 ft	*Charlet 1810 & T Charlet*	DUKE
T18N R18E S12	Hunter Creek Canyon	WA	7000 ft	*AA Heller 10372*	RENO
T18N R19E S18	head of Evans Creek, common	WA	7800 ft	*Nichols 641 & Lund*	RENO
T18N R19E S30	Thomas Creek	WA	7500 ft	*WA Archer 6585*	RENO
T19N R18E S20	1 mi S of Verdi	WA	—	*Roy Allen 28*	RENO
T19N R19E	off Hunter Lake Road	WA	—	*Bon Bonstead in 1968*	RENO
—	Mount Rose	WA	—	*Palmer Stockwell in 1942*	RSA

[DAC: (WA) T15N R19E, T16N R18E; (CC) T15N R18E; (DO) T14N R18E R19E, T13N R19E, T12N R19E]

Reported as present in the range by Little (1971).
DAC: abundant; dominant over large areas in the lower to middle elevations; probably the most abundant conifer on the range. *Johnson 76-192-3* is the high-elevation record for the species in the state (1971–1995).

Desert Creek Mountains (1)
elevation record: 6500 ft
1 record in 1 county (Lyon)

Towns/Range/Section	Location—Habitat	County	Elevation	Collector	Herbarium/a
T9N R24E S31	along Desert Creek, 5 mi S of Desert Creek Ranch	LY	6500 ft	*Billings 2083*	RENO

Reported as present in the range by Little (1971).
DAC: common along Desert Creek (1987–1994).

Flowery Range (1)
elevation record: 4760–6500 ft
3 records in 1 county (Storey)

Towns/Range/Section	Location—Habitat	County	Elevation	Collector	Herbarium/a
T16N R21E S3	3 mi E of Silver City	ST	6500 ft	*RJ Bandoni 9*	RENO
T17N R21E S23 S/2	N of Sixmile Canyon Creek	ST	4920 ft	*Charlet 1817 & T Charlet*	DUKE, RENO, UC
T17N R21E S25	N of Sixmile Canyon Creek	ST	4760 ft	*Charlet 1818 & T Charlet*	DUKE, RENO, UC

Not reported as present in the range by Little (1971).
DAC: a few drought-stressed individuals remain on altered andesite soils in Sixmile Canyon (1973–1994).

****Fort Sage Mountains (1)**
elevation record: 6640–6840 ft
2 records in 1 county (Washoe)

Towns/Range/Section	Location—Habitat	County	Elevation	Collector	Herbarium/a
T25N R18E S17 NE/4	below Stateline Peak	WA	6840 ft	*Charlet 2162*	DUKE, RENO, UC
T25N R18E S17 NE/4	below Stateline Peak	WA	6640 ft	*Charlet 2163*	DUKE, RENO, UC

Not reported as present in the range by Little (1971).
DAC: 2 stands of yellow pine occur on the north slope of Stateline Peak, just east of the California-Nevada border. *Pinus jeffreyi* occurs in the upper stand with aspen in a snow accumulation pocket; *P. ponderosa* occurs in the lower stand in a deep canyon. Local ranchers report that pines were much more common in the range before a fire that occurred around 1935 (1993–1994).

Peavine Mountain (1)
elevation record: 4600–8160 ft
9 records in 1 county (Washoe)

Towns/Range/Section	Location—Habitat	County	Elevation	Collector	Herbarium/a
T20N R18E S23	cluster of 5 trees below peak to W of Peavine Peak	WA	8160 ft	*Charlet 2167*	DUKE, RENO
T20N R18E S24	a few trees in mountain-mahogany, unaltered soil	WA	6600 ft	*Charlet 2166*	DUKE, RENO
T20N R18E S36	altered andesite, N of W altered knob, S side of Peavine	WA	6200 ft	*Billings 2065*	RENO
T20N R18E	altered andesite	WA	6000 ft	*Billings 2062*	RENO
T20N R18E	altered andesite	WA	6000 ft	*Billings 2063*	RENO
T20N R19E S20	isolated tree on altered rock, S aspect, 1 mi S of Black Springs	WA	5450 ft	*Billings 2052*	RENO
T20N R19E	hills north of Reno	WA	4600 ft	*AA Heller 10239*	RENO
T19N R18E S6	summit of Dog Valley grade near Verdi	WA	5500 ft	*JW Vetter in 1956*	RENO
T19N R18E S6	W of Verdi near CA	WA	5000 ft	*M Williams 74-64-1 & E McPherson*	RENO
[DAC: T20N R18E S4–S10, S14–S21]					

Reported as present in the range by Little (1971).
DAC: present in altered andesite stands at lower elevations, and in *Cercocarpus* woodlands at higher elevations. The most common conifer on the mountain; forms an extensive forest on the west slope (1971–1995).

Petersen Mountain (1)
elevation record: 5905 ft
2 records in 1 county (Washoe)

Towns/Range/Section	Location—Habitat	County	Elevation	Collector	Herbarium/a
T23N R18E	N end of Red Rock region (5000 ft) [DAC: elevation is in error]	WA	—	*RJ Bandoni 2*	RENO
T23N R18E S33	low end of stand, w/ *Cercocarpus*	WA	5905 ft	*Charlet 1050*	UCR

Reported as present in the range by Little (1971).
DAC: the lowest individual in the range that I have seen is *Charlet 1050.* More than 100 vigorous saplings are present in mountain-mahogany woodland on the east slope. In addition, about 40–50 very large trees (ca. 1 m dbh) are present; most of these were dying in 12/1993, but the saplings are doing well. This stand was burned in a fire in about 1975, and again in July 1994. I have not returned since the 1994 fire to see if there are survivors (1990–1994).

****Pine Grove Hills (1)**
elevations record: 7700 ft
1 record in 1 county (Lyon)

Towns/Range/Section	Location—Habitat	County	Elevation	Collector	Herbarium/a
T 9N R26E S17 N/2	above Rockland Mines, on ridge within pinyon woodlands	LY	7700 ft	*Charlet 2207 & T Charlet*	RENO, UC, DUKE

Not reported as present in the range by Little (1971).
DAC: rare; a stand of about 10 individuals occurs with *Pinus contorta* and *Pinus flexilis* above Rockland Mines, east of Pine Grove Summit. *Charlet 2207* is a large, fire-scarred individual, with a diameter at breast height in excess of 1m and is 14m tall (1989–1995).
New report for the species in the range.

Pine Nut Mountains (1)
elevation record: 6900–7600 ft
3 records in 1 county (Douglas)

Towns/Range/Section	Location—Habitat	County	Elevation	Collector	Herbarium/a
T11N R21E S1	near Ruby Hill Mine	DO	6900 ft	*M Williams 82-108-5 & Tiehm*	RENO
T11N R22E	E side Bald Mountain	DO	—	*RJ Bandoni 6*	RENO
T11N R21E	2 mi W of Bald Mountain, altered andesite	DO	7600 ft	*Billings 1675*	RENO

Not reported as present in the range by Little (1971).

****Seven Lakes Mountain (1)**
elevation record: 4600–5120 ft
3 records in 1 county (Washoe)

Towns/Range/Section	Location—Habitat	County	Elevation	Collector	Herbarium/a
T24N R18E S32	Red Rock Rd, near CA	WA	4600 ft	*Charlet 1783 & T Charlet*	DUKE
T24N R18E S32	Red Rock Rd, near CA	WA	4600 ft	*Charlet 1784 & T Charlet*	DUKE
T24N R18E S34 NE/4	Red Rock Rd, near CA	WA	5120 ft	*Charlet 1785 & T Charlet*	DUKE

Not reported as present in the range by Little (1971).
DAC: common in a canyon about 1 mi east of the California border, from the base to the ridge (1985–1994).
New records for the species in the range.

Sweetwater Mountains (1)
elevation record: 7200–7400 ft
2 records in 1 county (Lyon)

Towns/Range/Section	Location—Habitat	County	Elevation	Collector	Herbarium/a
T7N R25E S6	Sweetwater Creek, 1 mi from CA	LY	7400 ft	*Billings 1847*	RENO
T8N R24E S8	Desert Creek, S of Desert Creek Peak	LY	7200 ft	*Billings 1854*	RENO
[DAC: T7N R24E, T8N R25E (Cottonwood Creek)]					

Reported as present in the range by Little (1971) and Lavin (1981).
DAC: much more common in the California portion of the range, but present along streams in the Nevada portion (1985–1995).

Toiyabe Range (7)
Lander County

Robin Tausch: about 10 individuals were planted along Corral Creek above Big Creek campground, probably after the fire in 1976. This planting was reported by Kartesz (1988) as *Pinus ponderosa*. The individuals produced cones in 1994 but have not yet reproduced. This population may spread and naturalize.

Virginia Range (1)
elevation record: 4960–7000 ft
10 records in three 3 counties (Carson City, Storey, and Washoe)

Towns/Range/Section	Location—Habitat	County	Elevation	Collector	Herbarium/a
T16N R20E S31	Sugarloaf; SW end of range, in granite	CC	5220 ft	*Charlet 2066 & T Charlet*	DUKE, RENO
T17N R21E S19 NE/4	N arm of Christmas Tree Canyon	ST	7000 ft	*Billings 2068*	RENO
T18N R21E S32 E/4	Lousetown Rd at tributary of Lousetown Creek, altered andesite	ST	5780 ft	*Charlet 1819 & T Charlet*	DUKE
T19N R20E S25 SW/4	ridge above Hidden Valley, N aspect, altered andesite	ST	5680 ft	*Charlet 1820 & T Charlet*	DUKE
T19N R21E S33 S/4	W of Washington, altered andesite	ST	5100 ft	*Charlet 1779*	DUKE, RENO, UC
T17N R21E S6	head of old Geiger Grade	WA	6400 ft	*Billings 2018*	RENO
T18N R20E S26 S/4	Geiger Grade; "White Gold" stand, altered andesite	WA	5300 ft	*Charlet 1814 & T Charlet*	DUKE
T18N R20E S35	NE of Geiger Grade lookout, altered andesite	WA	5700 ft	*Billings 2058*	RENO
T19N R21E S30	along powerline road; alone among a small stand of *P. ponderosa*	ST	4960 ft	*Charlet 1808*	DUKE, RENO, UC
T19N R20E S35 NE/4	S of Hidden Valley Park, with *P. ponderosa;* only Jeffrey seen here	WA	5300 ft	*Charlet 2202 & James Morefield*	DUKE, RENO
[DAC: T17N R20E S12 (tributary of Newton Creek)]					

Reported as present in the range by Little (1971).
DAC: restricted to altered andesite soils scattered throughout the range, except for an individual found along a tributary of Newton Creek and one in a granite outcrop north of Carson City (*Charlet 2066*) (1973–1995).

Wassuk Range (1)
elevation record: 5740–8280 ft
4 records in 1 county (Mineral)

Towns/Range/Section	Location—Habitat	County	Elevation	Collector	Herbarium/a
T6N R29E S11	ridge N of Buller Mountain, w/*P. contorta* and *P. flexilis*	MI	8280 ft	*Charlet 1976*	DUKE, RENO, UC
T6N R29E S15	W slope of range, S of Lucky Boy Pass, on altered andesite	MI	7600 ft	*Charlet 1975*	DUKE, RENO
T7N R30E S28 NW/4	Alum Creek; lowest individual in the range	MI	5740 ft	*Charlet 1965*	DUKE, RENO
T7N R30E S29 E/4	Alum Creek, w/*P. ponderosa*	MI	5880 ft	*Charlet 1966*	DUKE, RENO

Reported as present in the range by Little (1971).
Reported by Billings (1954) and Cronquist et al. (1972) for Alum Creek.
DAC: mostly restricted to altered andesite soils. Observed on north slope of Buller Mountain, south of Lucky Boy Pass, with limber pine and lodgepole pine (1990–1995).

White Mountains (4)
Mineral and Esmeralda Counties

Not reported as present in the range by Little (1971).
DAC: present in the nearby California portion of the range, and may yet be found here.

OTHER FEATURES IN NEVADA, RECORDS (4)

Washoe County
elevation record: 4500 ft
1 record

Towns/Range/Section	Location—Habitat	Elevation	Collector	Herbarium/a
T18N R20E S32	1 mi W of Reno Hot Springs, isolated grove in sagebrush desert	4500 ft	*Billings 1206*	RENO

DAC: *Billings 1206* is the low-elevation record for the species in the state.

Douglas County
elevation record: 5280–6080 ft
3 records

Towns/Range/Section	Location—Habitat	Elevation	Collector	Herbarium/a
T11N R21E S32 S/2	1.1 mi from CA border	6080 ft	*Charlet 1906*	DUKE, RENO, UC, BRY
T11N R21E S25 W/4	Bryant Creek, altered andesite forest	5340 ft	*Charlet 1909*	DUKE, RENO, UC
T11N R20E S24	Bryant Creek, altered andesite forest	5280 ft	*Charlet 1910*	DUKE, RENO, UC
[Little: T12N R19E, T11N R20E, both sites W Fork Carson River]				

DAC: *Charlet 1906* is from a small population in pinyon woodland; *1909* & *1910* are from an extensive altered andesite area in the East Fork Carson River drainage (1994).

Pinus lambertiana Douglas

common name: sugar pine
elevation: 5200–7000 ft (1585–2135 m)
zones: Canadian, Arid Transition
17 Records for 3 counties (CC, DO, WA) in 2 mountain ranges

Sugar pine is common in central and southwestern Oregon, on the western slopes of the Sierra Nevada, and scattered along the Coast Ranges from southern Oregon south to extreme southern California. An isolated population occurs in northern Baja California.

In Nevada, sugar pine is found in Carson City, Douglas, and Washoe Counties in the Carson Range of the Sierra Nevada, and on Peavine Mountain (Washoe County). The species is not included in *Intermountain Flora* (Cronquist et al. 1972). Sugar pine is now a minor component of the Canadian Zone forests around Lake Tahoe but was probably much more widely distributed there before the intensive logging operations of the 1860s. It is present in altered andesite soils at lower elevations in both the Carson Range and on Peavine Mountain, and grows over a relatively narrow elevation range of 1800 ft (550 m).

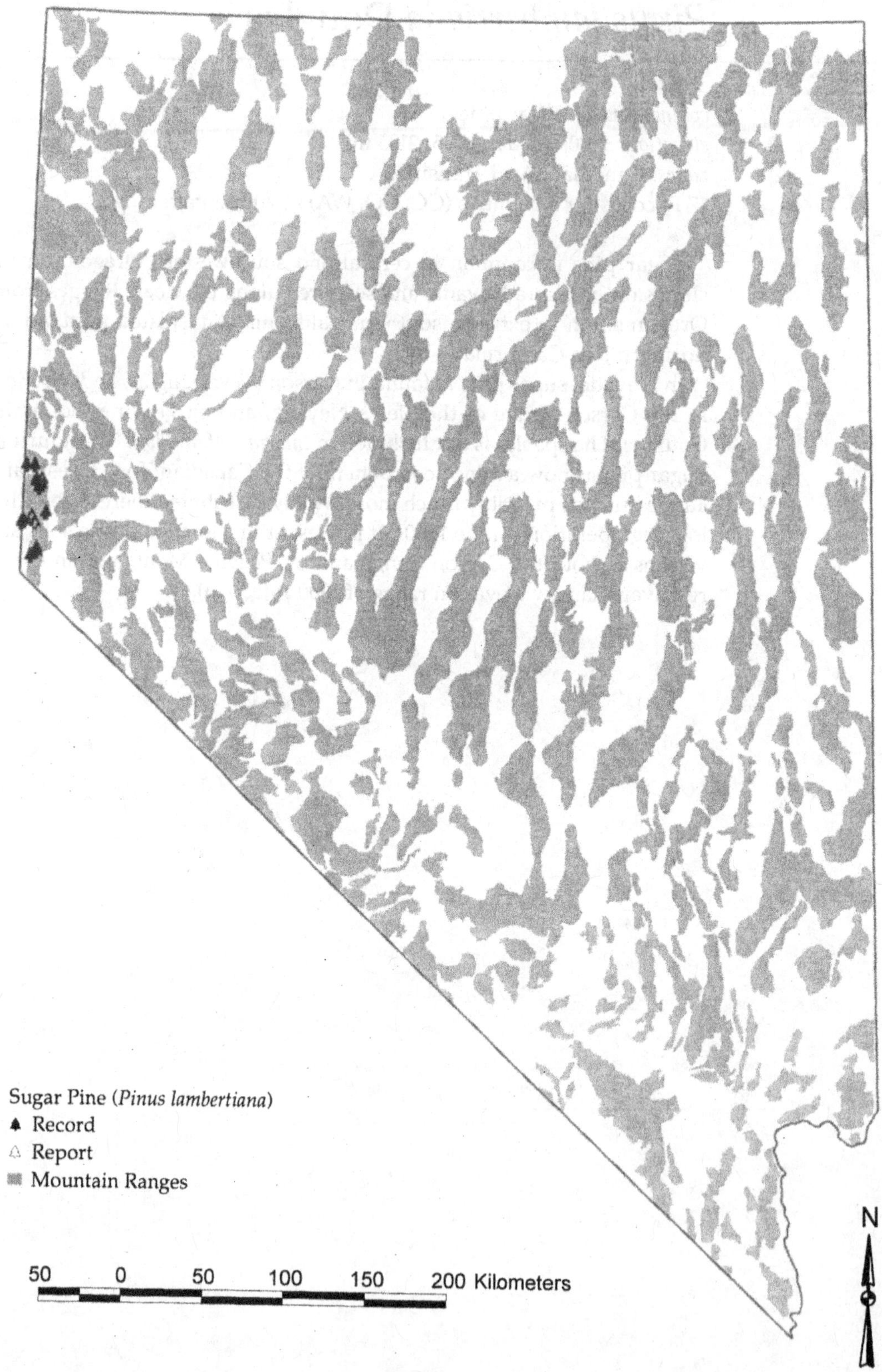
Sugar Pine (*Pinus lambertiana*)
Record
Report
Mountain Ranges
50 0 50 100 150 200 Kilometers
N

Carson Range (1)
elevation record: 5200–7000 ft
10 records in 3 counties (Carson City, Douglas, and Washoe)

Towns/Range/Section	Location—Habitat	County	Elevation	Collector	Herbarium/a
T14N R18E S15	N side of Shakespeare Rock, 1 mi S of Glenbrook	DO	6700 ft	*J Darling 22*	RENO
T14N R19E S6 NE/8	below US 50, E of Spooner Summit	DO	6500 ft	*Charlet 1917 & T Charlet*	DUKE, RENO, UC, BRY
T15N R19E S31	US 50 E of Spooner Summit	CC	6500 ft	*Ann Pinzl in 1973*	NSM
T16N R18E S18	3 mi E of Brockway, N end Tahoe	WA	6300 ft	*Nichols 685 & Lund*	RENO
T16N R18E S18	E [NE] shore Tahoe, near Kings Beach	WA	—	*L Sloan 2*	RENO
T17N R19E S27	above Bowers Mansion	WA	—	*PA Lehenbauer 1829*	RENO
T18N R19E S7	Alum Creek–Evans Creek divide, altered rock	WA	6000 ft	*TM Little in 1951*	RENO
T19N R18E S25	Hunter Creek	WA	5200 ft	*VK Johnson in 1949*	RENO
T19N R18E S36	Hunter Creek Canyon	WA	7000 ft	*AA Heller 10372*	RENO
T19N R18E	Hunter Creek, near sawmill	WA	—	*PA Lehenbauer in 1925*	RENO

[Billings (1992): T18N R18E R19E, in altered andesite soils along Evans, Alum, and Hunter Creeks]
[DAC: T16N R18E S16 (Incline Village, jct SR 28 & SR 431), T16N R18E S26 (Ponderosa Ranch)]

Reported as present in the range by Little (1971).
DAC: uncommon, but probably more widespread before the clear-cutting of the Carson Range during the mid-1860s. Observed in Hunter Creek Canyon (Washoe County) at junction of SR 431 and SR 28, and at Ponderosa Ranch (both Incline Village), and along the east shore of Lake Tahoe (1971–1995).

Peavine Mountain (1)
elevation record: 5500–6050 ft
7 records in 1 county (Washoe)

Towns/Range/Section	Location—Habitat	County	Elevation	Collector	Herbarium/a
T20N R18E	"yellow stand" altered andesite, SW facing, S side Peavine	WA	5500 ft	*Billings 1427*	RENO
T20N R18E S32	single tree, altered rock just E of Chokecherry Creek, S side Peavine	WA	5700 ft	*Billings 2061*	RENO
T20N R18E S35	westernmost altered patch, S side Peavine	WA	6000 ft	*Billings 2062*	RENO
T20N R18E S35	westernmost altered patch, S side Peavine	WA	6000 ft	*Billings 2063*	RENO
T20N R18E S35	S aspect, opposite Mogul exit	WA	5850 ft	*Loring Williams 72-55-5*	CAS, RENO
T20N R18E S35	on hillside in a gully	WA	6000 ft	*J Anderson in 1972*	RENO
T20N R18E S35 SE/4 NE/4	w/*P. ponderosa* & *P. jeffreyi* on altered andesite	WA	6050 ft	*Jan Nachlinger 1506 & Billings & Teri Knight*	DUKE

Reported as present in the range by Little (1971).
DAC: restricted to at most 3 locations on altered andesite (1971–1995).

Virginia Range (5)
Washoe and Storey Counties

Reported as present in the range by Little (1971).
DAC: Little's map (1971:59-W) is clearly in error. It shows a large bubble that covers half of Storey County and some of adjacent Washoe County, but I have seen no material in herbaria, nor have I seen the species after living in the range for 11 years and regularly visiting it for another 12 years (1973–1995).

Pinus longaeva D. K. Bailey

common name: Great Basin bristlecone pine
elevation: 6760–11,600 ft (2060–3535 m)
zones: Arctic-Alpine, Hudsonian, Canadian
72 Records for 8 counties (CL, EL, ES, EU, LI, MI, NY, WP)
Reports for 20 mountain ranges, Records for 17 of these
New Reports for 3 mountain ranges, New Records for 2 of these

Great Basin bristlecone pine is restricted to the high elevations of the hydrographic Great Basin of southeastern California, Utah, and Nevada. *Pinus longaeva* is closely related to the bristlecone pine of the Rocky Mountains (*P. aristata* Engelm.), and Little considered it to be a variety of that species: *P. aristata* Engelm. var. *longaeva* (D. K. Bailey) Little. A close relative of the bristlecone pines is the foxtail pine (*P. balfouriana* Grev. & Balf.) of northwestern California and the southern Sierra Nevada. Cronquist et al. (1972) list the common name of *P. longaeva* as "intermountain bristlecone pine," but this term is not used in the region.

After limber pine, Great Basin bristlecone pine is the most widely distributed high-elevation pine in Nevada, of which it is the state tree. It has been found in 20 of Nevada's mountain ranges in 8 counties (Clark, Elko, Esmeralda, Eureka, Lincoln, Mineral, Nye, and White Pine). I present new records from the Duck Creek Range (White Pine County) and the Goshute Mountains (Elko County), and a new report from Irish Mountain (Lincoln County). Little's map (1971) showing the species as present on Diamond Peak in the Diamond Mountains is in error; the only conifers on the peak are ancient limber pines. The species grows over a wide elevation range of 4840 ft (1475 m) in Nevada. The collection made at 6760 ft (2060 m) in Kalamazoo Creek in the Schell Creek Range (White Pine County) is the lowest one known anywhere in the native distribution. Grayson (1993) reported Great Basin bristlecone pine at 5800 ft (1770 m) in the Snake Range, but I have not been able to find a herbarium record for this tree. Until material is collected from this tree, the 6760-ft Kalamazoo Creek collection cited above is the lowest record in the native distribution of the species. Great Basin bristlecone pine may yet be found in other high mountains in eastern and central Nevada.

Cherry Creek Range (2)
White Pine County
[DAC: T25N R63E S7 S9]

Reported as present in the range by Loope (1969).
Not reported as present by Critchfield & Allenbaugh (1969) or Little (1971).
DAC: Loope (1969:189) reported *Pinus longaeva* to be common, but his distribution map (Loope 1969:182) does not show where in the mountain range it is. Perhaps it occurs only in the White Pine County portion. The TR assignments are mine, based on my observation of the species in the vicinity of Currie Summit (White Pine County) (1994).

Diamond Mountains (5)
Eureka and White Pine Counties

Reported as present in the range by Little (1971).
DAC: Little's map (1971) has a dot at Diamond Peak (T20N R55E) on the Eureka-White Pine County line. Although I have not visited this location, Alvin McLane saw very large, twisted trees at the summit of Diamond Peak. A look with binoculars led him to think the trees were *Pinus longaeva*, but when he climbed the peak to confirm it, he found that they were actually very old *P. flexilis* (McLane, pers. comm. 1994). I have seen no material in herbaria that verifies Little's report (1971); undoubtedly it is based on a long-distance observation of these *P. flexilis* on Diamond Peak. The rugged, inaccessible nature of the mountain range, the lack of botanizing here, and the regional distribution of the species all suggest that it may yet be found in the Diamond Mountains (1994).

****Duck Creek Range (1)**
elevation record: 9260 ft
1 record in 1 county (White Pine)

Towns/Range/Section	Location—Habitat	County	Elevation	Collector	Herbarium/a
T16N R64E S24 N/4	ridge N of Camel Peak, in limestone outcrops	WP	9260 ft	*Charlet 2020*	DUKE, RENO

Not reported as present in the range by Little (1971).
DAC: the presence of the species here is restricted to about 20–25 reproductive individuals growing on limestone outcrops along the ridge north of Camel Peak with *P. flexilis* (1994).
New record for the species in the range.

Egan Range (1)
elevation record: 8855–9600 ft
2 records in 1 county (White Pine)

Towns/Range/Section	Location—Habitat	County	Elevation	Collector	Herbarium/a
T15N R63E S31	Ward Mountain, W side	WP	9600 ft	*KH Thorne 1175 & BT Welsh*	BRY
T21N R62E S12	1 mi N of Telegraph Peak, NW-SE ridge	WP	8855 ft	*Ann Pinzl 10075 & Jim Morefield*	NSM, BRY

Reported as present in the range by Little (1971).
DAC: the species is probably present at many more high-elevation locations in the range, particularly in the northern part (1994).

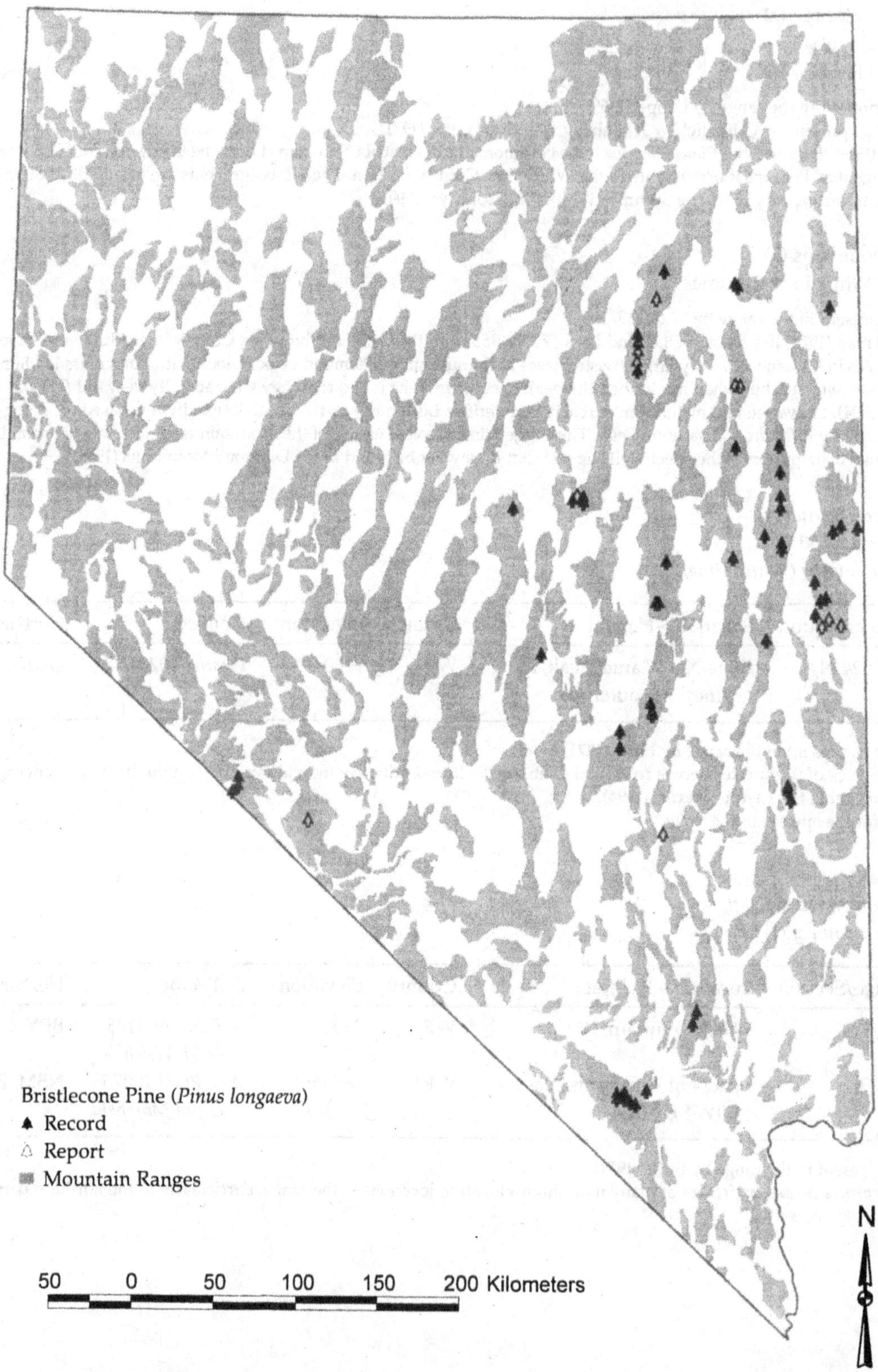
Bristlecone Pine (Pinus longaeva)
Record
Report
Mountain Ranges
N
50 0 50 100 150 200 Kilometers

Fish Creek Range (1)
elevation record: 7200–8060 ft
5 records in 1 county (Eureka)

Towns/Range/Section	Location—Habitat	County	Elevation	Collector	Herbarium/a
T18N R53E S11	Hamburg Ridge, w/*P. flexilis* & mountain-mahogany	EU	8060 ft	*Charlet 2084*	RENO
T18N R53E S20	Atrypa Peak	EU	8000 ft	*Percy Train 468*	NSM
T18N R53E S23	Secret Canyon, 10 mi S of US 50	EU	7200 ft	*Peter Herlan et al. in 1976*	NSM
T18N R53E	Hidden Canyon, SE of Prospect Peak	EU	—	*JG Gray in 1971*	NSM
T18N R53E	S of Eureka	EU	7200 ft	*M Williams in 1968*	RENO
[DAC: T18N R53E S3 (Prospect Mountain)]					

Not reported as present in the range by Little (1971).
DAC: I assume that "Hidden Canyon" is Secret Canyon. *Pinus longaeva* is abundant at high elevations, mostly as young individuals with narrow crowns. Given the extent of mining in this range, it seems likely that the original stands were logged (1994).

****Goshute Mountains (1)**
elevation record: 8800 ft
1 record in 1 county (Elko)

Towns/Range/Section	Location—Habitat	County	Elevation	Collector	Herbarium/a
T30N R68E S36	Hawk Ridge	EL	8800 ft	*Richard W. Rust 42*	UCR

Not reported as present in the range by Critchfield & Allenbaugh (1969), Loope (1969), or Little (1971).
DAC: I have seen no material from the Goshute Mountains other than *Rust* 42.
New record for the species in the range.

Grant Range (1)
elevation record: 9000–10,500 ft
4 records in 1 county (White Pine)

Towns/Range/Section	Location—Habitat	County	Elevation	Collector	Herbarium/a
T5N R58E S7	Troy Peak	WP	10,300 ft	*M Williams in 1967*	RENO
T5N R58E S7	Troy Peak	WP	10,500 ft	*Richard W. Rust 12*	UCR
T5N R58E S18	E face Troy Peak, Scofield Canyon	WP	9000 ft	*NK Johnson in 1953*	DUKE
T6N R57E S25	8.5 mi from valley road, Timber Mountain [DAC: label says "Seaman Range"]	WP	10,000 ft	*M Williams 79-110-8 et al.*	RSA, RENO

Reported as present in the range by Little (1971).

Highland Range (1)
elevation record: 8580–9200 ft
5 records in 1 county (Lincoln)

Towns/Range/Section	Location—Habitat	County	Elevation	Collector	Herbarium/a
T1N R66E S33	below Highland Peak	LI	9200 ft	*M Williams 83-75-1*	RENO
T1N R66E S33	Highland Ridge	LI	8580 ft	*Charlet 1995*	DUKE, RENO
T1N R66E S33	Highland Ridge	LI	—	*Jeff Knight & Scott Cichowlaz in 1993*	DUKE
T0 R66E	Highland Peak Barbara Ertter annotation: "In Purpus' field notes, 6123 was collected in the Charleston Mts of Clark County, NV"	LI	—	*CA Purpus 6123*	UC
T1S R66E S26	Comet Peak, Pioche	LI	—	*ME Jones in 1912*	RSA

Reported as present in the range by Little (1971).
DAC: abundant, and sometimes dominant, at high elevations, forming forest stands at the head of Water Canyon (1994).

Hot Creek Range (1)
elevation record: 8800 ft
1 record in 1 county (Nye)

Towns/Range/Section	Location—Habitat	County	Elevation	Collector	Herbarium/a
T9N R51E S24	Morey Peak	NY	8800 ft	*V Bostick 6412*	UNLV

Not reported as present in the range by Little (1971).

****Irish Mountain (3)**
Lincoln County
[McLane: T3S R59E]

Not reported as present in the range by Little (1971).
DAC: rare; Alvin McLane (pers. comm. 1993) told me of a stand of about 6 trees at the summit.
New report for the species in the range.

Monitor Range (1)
1 record in 1 county (Eureka)

Towns/Range/Section	Location—Habitat	County	Elevation	Collector	Herbarium/a
T18N R49E S33	N end, probably Antelope Peak (determined Billings & J Darling 1951)	EU	—	*Art Cusick in 1949*	RENO

Reported as present in the range by Little (1971).

Pequop Mountains (1)
elevation record: 8000–11,500 ft
5 records in 1 county (Elko)

Towns/Range/Section	Location—Habitat	County	Elevation	Collector	Herbarium/a
T31N R63E S20	top of Spruce Mountain	EL	10,260 ft	*M Williams 81-84-1 & Tiehm*	RSA, UT, BRY, UTC, RENO
T31N R63E	top of Spruce Mountain, N aspect	EL	11,500 ft	*A Holmgren 1273*	UT, UTC
T31N R63E S12	N end of Spruce Mountain, ca. 25 mi S of Latham Springs, W slope	EL	8000 ft	*R Foster 7860*	BRY
T31N R63E	Spruce Mountain, at lower forest border	EL	8800 ft	*Lloyd Loope 673*	DUKE
—	Spruce Mountain	EL	10,200 ft	*Peter Herlan & Ann Pinzl in 1976*	NSM

Reported as present in the range by Loope (1969) and Little (1971).

Quinn Canyon Range (1)
elevation record: 8000 ft
2 records in 1 county (Nye)

Towns/Range/Section	Location—Habitat	County	Elevation	Collector	Herbarium/a
T3N R56E	Quinn Canyon Range, slopes & ridges at head of Cherry Creek	NY	—	*B Maguire 25596 & A Holmgren*	UC, UT, UTC
T4N R56E	above Cherry Creek summit	NY	8000 ft	*M Williams 6-11-67*	RENO

Reported as present in the range by Little (1971).
DAC: no additional location information was available on Maguire & Holmgren's label. This TR assignment is mine, centered on the highest portion of the range (1994).

Ruby Mountains (1)
elevation record: 8800–9800 ft
6 records in 2 counties (Elko and White Pine)

Towns/Range/Section	Location—Habitat	County	Elevation	Collector	Herbarium/a
—	Ruby Range	—	—	*CL Hitchcock & JS Martin 5686*	CAS
T26N R57E S19	Sherman Ridge, head of Sherman Creek Canyon, near EL County line	WP	9100 ft	*N Holmgren 3892 & B Bethers*	UT, UTC
T26N R57E S19	N slope, Sherman Mountain, 60 ft tall	WP	9800 ft	*CL Hitchcock 5686 & JS Martin*	UT
T28N R57E ca. S35	Ruby Mountains, S of Harrison Pass	EL	9100 ft	*J Gentry, Jr. & G Davidse 1832*	BRY, UT, UTC
T32N R58E S23	Thomas Creek Canyon, W slope of canyon near mouth, 22 mi SE of Elko	EL	8800 ft	*N Holmgren 6273 & P Holmgren*	BRY
T32N R58E S23	Thomas Canyon, farthest N location of the species in the Ruby Mts	EL	8800 ft	*Lloyd Loope 1422*	DUKE

[Loope: (EL) T30N R58E, T27N R57E S21 (Cass House Peak) S10 (Pearl Peak), T26N R57E]
[Critchfield & Allenbaugh: (WP) T26N R57E]

Reported as present in the range by Critchfield & Allenbaugh (1969), Loope (1969), and Little (1971).
DAC: The Cass House Peak and Pearl Peak locations were verified by Alvin McLane (pers. comm. 1994).

Schell Creek Range (1)
elevation record: 6760–11,000 ft
8 records in 1 county (White Pine)

Towns/Range/Section	Location—Habitat	County	Elevation	Collector	Herbarium/a
T10N R65E S31	Mount Grafton, E slope	WP	9000 ft	*BF Harrison 13390 & KH Thorne*	BRY
T10N R65E S31	Mount Grafton, E slope	WP	10,000 ft	*N Holmgren 3848 & B Bethers*	UT
T15N R66E S18	Bastian Peak [Cave Mountain]	WP	10,000 ft	*Dimaggio in 1967*	RENO
T15N R66E S6	15 mi SE of Ely, ca. 2 mi N of Cave Mountain summit	WP	10,600 ft	*E Neese 10769 & Goodrich & B Welsh*	BRY
T18N R65E S12	above N Fork Timber Creek	WP	11,000 ft	*Ann Pinzl 4362 & Austin*	NSM
T18N R65E S36	S Fork Timber Creek; rare in spruce forest	WP	10,200 ft	*Lloyd Loope 1200*	DUKE
T20N R65E S25 S/4	Kalamazoo Creek, w/*A. concolor, J. scopulorum,* and *Betula*	WP	6760 ft	*Charlet 2024*	DUKE, RENO
T21N R65E S10	Aurum	WP	8000 ft	*ME Jones in 1905*	RSA

Reported as present in the range by Little (1971).
DAC: mostly restricted to limestone cliffs and outcrops at high elevations, but is occasional at lower elevations in canyons. *Charlet 2024* from along Kalamazoo Creek is the lowest elevation record anywhere in the range of *P. longaeva.* Grayson (1993:27) observed the species at 5800 ft (1770 m) in the Snake Range, but I have not encountered material in herbaria collected from that tree (1994).

Seaman Range (4)
Lincoln and Nye Counties

Not reported as present in the range by Little (1971).
DAC: possibly occurs in the highest portions of the range (see also note under Grant Range, collection no. *M Williams 79-110-8 et al.*).

Sheep Range (1)
elevation record: 9000 ft
3 records in 1 county (Clark)

Towns/Range/Section	Location—Habitat	County	Elevation	Collector	Herbarium/a
—	Bald Peak	CL	—	*Frandsen 1081*	RENO
T14S R61E S20	0.5 mi S of Timber Canyon	CL	9000 ft	*T Ackerman 6711*	UNLV
T15S R61E S1	base of Hayford Peak	CL	—	*BH Banta in 1950*	RENO

Reported as present in the range by Little (1971).

Silver Peak Range (3)
Esmeralda County
[Little: T2S R37E]

Reported as present in the range by Little (1971).
DAC: the single dot on Little's map (1971) corresponds to Piper Peak at the above coordinates (the TR assignment is mine). I have seen no material of *P. longaeva* from this range (1993).

Snake Range (1)
elevation record: 7060–11,600 ft
12 records in 1 county (White Pine)

Towns/Range/Section	Location—Habitat	County	Elevation	Collector	Herbarium/a
T11N R68E	saddle, S side of Mount Washington	WP	11,300 ft	*I Ting in 1959*	RENO
T12N R68E S21	W side of range on slopes between Pole & Lincoln Canyons	WP	7500 ft	*N Holmgren 935 & J Reveal*	BRY, UTC
T12N R68E S14	S slopes of Mount Washington; Liberty Mine	WP	10,400 ft	*N Holmgren 3762 & A Holmgren*	UTC
T13N R68E	Wheeler Peak area	WP	10,500 ft	*Mozingo 74-41b*	RENO
T13N R68E	Wheeler Peak scenic area, holotype, 39°N, 114°18′W	WP	10,620 ft	*DK Bailey 7001 & JE Whitson*	UC, IFGP, UT
T13N R68E	Mount Jeff Davis; above timberline, N slope	WP	—	*B Maguire 2470 & AG Richards*	UTC
T16N R69E	Nigger Creek head, Mount Moriah, rocky ridges	WP	10,000 ft	*Percy Train 1184*	RENO, DUKE
T16N R69E S3 NE/4	Mount Moriah	WP	11,600 ft	*JC Lyman 430B*	RSA
T16N R69E	Henry [Hendrys] Creek	WP	—	*T Frantz in 1952*	DUKE
T16N R70E S10	Mount Moriah, Hampton Creek	WP	7060 ft	*Charlet 2114 & T Charlet*	RENO
—	Snake Range	WP	—	*JH Thomas 16250*	CAS
—	Mount Wheeler, slopes	WP	10,000 ft	*WP Cottam 8130*	UT
[Hendricks: T11N R69E NW/4 SE/4, T11N R68E E/4]					

Reported as present in the range by Little (1971).
DAC: Albert J. Hendricks, the Superintendent of Great Basin National Park, reported the dominance of *Pinus longaeva* in the southern portion of the range above 9000 ft (pers. comm. 1994). Grayson (1993) reported the species in a canyon on the east slope at 5800 ft (1994).

Spring Mountains (1)
elevation record: 8200–10,920 ft
9 records in 1 county (Clark)

Towns/Range/Section	Location—Habitat	County	Elevation	Collector	Herbarium/a
T19S R56E	Charleston Peak	CL	10,900 ft	*IW Clokey 5378*	RENO, UTC
T19S R56E	Mount Charleston	CL	9000 ft	*G Muennig 3230*	RENO
T19S R56E S15	rimrock, head of Lee Canyon	CL	10,000 ft	*Percy Train 2153*	RENO
T19S R56E S21	ridge to Charleston Peak	CL	10,920 ft	*IW Clokey 7817*	UT, BRY, UTC, RENO, DUKE
T19S R56E S27	E face of Charleston Peak, steep slopes with *P. flexilis*	CL	9445 ft	*IW Clokey 5377*	UT, DUKE
T19S R56E S36	Charleston Resort	CL	—	*ME Jones in 1927*	RSA
T19S R57E S12	Deer Creek, 5–6 mi NE of Charleston Peak, Canadian Zone	CL	8500 ft	*LaRivers 502 & Hancock*	RENO
T19S R57E S12	head, Deer Creek	CL	8200 ft	*Beatley 9519 & Dainty & Jeffries*	RENO
T20S R57E S5	Harris Mountain Rd, E side of Charleston Mountain	CL	—	*D Atwood 7981*	BRY

Reported as present in the range by Little (1971).

White Mountains (1)
elevation record: 9500–11,000 ft
4 records in 2 counties (Esmeralda and Mineral)

Towns/Range/Section	Location—Habitat	County	Elevation	Collector	Herbarium/a
T1S R33E S7	N slope of Boundary Peak	ES	11,000 ft	*Billings 1682*	RENO
T1S R33E S7	N ridge Boundary Peak	ES	10,800 ft	*Ann Pinzl 2387 & Pinzl*	NSM
T1N R33E S33	Mustang Peak	ES	9500 ft	*John Roos 1032*	RSA
T1N R33E S16	ca. 7.5 mi up Mustang Mountain Rd from US 6	MI	9800 ft	*Vasek 640928-1*	UCR

Reported as present in the range by Little (1971).

White Pine Range (1)
elevation record: 7600–7800 ft
3 records in 2 counties (Nye and White Pine)

Towns/Range/Section	Location—Habitat	County	Elevation	Collector	Herbarium/a
T12N R58E S20	Broom Canyon	NY	7600 ft	*P Herlan et al. in 1976*	NSM
T12N R58E S21 NE/4	9.8 mi E on, and end of, Broom Creek Rd, N of White Pine Peak	NY	7800 ft	*Mozingo 76-54-1*	BRY, RENO
T14N R58E S1	open ridge S of pass on Currant-Hamilton Rd, 14 mi SE of Hamilton	WP	—	*Helen Sharsmith 4802a*	UC, UTC

Reported as present in the range by Little (1971).

Pinus monophylla Torrey & Frémont

common name: singleleaf pinyon, nut pine, pinyon
elevation: 2800–8740 ft (855–2665 m)
zones: Canadian, Arid Transition, Upper Sonoran
Reports for all Nevada counties except Humboldt; Unverified Reports for Pershing County
131 Records for all counties except Pershing and Humboldt
Reports for 126 mountain ranges, Records for 70 of these
New Reports for 6 mountain ranges, New Records for 3 of these

Pinus monophylla is distributed mainly in Nevada, but its range extends into eastern and southern California south to Baja California and western Utah. Isolated occurrences are known from southern Idaho, western Arizona, and central and northern Utah.

Singleleaf pinyon, the original state tree of Nevada, is by far the most common and widely distributed pine in the state, covering an impressive elevation range of 5940 ft (1810 m). The only conifer with a wider distribution in Nevada is Utah juniper. Throughout most of its range in Nevada, singleleaf pinyon is codominant, or nearly so, with Utah juniper in pinyon-juniper woodlands. Singleleaf pinyon usually dominates and often overwhelms Utah juniper when the species occur together in the southwestern Lahontan Basin (Esmeralda and Mineral Counties). At the northern limits of its range in Nevada, singleleaf pinyon is uncommon and appears to be invading open Utah juniper woodlands.

Although there are reports from 126 mountain ranges in most Nevada counties, singleleaf pinyon is very rare north of the Truckee and Humboldt Rivers. New reports are made here for the Candelaria Hills (Mineral County), the East Gate Range (Churchill County), the Golden Gate Range (Lincoln and Nye Counties), the Horse Range (Nye County), and the Junction House Range and Virginia Mountains (both of these locations are north of the Truckee River in Washoe County).

Anchorite Hills (1)
elevation record: 7620 ft
1 record in 1 county (Mineral)

Towns/Range/Section	Location—Habitat	County	Elevation	Collector	Herbarium/a
T4N R29E S13 SE/4	Anchorite Pass	MI	7620 ft	*Charlet 1964*	RENO
[DAC: T5N R30E, T4N R29E R30E, T3N R30E]					

Reported as present in the range by Little (1971).
DAC: dominant in pinyon woodlands (1991–1995).

Antelope Range (2)
Elko and White Pine Counties
[DAC: (EL) T26N R67E; (WP) T25N R67E, T24N R67E, T23N R67E]

Reported as present in the range by Little (1971).
DAC: common in pinyon-juniper woodlands in the middle elevations (1994).

Antelope Range (3)
Eureka and Nye Counties
[Little: (NY) T15N R51E, T14N R51E, T13N R51E; (EU) T15N R51E]

Reported as present in the range by Little (1971).

Augusta Mountains (4)
Lander and Pershing Counties

Not reported as present in the range by Little (1971).
DAC: the species is almost certain to occur from Cain Mountain, in the center of the range, to the south. Definitely absent north of Cain Mountain to Cedar Canyon (1993–1994).

Battle Mountain (5)
Humboldt County
[Little: T32N R43E]

Reported as present in the range by Little (1971).
DAC: *Pinus monophylla* is the only conifer reported from Battle Mountain by Little (1971), and the location is given as Humboldt County. However, the only conifer I have found on the mountain is *Juniperus osteosperma*, and this only in Lander County. The species may yet be found in Timber Canyon in Lander County, but the conifer stands there are more likely composed solely of *J. osteosperma* (1994).

Belted Range (1)
elevation record: 6500–7000 ft
2 records in 1 county (Nye)

Towns/Range/Section	Location—Habitat	County	Elevation	Collector	Herbarium/a
T6S R53E S8	E slope near Johnnies Water [ca. 1.2 mi SSW of Wheelbarrow Peak]	NY	6500 ft	*Beatley 4407 & Rhoads*	RENO
T8S R52E S31	Central Rainier Mesa	NY	7000 ft	*Beatley 3789*	RENO
[Little: T8S R52E, T7S R52E, T6S R52E]					

Reported as present in the range by Little (1971).

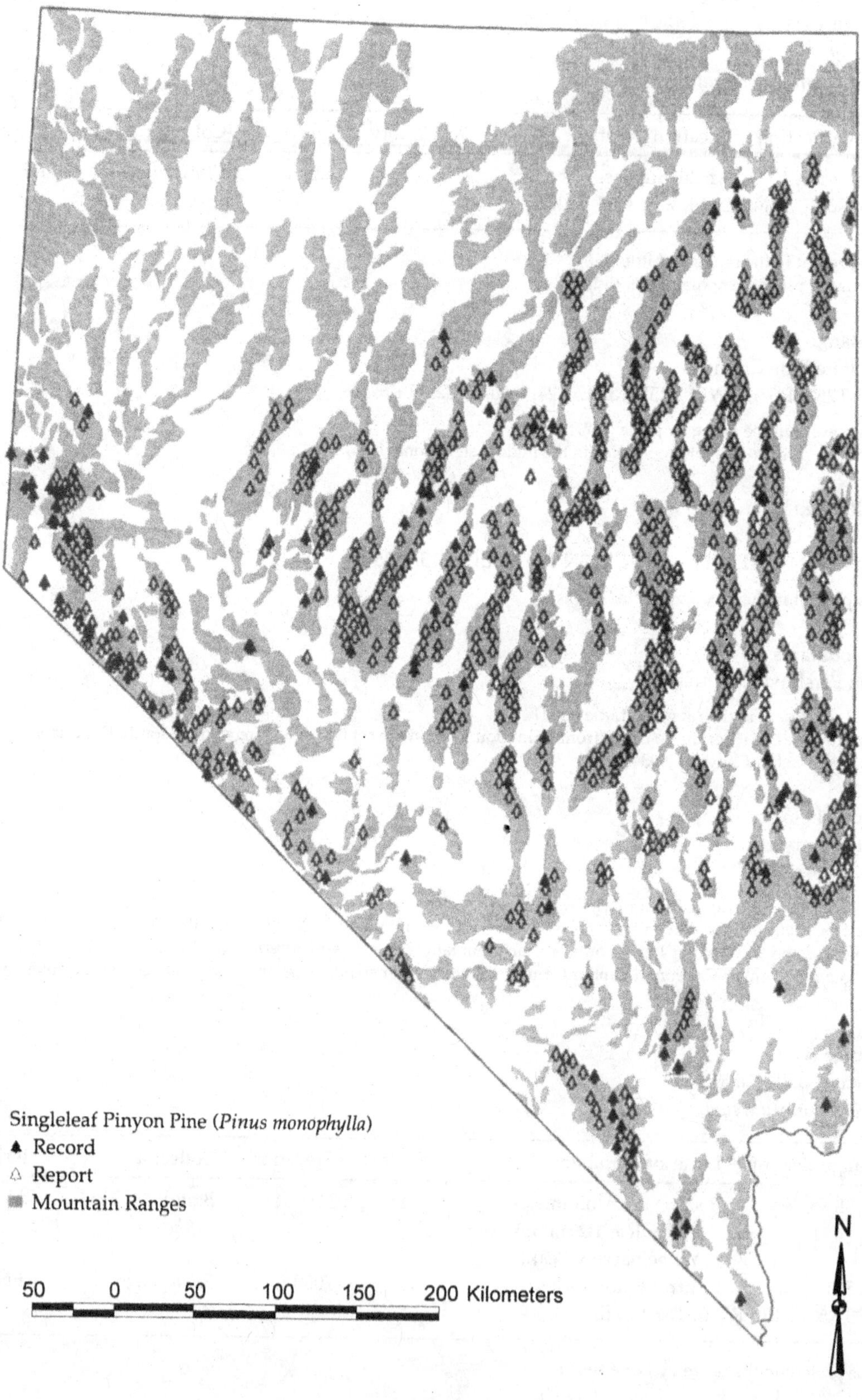
Singleleaf Pinyon Pine (*Pinus monophylla*)
Record
Report
Mountain Ranges
50 0 50 100 150 200 Kilometers
N

Bird Spring Range (1)
elevation record: 4800 ft
1 record in 1 county (Clark)

Towns/Range/Section	Location—Habitat	County	Elevation	Collector	Herbarium/a
T23S R58E S11	2.5 mi S of Mount Spring Pass Rd (SR 160) on Cottonwood Pass Rd	CL	4800 ft	*Ann Pinzl 2692*	NSM

Not reported as present in the range by Little (1971).

Black Canyon Range (3)
Lincoln County
[Little: T1S R65E]

Reported as present in the range by Little (1971).

Bodie Mountains (1)
elevation record: 7500–7600 ft
2 records in 1 county (Mineral)

Towns/Range/Section	Location—Habitat	County	Elevation	Collector	Herbarium/a
T6N R26E S16	6 mi N of Masonic Lower Town	MI	7600 ft	*J & Y Robertson in 1979*	RENO
T5N R28E S17	Aurora, in graveyard on hillside	MI	7500 ft	*Cooper 2540*	RSA
[DAC: T7N R26E, T6N R27E, T5N R27E, T4N R28E]					

Reported as present in the range by Little (1971).
DAC: dominant in the widespread pinyon woodlands of the range (1991–1995).

Bristol Range (1)
elevation record: 6140 ft
1 record in 1 county (Lincoln)

Towns/Range/Section	Location—Habitat	County	Elevation	Collector	Herbarium/a
T3N R65E S13	Bristol Pass	LI	6140 ft	*Charlet 1992*	RENO
[DAC: T2N R66E]					

Reported as present in the range by Little (1971).
DAC: codominant with *Juniperus osteosperma* in pinyon-juniper woodlands throughout the range (1994).

Burnt Spring Range (3)
Lincoln County
[Little: T3S R65E]

Reported as present in the range by Little (1971).

Butte Mountains (2)
White Pine County
[DAC: T25N R60E, T18N R60E]
[Little: T24N R60E, T23N R59E, T22N R59E R60E, T21N R59E R60E, T20N R59E R60E, T19N R60E]

Reported as present in the range by Little (1971).
DAC: abundant in the pinyon-juniper woodlands scattered throughout the range (1994).

Cactus Range (3)
Nye County
[McLane: T3S R46E]

Not reported as present in the range by Little (1971).
Reported as present by McLane (1978).

****Candelaria Hills (2)**
Mineral and Esmeralda Counties
[RWR: (MI) T2N R34E; (ES) T2N R34E]

Not reported as present in the range by Little (1971).
New report for the species in the range.

Carson Range (1)
elevation record: 4500–5500 ft
3 records in 1 county (Washoe) and reports for 2 others (Carson City and Douglas)

Towns/Range/Section	Location—Habitat	County	Elevation	Collector	Herbarium/a
T17N R19E	near foot of Mount Rose	WA	5500 ft	*WP Stockwell 1454*	UTC
T18N R19E	2 mi W of Reno Hot Springs	WA	4600 ft	*KC Williams 84*	RENO
T18N R19E	———	WA	4500 ft	*Billings 1207*	RENO
[DAC: (DO) T12N R19E; (CC) T15N R19E (Clear Creek Canyon)]					

Reported as present in the range by Little (1971).
DAC: very uncommon, restricted primarily to the extreme southeastern portion of the range (1971–1995).

Cedar Mountains (3)
Mineral County
[Little: T8N R37E]

Reported as present in the range by Little (1971).

Cedar Range (1)
elevation record: 6500 ft
2 records in 1 county (Lincoln)

Towns/Range/Section	Location—Habitat	County	Elevation	Collector	Herbarium/a
T4S R69E	12 mi E of Caliente	LI	—	*LaRivers & Hancock 757*	RENO
T1S R70–71E	Mahogany Mts near Big Summit, 16 mi NE of Panaca	LI	6500 ft	*Leila Shultz 6242 & J Shultz*	UTC

[RWR: T2S R69E]
[Little: T1S R70E R71E, T2S R69E R70E R71E, T3S R69E R70E, T4S R70E]

Reported as present in the range by Little (1971).

Cherry Creek Range (2)
Elko and White Pine Counties
[RWR: (EL) T27N R63E, T26N R62E R64E]
[Little: (EL) T28N R63E]
[DAC: (WP) T25N R63E R64E, T24N R63E R64E, T23N R62E R63E]

Reported as present in the range by Little (1971).
DAC: abundant in pinyon-juniper woodlands at lower elevations throughout the range (1994).

Chief Range (3)
Lincoln County
[Little: T2S R66E]

Reported as present in the range by Little (1971).

Clan Alpine Mountains (1)
elevation record: 5800–6000 ft
3 records in 1 county (Churchill)

Towns/Range/Section	Location—Habitat	County	Elevation	Collector	Herbarium/a
T20N R37E	22 mi N of East Gate	CH	6000 ft	*PA Allen 400*	RENO
T20N R37E	War Canyon	CH	5800 ft	*Ann Pinzl in 1974*	NSM
—	E side of mountains, SE aspect; high incidence of 2-needled fascicles	CH	—	*RJ Tausch & KH Rea in 1974*	UTC

[DAC: T20N R36E, T19N R36E R37E]
[Little: T22N R37E R38E, T21N R36E R37E]

Reported as present in the range by Little (1971).
DAC: common in the pinyon-juniper woodlands in the southern and central portions of the range (1991–1994).

Clover Mountains (1)
elevation record: 5875 ft
1 record in 1 county (Lincoln)

Towns/Range/Section	Location—Habitat	County	Elevation	Collector	Herbarium/a
T5S R66E	1 mi S of Sawmill Canyon turnoff	LI	5875 ft	*Ann Pinzl 8618*	UNLV, NSM
[Little: T5S R67E R68E R71E, T6S R67E R68E R69E R70E R71E, T7S R68E R69E]					

Reported as present in the range by Little (1971).

Cortez Mountains (1)
1 record in 1 county (Eureka)

Towns/Range/Section	Location—Habitat	County	Elevation	Collector	Herbarium/a
T26N R48E	above Horse Ranch	EU	—	*Goodner 515 & Henning*	RENO, RSA

Not reported as present in the range by Little (1971).
DAC: restricted to the extreme southern tip of the range in pinyon-juniper woodlands (1993).

Deep Creek Mountains (2)
White Pine County
[Little: T23N R70E]
[DAC: T22N R69E R70E]

Reported as present in the range by Little (1971).
DAC: codominant with *Juniperus osteosperma* in pinyon-juniper woodlands throughout the middle elevations in most of the Nevada portion of the range (1994).

De Lamar Mountains (1)
elevation record: 4200 ft
1 record in 1 county (Lincoln)

Towns/Range/Section	Location—Habitat	County	Elevation	Collector	Herbarium/a
T5S R66E	Meadow Valley Wash, 8 mi S of Caliente	LI	4200 ft	*Billings 2202*	RENO
[Little: T6S R65E, T5S R65E, T4S R65E R66E]					

Reported as present in the range by Little (1971).
DAC: *Billings 2202* was determined to be *P. monophylla* by Billings, and *P. edulis* Engelm. var. *fallax* Little by DK Bailey. The material has a few double needles, but there seems to be no basis for calling it *P. edulis.*

Desatoya Mountains (2)
Churchill and Lander Counties
[DAC: (CH) T18N R37E R38E, T17N R37E, T16N R37E R38E, T15N R37E; (LA) T17N R37E]
[Little: (CH) T19N R38E R39E, T15N R37E]

Reported as present in the range by Little (1971).
DAC: abundant in pinyon-juniper woodland throughout the range (1992–1994).

Desert Creek Mountains (2)
Lyon County
[DAC: T9N R24E, T8N R24E]

Reported as present in the range by Little (1971).
DAC: dominant in pinyon woodlands throughout the range, with *Juniperus osteosperma, J. occidentalis,* and *Cercocarpus* (1988–1994).

Diamond Mountains (1)
elevation record: 7500 ft
1 record in 1 county (Eureka) and reports in another (White Pine)

Towns/Range/Section	Location—Habitat	County	Elevation	Collector	Herbarium/a
T19N R54E S12	Newark Summit, w/*Cercocarpus*	EU	7500 ft	*P & H Vreeland III-28*	RENO

[DAC: (EU) T26N R54E, T25N R54E, T24N R54E, T22N R54E, T22.5N R54E, T21N R54E, T20N R54E, T19N R54E, T18N R54E; (WP) T26N R55E, T25N R55E, T23N R55E, T19N R54E R55E, T18N R54E]

Reported as present in the range by Little (1971).
DAC: present in widely scattered pinyon-juniper woodlands in the lower and middle elevations. In the north at lower elevations, the woodlands are dominated by juniper, and pinyon is only a minor presence. Singleleaf pinyon is a major component of woodlands in the center of the range, and it dominates in the south (1994).

Dolly Varden Mountains (1)
elevation record: 6230 ft
1 record in 1 county (Elko)

Towns/Range/Section	Location—Habitat	County	Elevation	Collector	Herbarium/a
T29N R66E S35	Dolly Varden Canyon	EL	6230 ft	*Ann Pinzl 9301*	NSM, UTC

[RWR: T28N R65E R66E, T29N R65E]
[Little: T27N R66E]

Reported as present in the range by Little (1971).

Duck Creek Range (1)
elevation record: 7300 ft
1 record in 1 county (White Pine)

Towns/Range/Section	Location—Habitat	County	Elevation	Collector	Herbarium/a
T15N R65E S5	Elk Flat Campground, in pinyon-juniper woodlands	WP	7300 ft	*Charlet 2102 & T Charlet*	RENO

[DAC: T18N R64E, T17N R64E, T16N R64E]

Reported as present in the range by Little (1971).
DAC: codominant in pinyon-juniper woodlands in the lower elevations, particularly in the north (1994).

****East Gate Range (1)**
elevation record: 6140 ft
1 record in 1 county (Churchill)

Towns/Range/Section	Location—Habitat	County	Elevation	Collector	Herbarium/a
T16N R36E S24	low end pinyon-juniper woodlands	CH	6140 ft	*Charlet 2052 & T Charlet*	RENO

Not reported as present in the range by Little (1971).
DAC: abundant in the pinyon-juniper woodlands widespread in the central and southern portions of the range (1991–1994).
New record for the species in the range.

East Humboldt Range (1)
elevation record: 6680 ft
1 record in 1 county (Elko)

Towns/Range/Section	Location—Habitat	County	Elevation	Collector	Herbarium/a
T36N R61E S2	along road to Angel Lake in juniper woodlands	EL	6680 ft	*Charlet 2123 & T Charlet*	RENO
[DAC: T34N R61E, T35N R61E]					

Reported as present in the range by Little (1971).
DAC: observed commonly throughout the pinyon-juniper woodlands on the eastern slope (1988–1994).

East Range (5)
Pershing County

Reported as present in the range by Little (1971).
DAC: the only conifer species I have been able to find here is *Juniperus osteosperma* (1994).

Egan Range (2)
Lincoln, Nye, and White Pine Counties
[DAC: (WP) T23N R62E R63E, T22N R62E R63E, T21N R62E R63E, T20N R62E R63E, T19N R62E, T18N R61E R62E R63E, T17N R62E R63E, T16N R62E R63E, T15N R62E R63E, T14N R62E R63E, T13N R63E, T12N R63E, T11N R63E, T10N R62E R63E; (LI) T9N R62E R63E, T8N R62E R63E, T7N R62E R63E, T6N R62E R63E; (NY) T9N R62E, T6N R62E]

Reported as present in the range by Little (1971).
DAC: abundant in pinyon-juniper woodlands; at lower elevations in the north and middle, and at higher elevations in the south (1994).

Eleana Range (3)
Nye County
[Little, Beatley: T9S R51E]

Reported as present in the range by Little (1971) and Beatley (1976).

Excelsior Mountains (2)
Mineral County
[DAC: T3N R29E R30E, T4N R30E R31E R32E, T5N R31E R32E R33E R34E, T6N R34E]

Reported as present in the range by Little (1971).
DAC: dominant in the pinyon woodlands; with *Juniperus osteosperma* and, rarely, *J. occidentalis* (1993).

Fairview Range (2)
Lincoln County
[DAC: T5N R65E]

Reported as present in the range by Little (1971).
DAC: abundant in the pinyon-juniper woodlands at the north end of the range (1994).

Fairview Range (1)
elevation record: 6620 ft
1 record in 1 county (Churchill)

Towns/Range/Section	Location—Habitat	County	Elevation	Collector	Herbarium/a
T16N R34E S34 NW/4	on earthquake fault in pinyon-juniper woods	CH	6620 ft	*Charlet 2054*	RENO
[DAC: T15N R34E]					

Reported as present in the range by Little (1971).
DAC: abundant in pinyon-juniper woodlands throughout the middle and upper elevations (1994).

Fish Creek Mountains (5)
Lander County
[Little: T26N R40E R41E, T27N R40E R41E, T26N R40E]

Reported as present in the range by Little (1971).
DAC: I have not seen *Pinus monophylla* anywhere in this range. The only conifers present appear to be *Juniperus osteosperma* and an occasional *J. occidentalis.* Little's report (1971) is almost certainly in error (1993–1994).

Fish Creek Range (2)
Eureka County
[DAC: T19N R53E, T18N R52E R53E R54E, T17N R52E R53E, T16N R52E, T15N R52E]

Reported as present in the range by Little (1971).
DAC: abundant in pinyon-juniper woodlands in the lower elevations, extending to above 8000 ft with *Cercocarpus* and *P. flexilis* on Prospect Mountain (1994).

Flowery Range (2)
Storey County
[DAC: T17N R21E R22E, T18N R22E]

Reported as present in the range by Little (1971).
DAC: abundant in pinyon-juniper woodlands; most common in the southern and western parts of the range (1973–1994).

Fortification Range (3)
Lincoln and White Pine Counties
[Little: (WP) T10N R66E; (LI) T9N R67E, T8N R67E]

Reported as present in the range by Little (1971).

Gabbs Valley Range (1)
elevation record: 7500 ft
1 record in 1 county (Mineral)

Towns/Range/Section	Location—Habitat	County	Elevation	Collector	Herbarium/a
T9N R34E	1 mi S of Alkali Lake, in Alkali Valley, dry W-facing slopes w/juniper [DAC: this collection must be > 1 mi S at this elevation]	MI	7500 ft	*JL Reveal 402*	UC, UTC

Not reported as present in the range by Little (1971).
DAC: common in the pinyon-juniper woodlands of the range (1992–1993).

****Golden Gate Range (2)**
Lincoln County
[RWR: T1S R58E (above Sharp Spring in juniper woodland)]

Not reported as present in the range by Little (1971).
New report for the species in the range.

Gold Mountain Range (3)
Esmeralda County
[Little: T7S R42E, T8S R41.5E]

Reported as present in the range by Little (1971).

Goshute Mountains (2)
Elko County
[RWR: T30N R68E R69E, T29N R68E]
[Little: T35N R67E R68E, T34N R67E R68E, T33N R68E, T32N R68E R69E, T31N R68E]

Reported as present in the range by Little (1971).

Grant Range (1)
elevation record: 7300 ft
1 record in 1 county (Nye)

Towns/Range/Section	Location—Habitat	County	Elevation	Collector	Herbarium/a
T4N R57E S20	Quinn River [DAC: cannot be the Quinn River]	NY	7300 ft	*J Robertson BD 4-74*	RENO
[DAC: T3N R57E, T4N R58E, T5N R57E R58E, T6N R57E R58E, T7N R58E R59E, T8N R57E R58E R59E, T9N R58E R59E, T10N R58E]					

Reported as present in the range by Little (1971).
DAC: abundant in the pinyon-juniper woodlands that begin above the base of the mountain and extend to above 8000 ft (1994).

Grapevine Mountains (1)
elevation record: 6200 ft
1 record in county (Nye) and reports in another (Esmeralda)

Towns/Range/Section	Location—Habitat	County	Elevation	Collector	Herbarium/a
T11S R44E S32	———	NY	6200 ft	*M Kurzius 1018 & K Kingsley*	UNLV
[Little: (NY) T11S R43E R44E, T12S R44E; (ES) T11S R43E]					

Reported as present in the range by Little (1971) and Beatley (1976).

Groom Range (3)
Lincoln County
[Little: T5S R55E, T6S R55E R56E]

Reported as present in the range by Little (1971) and Beatley (1976).

Highland Range (1)
elevation record: 7200 ft
1 record in 1 county (Lincoln)

Towns/Range/Section	Location—Habitat	County	Elevation	Collector	Herbarium/a
T1N R66E S35 [DAC: T1S R66E]	Pine Spring	LI	7200 ft	*Ann Pinzl 3365*	NSM

Reported as present in the range by Little (1971).
DAC: abundant in pinyon-juniper-oak (*Quercus gambellii*) woodlands at low and middle elevations (1994).

****Horse Range (1)**
elevation record: 6300 ft
1 record in 1 county (Nye)

Towns/Range/Section	Location—Habitat	County	Elevation	Collector	Herbarium/a
T11N R59E S16	N end of mountain	NY	6300 ft	*Charlet 1985*	RENO
[DAC: T9N R59E, T10N R59E R60E]					

Not reported as present in the range by Little (1971).
DAC: abundant in the pinyon-juniper woodlands that occur over most of the range (1994).
New record for the species in the range.

Hot Creek Range (2)
Nye County
[Little: T12N R51E, T11N R50E R51E, T10N R50E, T9N R50E R51E, T8N R49E R50E, T7N R49E R50E, T6N R49E R50E]

Reported as present in the range by Little (1971).
DAC: abundant in the pinyon-juniper woodlands that occur throughout the range, but absent from the extreme south (1994).

Humboldt Range (5)
Pershing County
[Little: T28N R34E]

Reported as present in the range by Little (1971).
DAC: I have been unable to verify the occurrence of *Pinus monophylla* here (1990–1994).

Hunter Point Mountains (1)
elevation record: 6720 ft
1 record in 1 county (White Pine)

Towns/Range/Section	Location—Habitat	County	Elevation	Collector	Herbarium/a
T25N R62E S28	in pinyon-juniper woodland	WP	6720 ft	*Charlet 2117 & T Charlet*	RENO
[DAC: T24N R62E, T23N R61E, T22N R61E]					

Reported as present in the range by Little (1971).
DAC: codominant with *Juniperus osteosperma* throughout most of the range (1994).

Huntoon Mountains (2)
Mineral County
[DAC: T1N R31E R32E R33E, T2N R30E R31E R32E R33E]

Reported as present in the range by Little (1971).
DAC: dominant in the pinyon woodlands of the range, with *Juniperus osteosperma* (1992–1994).

Irish Mountain (2)
Lincoln County
[RWR: T3S R58E R59E R60E, T4S R59E]

Reported as present in the range by Little (1971).

****Junction House Range (1)**
elevation record: 4700 ft
1 record in 1 county (Washoe)

Towns/Range/Section	Location—Habitat	County	Elevation	Collector	Herbarium/a
T20N R20E S18 W/8	cluster of 6 individuals	WA	4700 ft	*Charlet 1799*	RENO, DUKE

Not reported as present in the range by Little (1971).
DAC: this is one of the few locations known for pinyon pine north of the Truckee and Humboldt Rivers (1971–1995).
New record for the species in the range.

Kawich Range (1)
elevation record: 7600 ft
1 record in 1 county (Nye)

Towns/Range/Section	Location—Habitat	County	Elevation	Collector	Herbarium/a
T1N R50E S5	Eden Creek Canyon	NY	7600 ft	*Beatley 12952*	RENO
[Little: T3N R49E R50E, T2N R49E R50E, T1N R49E R51E, T1S R50E, T2S R50E]					

Reported as present in the range by Little (1971) and Beatley (1976).

Kern Mountains (1)
elevation record: 6840 ft
1 record in 1 county (White Pine)

Towns/Range/Section	Location—Habitat	County	Elevation	Collector	Herbarium/a
T22N R69E S16	Tippett Canyon, in pinyon-juniper woodlands	WP	6840 ft	*Charlet 2106 & T Charlet*	RENO
[DAC: T21N R69E R70E]					

Reported as present in the range by Little (1971).
DAC: codominant with *Juniperus osteosperma* in pinyon-juniper throughout the middle elevations in the Nevada portion of the range (1994).

Kinsley Mountains (2)
Elko County
[RWR: T26N R67E, T27N R68E]

Reported as present in the range by Little (1971).

Las Vegas Range (1)
elevation record: 5800 ft
1 record in 1 county (Clark)

Towns/Range/Section	Location—Habitat	County	Elevation	Collector	Herbarium/a
T17S R61E	second drainage N of Quail Spring	CL	5800 ft	*T Ackerman 6723*	UNLV

Not reported as present in the range by Little (1971).

Leach Range (3)
Elko County
[Little: T40N R68E, T39N R67E, T38N R67E]

Reported as present in the range by Little (1971).

Leppy Range (3)
Elko County
[McLane: T34N R70E]

Reported as present in the range by McLane (1978).
Not reported as present by Little (1971).

Lone Mountain (2)
Esmeralda County
[RWR: T2N R40E]

Reported as present in the range by Little (1971).
RWR: dominant in pinyon woodland; ca. 99.9% pinyon, 0.1% juniper.

Magruder Mountain (1)
elevation record: 6900 ft
1 record in 1 county (Esmeralda)

Towns/Range/Section	Location—Habitat	County	Elevation	Collector	Herbarium/a
T6S R40E	SR 3, 1.5 mi E of Lida Summit	ES	6900 ft	*Christopher Davidson 2336*	RSA

Reported as present in the range by Little (1971).
DAC: observed in pinyon-juniper woodland (12/1993).

Mahogany Hills (2)
Eureka County
[DAC: T19N R52E, T18N R52E]

Reported as present in the range by Little (1971).
DAC: abundant in pinyon-juniper woodlands throughout the range (1994).

Mahogany Mountains (1)
elevation record: 7200–7500 ft
2 records in 1 county (Lincoln)

Towns/Range/Section	Location—Habitat	County	Elevation	Collector	Herbarium/a
T1N R71E	Deer Lodge	LI	7200 ft	*Desma Hall in 1935*	UC, BRY
T1N R71E	Deer Lodge	LI	7500 ft	*Desma Hall in 1935*	RENO
[Little: T1N R70E R71E, T2N R71E]					

Reported as present in the range by Little (1971).

Maverick Springs Range (3)
Elko and White Pine Counties
[Little: (EL) T26N R59E; (WP) T26N R59E, T25N R59E, T24N R58E R59E, T23N R57E R58E]

Reported as present in the range by Little (1971).

McCullough Range (1)
elevation record: 4500–5600 ft
3 records in 1 county (Clark)

Towns/Range/Section	Location—Habitat	County	Elevation	Collector	Herbarium/a
T26S R61E	16 mi NW of Bell Ranch [near Searchlight]	CL	4500 ft	*Percy Train 1489*	RENO, UTC
T27S R61E	15 mi W of Searchlight	CL	5000 ft	*Mozingo in 1960*	RENO
T27S R61E S3	——	CL	5600 ft	*Ann Pinzl in 1976*	NSM

Reported as present in the range by Little (1971).
DAC: *Mozingo in 1960* was determined to be *Pinus edulis* Engelm. var. *fallax* Little by DK Bailey at RENO, while *Ann Pinzl in 1976* was determined to be *P. edulis* Engelm. var. *fallax* Little by DK Bailey at NSM. This material is more than 95% single-needled.

Medicine Range (1)
elevation record: 6520 ft
1 record in 1 county (Elko)

Towns/Range/Section	Location—Habitat	County	Elevation	Collector	Herbarium/a
T28N R60E S5 N/8	above Medicine Spring, in juniper woodland	EL	6520 ft	*Charlet 2119 & T Charlet*	RENO
[DAC: T28N R61E, T27N R60E R61E, T26N R61E]					

Reported as present in the range by Little (1971).
DAC: codominant with *Juniperus osteosperma* in woodlands in the middle elevations (1994).

Monitor Range (1)
1 record in 1 county (Nye) and reports in another (Eureka)

Towns/Range/Section	Location—Habitat	County	Elevation	Collector	Herbarium/a
T7N R47E S6	——	NY	–	*Ann Pinzl 8566*	UNLV, NSM
[Little (EU): T18N R48E R49E, T17N R48E R49E, T16N R48E R49E R50E, T15N R48E R49E R50E; (NY) T15N R48E R49E, T14N R48E R49E, T13N R48E R49E, T12N R48E, T11N R47E R48E, T10N R47E R48E, T9N R47E R48E, T8N R47E R48E, T7N R46E, T6N R46E R47E, T5N R45E R46E, T4N R45E R46E]					

Reported as present in the range by Little (1971).
DAC: abundant in pinyon-juniper woodland at moderate elevations (1993–1994).

Montezuma Range (2)
Esmeralda County
[RWR: T3S R41E]

Reported as present in the range by Little (1971).

Mormon Mountains (1)
elevation record: 4000–4800 ft
2 records in 1 county (Lincoln)

Towns/Range/Section	Location—Habitat	County	Elevation	Collector	Herbarium/a
T12S R67E	S end of range	LI	4800 ft	*WE Niles 3676, Landau & Durham*	UNLV
—	Mormon Mountains	LI	4000 ft	*JD Armstrong in 1966*	UNLV

Reported as present in the range by Little (1971).

Mountain Boy Range (1)
elevation record: 7500 ft
1 record in 1 county (Eureka)

Towns/Range/Section	Location—Habitat	County	Elevation	Collector	Herbarium/a
T20N R52E S26	Devils Gate, 10 mi W of Eureka	EU	7500 ft	*C Condit 34*	UC

Reported as present in the range by Little (1971).
DAC: abundant in pinyon-juniper woodlands throughout the range (1994).

Needle Range (3)
Nye County
[Little: T10N R52E]

Reported as present in the range by Little (1971).

Newberry Mountains (1)
elevation record: 3700 ft
1 record in 1 county (Clark)

Towns/Range/Section	Location—Habitat	County	Elevation	Collector	Herbarium/a
T31S R65E	1.4 mi W of Christmas Tree Pass	CL	3700 ft	*Ann Pinzl 2653*	NSM

Reported as present in the range by Little (1971).

New Pass Range (2)
Churchill and Lander Counties
[Little: (CH) T20N R40E; (LA) T22N R40E, T21N R40E]

Reported as present in the range by Little (1971).
DAC: abundant in the limited pinyon-juniper woodland (1991–1994).

North Pahroc Range (2)
Lincoln County
[RWR: T1S R63E, T2S R63E, T3S R63E]

Reported as present in the range by Little (1971).
RWR: in extensive pinyon-juniper woodlands.

Pah Rah Range (1)
elevation record: 5000 ft
1 record in 1 county (Washoe)

Towns/Range/Section	Location—Habitat	County	Elevation	Collector	Herbarium/a
T23N R22E S23	"Pinyon Canyon" 1 mi S of Pyramid Lake	WA	5000 ft	*Billings 2041*	RENO
[Robin Tausch: T22N R22E S20 S/8, at 5345 ft]					

Not reported as present in the range by Little (1971).
DAC: rare; these are the only known locations of the species on the mountain; most of the woodlands present are composed solely of *Juniperus osteosperma*. Tausch cored the individual at T22N R22E and determined it to be 250 years old at 2 ft above the ground; circumference = 139 in, height = 53 ft, spread = 66 ft. This is the largest singleleaf pinyon ever found. It appears in the National Register of Champion Trees (1975–1994).

Pahranagat Range (2)
Lincoln County
[Little: T7S R59E]
[RWR: T4S R58E]

Reported as present in the range by Little (1971).

Pahute Mesa (3)
Nye County
[Little: T8S R51E, T9S R50E R51E]

Reported as present in the range by Little (1971) and Beatley (1976).

Palmetto Mountains (1)
elevation record: 6890 ft
1 record in 1 county (Esmeralda)

Towns/Range/Section	Location—Habitat	County	Elevation	Collector	Herbarium/a
T6S R40E	mountains W of Lida	ES	6890 ft	*Ivar Tidestrom 9847*	RSA
[RWR: T5S R39E R40E]					

Reported as present in the range by Little (1971).

Pancake Range (2)
Nye and White Pine Counties
[RWR: (WP) T15N R55E, T16N R55E]
[Little: (NY) T14N R54E, T13N R54E, T12N R55E, T11N R55E, T10N R55E]

Reported as present in the range by Little (1971).

Paradise Range (1)
elevation record: 6900 ft
2 records in 1 county (Nye)

Towns/Range/Section	Location—Habitat	County	Elevation	Collector	Herbarium/a
T12N R37E	just E of Green Springs Summit	NY	6900 ft	*Ann Pinzl 1189*	NSM
T14N R37E	15 mi NW of Ione, 2 mi NW of Penelas Mill	NY	—	*KH Beach 1042*	UC
[Little: T13N R37E]					

Reported as present in the range by Little (1971).
DAC: observed in pinyon-juniper woodland throughout the range (1992–1993).

Park Range (3)
Nye County
[Little: T12N R51E]

Reported as present in the range by Little (1971).

Peavine Mountain (1)
elevation record: 5200–6450 ft
3 records in 1 county (Washoe)

Towns/Range/Section	Location—Habitat	County	Elevation	Collector	Herbarium/a
T20N R18E	NW side	WA	6450 ft	*M Williams 72-10*	RENO
T20N R19E S20	Ecology Canyon	WA	5200 ft	*M Williams 72-6*	RENO
T20N R18E S8	3 mi W, 2 mi N of Peavine Peak; S aspect	WA	6000 ft	*NK Johnson & JS Spencer in 1953*	DUKE

Reported as present in the range by Little (1971).
DAC: the population from which *M Williams 72-10* was collected was extirpated in 1984 in a brush fire (Williams et al. 1992). The species still occurs at another site in T20N R18E (1971–1995).

Pequop Mountains (1)
elevation record: 6200–6600 ft
2 records in 1 county (Elko)

Towns/Range/Section	Location—Habitat	County	Elevation	Collector	Herbarium/a
T37N R65E S24	Pequop Summit, 35 mi W of Wendover	EL	6200 ft	*CL Porter 7698 & MW Porter*	UC
T37N R65E S23	[Maverick Canyon]	EL	6600 ft	*N Holmgren 3518 & E Jensen*	UTC

[Little: T36N R65E R66E, T35N R65E, T34N R65E, T33N R65E, T32N R63E R64E R65E, T31N R63E R64E R65E, T30N R63E R64E]

Reported as present in the range by Little (1971).

Pilot Mountains (5)
Mineral County

Reported as present in the range by Little (1971).
DAC: I have not yet visited this range, but both Rust and Tausch have, and neither found a *P. monophylla* individual (RW Rust, pers. comm. 1993; RJ Tausch, pers. comm. 1994).

Pilot Range (1)
elevation record: 5800 ft
1 record in 1 county (Elko)

Towns/Range/Section	Location—Habitat	County	Elevation	Collector	Herbarium/a
T36N R70E	near Danger Cave, Bonneville level, E of Pilot Peak	EL	5800 ft	*WP Cottam 13821*	UT
[RWR: T37N R70E]					

Reported as present in the range by Little (1971).

Pine Grove Hills (1)
elevation record: 5000–8460 ft
2 records in 1 county (Lyon)

Towns/Range/Section	Location—Habitat	County	Elevation	Collector	Herbarium/a
T10N R25E	15 mi S of Yerington	LY	5000 ft	*E Koch, Holstein & Carlson N-119*	UC
T8N R26E S18 E/4	N slope Bald Mountain, upper pinyon limit	LY	8460 ft	*Charlet 1952*	RENO
[DAC: T9N R24E R25E R26E, T8N R25E R26E R27E, T7N R25E R26E]					

Reported as present in the range by Little (1971).
DAC: dominant in extensive pinyon woodlands throughout the range, with *Juniperus osteosperma* (1989–1995).

Pine Nut Mountains (1)
elevation record: 5400 ft
2 records in 2 counties (Carson City and Douglas) and reports in another (Lyon)

Towns/Range/Section	Location—Habitat	County	Elevation	Collector	Herbarium/a
T11N R21E	10 mi S of Gardnerville	DO	—	*Breene 80 & Sampson*	RENO
T15N R21E S12	0.75 mi SW of Toll (Ormsby)	CC	5400 ft	*V Stackhouse 11*	UC
[DAC: (CC) T14N R21E; (DO) T14N R21E R22E R23E, T13N R21E R22E R23E, T12N R21E R22E R23E, T11N R22E R23E, T10N R22E; (LY) T16N R22E R23E, T15N R22E R23E, T14N R22E R23E]					

Reported as present in the range by Little (1971).
DAC: dominant in extensive pinyon woodlands throughout the range; extends nearly to the highest summits (1991–1994).

Piñon Range (2)
Elko County
[RWR: T32N R52E R53E, T31N R52E R53E, T30N R53E]

Reported as present in the range by Little (1971).

Pioche Hills (1)
elevation record: 5800 ft
1 record in 1 county (Lincoln)

Towns/Range/Section	Location—Habitat	County	Elevation	Collector	Herbarium/a
T1N R67E [RWR: T1S R68E]	E-facing slope S of Pioche	LI	5800 ft	*Billings 2196*	RENO

Reported as present in the range by Little (1971).

Quinn Canyon Range (2)
Lincoln and Nye Counties
[Little (LI): T1N R54E R55E R56E, T2N R54E R55E R56E; (NY) T2N R55E R56E R57E, T3N R54E R57E, T4N R55E R56E]

Reported as present in the range by Little (1971).
DAC: abundant in the pinyon-juniper woodlands that extend from above the base of the mountain to above 8000 ft (1994).

Reveille Range (1)
elevation record: 6700 ft
1 record in 1 county (Nye)

Towns/Range/Section	Location—Habitat	County	Elevation	Collector	Herbarium/a
T4N R52E	lower canyon below Echo Peak, 38°9'1"N, 116°6'17"W, 40-mile cable access road from Pahute CP [?] to Echo Peak	NY	6700 ft	*Beatley 6302*	RENO, UTC
[Little: T1N R52E, T2N R52E, T3N R52E]					

Reported as present in the range by Little (1971).

Roberts Mountains (1)
elevation record: 8740 ft
1 record in 1 county (Eureka)

Towns/Range/Section	Location—Habitat	County	Elevation	Collector	Herbarium/a
T23N R50E S12 E/8	ridge E of high peaks, shrub 0.5 m tall	EU	8740 ft	*Charlet 2089*	RENO
[DAC: T24N R50E R51E, T23N R49E R51E, T22N R50E R51E]					

Reported as present in the range by Little (1971).
DAC: abundant in the extensive pinyon-juniper woodlands. *Charlet 2089* is the high-elevation record for the species in the state (1993–1994).

Ruby Mountains (1)
elevation record: 6150 ft
3 records in 1 county (Elko) and reports in another (White Pine)

Towns/Range/Section	Location—Habitat	County	Elevation	Collector	Herbarium/a
T28N R57E	Harrison Pass	EL	—	*EVA Murphy 424*	RENO
T27N R57E S31	E base Ruby Mountains	EL	6150 ft	*Ann Pinzl 6807*	NSM
T27N R57E	Hills about Cave Creek PO	EL	—	*HL Mason 4746*	UC

[Little: (EL) T34N R59E, T33N R58E R59E, T32N R57E, T31N R58E R59E, T30N R58E, T29N R58E, T26N R55E R56E; (WP) T26N R56E R57E, T25N R56E R57E, T24N R57E, T23N R57E, T22N R57E, T21N R56E R57E]
[DAC: (EL) T28N R57E (Longhair Smith Canyon)]

Reported as present in the range by Little (1971).
DAC: rare to uncommon in the north, abundant in pinyon-juniper woodlands in the south (1988–1994).

San Antonio Mountains (3)
Nye County
[Little: T5N R42E]

Reported as present in the range by Little (1971).

Schell Creek Range (1)
elevation record: 6600–7800 ft
6 records in 2 counties (White Pine and Lincoln)

Towns/Range/Section	Location—Habitat	County	Elevation	Collector	Herbarium/a
T9N R65E S33	S Schell Creek Range	LI	6600 ft	*N Holmgren 3551 & E Jensen*	UTC
T14N R65E S35	Connor's Pass	WP	7700 ft	*Olga Reifschneider in 1950*	RENO
T14N R65E S35	Connor's Summit	WP	7800 ft	*N Holmgren 3761 & A Holmgren*	UTC
T19N R65E	E Fork Duck Creek	WP	—	*Percy Train 1112*	RENO, DUKE
T19N R65E	Bird Creek, 15 mi E of McGill	WP	7000 ft	*BO Boone & GE Franklin, Jr. 673*	UC, RENO
T22N R65E	Schellbourne	WP	—	*ME Jones in 1892*	RSA

[DAC: (WP) T25N R65E, T24N R65E, T23N R65E, T22N R65E, T21N R65E R66E, T20N R65E R66E, T19N R65E R66E, T18N R65E R66E, T17N R65E R66E, T15N R65E R66E, T14N R65E R66E, T13N R65E R66E, T12N R64E R65E R66E, T11N R64E R65E, T10N R64E R65E; (LI) T9N R64E R65E, T8N R64E, T7N R64E, T6N R64E, T5N R64E]

Reported as present in the range by Little (1971).
DAC: abundant in pinyon-juniper woodlands at low elevations in the north, and at middle to high elevations in the south (1994).

Seaman Range (2)
Lincoln and Nye Counties
[RWR: (LI) T1N R62E, T2N R61E]
[Little: (NY) T3N R61E, T2N R61E]

Reported as present in the range by Little (1971).

Sheep Range (1)
elevation record: 6200 ft
2 records in 1 county (Clark)

Towns/Range/Section	Location—Habitat	County	Elevation	Collector	Herbarium/a
T15S R60E	Wiregrass Spring, Hidden Forest	CL	—	*BH Banta in 1950*	RENO
T16S R60E	Above Cow Camp Springs	CL	6200 ft	*WG Spaulding in 1972*	UNLV

[Little: T11S R61E, T12S R61E, T13S R61E, T14S R61E, T15S R61E]

Reported as present in the range by Little (1971).

Shoshone Mountain (3)
Nye County
[Paul Tueller: T11S R51E, T12S R50E R51E]

Not reported as present in the range by Little (1971).
Reported as present by Paul Tueller (Dept. of Environmental and Resource Sciences, Univ. Nevada, pers. comm. 1994).

Shoshone Mountains (2)
Lander and Nye Counties
[DAC: (LA) T19N R41E, T18N R41E, T16N R40E R41E; (NY) T9N R39E, T10N R39E R40E, T11N R39E R40E, T12N R39E R40E, T13N R39E R40E, T14N R38E R39E, T15N R40E]

Reported as present in the range by Little (1971).
DAC: abundant in pinyon-juniper woodlands throughout the range (1991–1994).

Shoshone Range (1)
elevation record: 5840 ft
1 record in 1 county (Lander)

Towns/Range/Section	Location—Habitat	County	Elevation	Collector	Herbarium/a
T29N R45E S31	Mill Creek, in pinyon juniper woodland	LA	5840 ft	*Charlet 1893*	UCR, RENO, UC, BRY

[DAC: T29N R45E, T28N R45E, T27N R44E]

Reported as present in the range by Little (1971).
DAC: abundant as far north as Mount Lewis, apparently invading juniper woodland and sagebrush steppe (1993–1994).

Silver Peak Range (1)
elevation record: 7200 ft
1 record in 1 county (Mineral)

Towns/Range/Section	Location—Habitat	County	Elevation	Collector	Herbarium/a
T2S R38E S10 NW/4	0.25 mi NW of Coyote Summit	MI	7200 ft	*Ann Pinzl 9923*	NSM

[RWR: T1S R38E, T2S R37E, T4S R37E]
[Little: T3S R37E, T1S R37E, T4S R38E]

Reported as present in the range by Little (1971).

Simpson Park Range (1)
elevation record: 6600–6800 ft
2 records in 2 counties (Eureka and Lander)

Towns/Range/Section	Location—Habitat	County	Elevation	Collector	Herbarium/a
T24N R48E	Gund Ranch	EU	6600 ft	*Rowan Sage in 1980*	RENO
T19N R46E	1.7 mi above Hickison petroglyphs	LA	6800 ft	*J Robertson NS 115-74*	RENO

[DAC: (EU) T24N R48E]
[Little: (EU) T23N R48E; (LA) T22N R47E R48E, T21N R47E, T20N R46E R47E]

Reported as present in the range by Little (1971).
DAC: abundant in extensive pinyon-juniper woodlands (1993–1994).

Singatse Range (3)
Lyon County
[Little: T12N R25E]

Reported as present in the range by Little (1971).

Snake Range (1)
elevation record: 6000–7000 ft
5 records in 1 county (White Pine)

Towns/Range/Section	Location—Habitat	County	Elevation	Collector	Herbarium/a
T13N R68/69E	Mount Wheeler	WP	6500 ft	*Peter Herlan & Ann Pinzl in 1973*	NSM
T13N R69E	Baker Creek	WP	7000 ft	*George Larson in 1919*	OGDF
T13N R69E	Baker Canyon, 5 mi SW of Baker, canyon bottom	WP	6500 ft	*SJ Preece, Jr 1456*	UT
T13N R69E	3 mi W of Baker	WP	6500 ft	*B Maguire 2469 & RJ Becraft*	UTC
T17N R70E	Horse Canyon, 1 mi W of mouth on N-facing slope	WP	6000 ft	*BF Harrison 11654*	BRY

[DAC: T15N R68E]
[Little: T20N R69E, T19N R68E R69E R70E, T18N R68E R69E R70E, T17N R68E R70E, T15N R68E R69E R70E, T14N R68E R69E, T13N R68E R69E R70E, T12N R68E R69E R70E, T11N R68E R69E R70E, T10N R68E R69E R70E]

Reported as present in the range by Little (1971).
DAC: abundant in pinyon-juniper woodlands at lower elevations. Some populations have a high incidence of double needles (ca. 5%), reflecting a possible zone of hybridization with *P. edulis* (RJ Tausch, pers. comm. 1993) (1994).

South Pahroc Range (3)
Lincoln County
[Little: T5S R62E, T6S R62E]

Reported as present in the range by Little (1971).

South Virgin Mountains (1)
elevation record: 2800 ft
1 record in 1 county (Clark)

Towns/Range/Section	Location—Habitat	County	Elevation	Collector	Herbarium/a
T19S R70E	2 mi in on road to Devils Cove, from fork in Gold Butte Rd	CL	2800 ft	*Ann Pinzl 182 Peter Herlan & Ryser*	NSM

Not reported as present in the range by Little (1971).
DAC: *Pinzl 182* is the low-elevation record for the species in the state.

Spring Mountains (1)
elevation record: 5000–7950 ft
5 records in 1 county (Clark) and reports in another (Nye)

Towns/Range/Section	Location—Habitat	County	Elevation	Collector	Herbarium/a
T18S R55E S1	Cold Creek Spring, NE slope	CL	6200 ft	*Beatley & Reveal 10834*	RENO
T19S R57E	Kyle Canyon, Deer Creek, hillside	CL	7950 ft	*IW Clokey 7819*	UC, UT, RENO, UTC, DUKE
T19S R57E	Lee Canyon, flat top of ridge	CL	—	*IW Clokey 7025*	UT
T20S R57E	Harris Mountain Rd, E side of Charleston Mts	CL	—	*D Atwood 7962*	BRY
T22S R57E	1 mi E of Mountain Pass summit	CL	5000 ft	*JS Harris*	UNLV

[Little: (CL) T25S R58E, T24S R58E, T23S R58E, T22S R58E, T21S R57E R58E, T18S R54E, 17S R53E R54E; (NY) T17S R53E R54E, T18S R54E]
[DAC: (CL) T18S R57E R58E, T19S R55E R58E, T20S R55E R56E R57E R58E, T21S R56E, T22S R58E, T23S R57E]

Reported as present in the range by Little (1971).
DAC: abundant, with a wide elevation range (4500 ft to at least 8500 ft [Kyle Canyon]); in pinyon-juniper woodland with *Juniperus osteosperma* at middle elevations and *J. scopulorum* and *P. ponderosa* at higher elevations. At lower elevations, sometimes found with *Yucca brevifolia* (1982 & 1993).

Stillwater Range (2)
Churchill County
[Little: T19N R33E, T20N R33E, T21N R33E, T22N R33E R34E]
[DAC: T24N R34E R35E, T23N R34E R35E]

Reported as present in the range by Little (1971).
DAC: with *Juniperus osteosperma* in pinyon-juniper woodland in the southern and central portions of the range only; appears to be absent from the juniper woodlands that extend north from Copper Kettle Canyon to McKinney Pass at the juncture with the East Range. Little reported the species as occurring in the Pershing County portion of the range, but there are no records, and I have not found it there (1991–1994).

Stonewall Mountain (1)
elevation record: 5800–8000 ft
1 record in 1 county (Nye)

Towns/Range/Section	Location—Habitat	County	Elevation	Collector	Herbarium/a
T5S R44E	Stonewall Flat [Mountains]	NY	5800–8000 ft	*Beatley & Smith 13113*	RENO

Reported as present in the range by Little (1971) and Beatley (1976).

Sulphur Spring Range (1)
1 record in 1 county (Eureka) and reports in another (Elko)

Towns/Range/Section	Location—Habitat	County	Elevation	Collector	Herbarium/a
T23N R52E	Garden Pass, SR 51, Suphur Springs, high incidence of 2-needled fascicles	EU	—	*RJ Tausch & KH Rea in 1974*	UTC
[Little: (EL) T28N R53E, T27N R53E; (EU) T27N R52E R53E, T26N R52E R53E, T25N R52E, T24N R52E, T23N R52E]					

Reported as present in the range by Little (1971).
DAC: abundant in pinyon-juniper woodlands throughout the range (1993–1994).

Sweetwater Mountains (2)
Douglas and Lyon Counties
[DAC: (LY) T8N R24E R25E, T7N R24E; (DO) T9N R23E, T8N R23E]

Reported as present in the range by Little (1971).
DAC: dominant in pinyon woodlands throughout the range at lower and middle elevations (1985–1995).

Sylvania Mountains (2)
Esmeralda County
[RWR: T6S R39E]

Reported as present in the range by Little (1971).

Timber Mountain (3)
Nye County
[McLane: T10S R49E]

Reported as present in the range by McLane (1978).
Not reported as present by Little (1971).

Timpahute Range (2)
Lincoln County
[RWR: T3S R58E, T2S R58E]

Reported as present in the range by Little (1971).

Toana Range (1)
elevation record: 6400 ft
1 record in 1 county (Elko)

Towns/Range/Section	Location—Habitat	County	Elevation	Collector	Herbarium/a
T37N R68E S9	11 mi S of Montello on SR 30	EL	6400 ft	*N Holmgren 3621 & P Kern*	UTC
[RWR: T35N R68E, T36N R68E]					

Reported as present in the range by Little (1971).
DAC: *Holmgren 3621* cannot be on SR 30 at the TRS recorded on the label.

Toiyabe Range (1)
elevation record: 6200–7200 ft
7 records in 2 counties (Nye and Lander)

Towns/Range/Section	Location—Habitat	County	Elevation	Collector	Herbarium/a
T9N R40E	6 mi N of USFS-BLM line, along road 018	NY	—	*D Atwood 15293*	BRY
T17N R43E	Big Creek campground	LA	6700 ft	*P & H Vreeland IV-16*	RENO
T18N R44E S19 NE/4	along US 50, 8.5 mi SE of Austin at Forest Service boundary	LA	6200 ft	*S Goodrich 10907*	BRY
T19N R44E	Toiyabe NF, just below Scott's Summit Pass	LA	—	*Jeffrey Huber 309*	BRY
T19N R44E	along SR 8A, E of Austin Summit	LA	7000 ft	*S Goodrich 6704a*	BRY
T19N R44E	Bob Scott Campground, US 50	LA	6500 ft	*M Williams 76-99-1 & Tiehm*	RENO
T19N R44E	Bob Scott Campground	LA	7200 ft	*Ann Pinzl 885*	NSM

[DAC: (LA) T26N R47E, T25N R46E R47E, T24N R46E, T23N R45E R46E, T22N R44E R45E R46E, T21N R44E R45E, T20N R44E, T19N R43E R44E, T18N R43E R44E, T17N R44E, T16N R43E, T15N R42E R43E; (NY) T15N R52E R43E, T14N R42E R43E, T13N R41E R42E, T12N R40E R41E R42E, T11N R40E R42E, T10N R40E R41E R42E, T9N R40E R41E R42E]

Reported as present in the range by Little (1971).
DAC: dominant throughout the pinyon-juniper woodlands on the mountain; much more common than any of the juniper species (*J. occidentalis, J. osteosperma,* and *J. scopulorum*) present.

Toquima Range (1)
elevation record: 7860–8350 ft
4 records in 2 counties (Lander and Nye)

Towns/Range/Section	Location—Habitat	County	Elevation	Collector	Herbarium/a
T8N R44E	N of Manhattan, common	NY	—	*Goodner 1200 & Henning*	RENO
T8N R44E	Manhattan Station	NY	—	*Steele in 1917*	AUSTIN
T12N R45E	Moores Creek Summit	NY	8350 ft	*Charlet 904 & S Vander Wall*	RENO
T16N R46E	picnic area at Petes Summit	LA	7860 ft	*BD Parfitt 3813 & K Roberts*	RSA

[DAC: (NY) T15N R45E R46E, T14N R45E R46E R47E, T13N R45E R46E, T12N R46E, T11N R44E R45E, T10N R44E R45E, T9N R44E R45E, T7N R43E]
[Little: (LA) T17N R46E, T16N R46E]

Reported as present in the range by Little (1971).
DAC: dominant in the middle and upper elevations of the pinyon-juniper woodlands in the range. These woodlands occupy a relatively narrow elevational band (ca. 7200–8400 ft) compared with other Nevada mountain ranges. The species feebly extends through the mountain brush zone to the mountain-mahogany and *Pinus flexilis* upper montane woodlands (1991–1993).

****Virginia Mountains (3)**
Washoe County
[Robin Tausch: T24N R21E S32 S/2, S33]

Not reported as present in the range by Little (1971).
DAC: rare; a small stand is present near and below Needle Rock at the south end of the range (5250 ft), reported by Robin Tausch (pers. comm. 1994).
New report for the species in the range.

Virginia Range (1)
elevation record: 5000–6500 ft
7 records in 3 counties (Lyon, Storey, and Washoe)

Towns/Range/Section	Location—Habitat	County	Elevation	Collector	Herbarium/a
T16N R21E	1 mi W of Silver City, American Flat Canyon	LY	5000 ft	*RA Allen 137*	RENO
T17N R21E	3 mi E of WA County line, Geiger	ST	—	*TC Frantz in 1951*	RENO
—	in mined area below Virginia City	ST	—	*C Ritchie Bell 17656*	BRY
—	Geiger Grade, 17 mi from Reno	ST	—	*BO Moore 47 & GE Franklin, Jr*	UTC, DUKE
T17N R21E	Mount Davidson	WA	6500 ft	*Frank C Gregory sn*	UC
T17N R21E	Old Geiger Grade summit	WA	6300 ft	*Ann Pinzl in 1975*	NSM
T18N R20E	11 mi SE of Reno	WA	—	*WA Archer 6487*	RENO
[DAC: (ST) T19N R21E R22E, T18N R20E R21E R22E]					

Reported as present in the range by Little (1971).
DAC: dominant in the middle and upper elevations of the pinyon-juniper woodlands (1973–1995).

Virgin Mountains (1)
elevation record: 4400–6000 ft
2 records in 1 county (Clark)

Towns/Range/Section	Location—Habitat	County	Elevation	Collector	Herbarium/a
T15S R71E	Cabin Canyon	CL	6000 ft	*JD Armstrong 696*	UNLV
T14S R71E	Cabin Canyon	CL	4400 ft	*Ann Pinzl in 1974*	NSM

Reported as present in the range by Little (1971).

Wassuk Range (1)
elevation record: 6000–8340 ft
6 records in 1 county (Mineral)

Towns/Range/Section	Location—Habitat	County	Elevation	Collector	Herbarium/a
T6N R29E S25	Hilltop S of Powell Canyon	MI	8340 ft	*Charlet 1973*	RENO
T7N R29E	Cory Creek, 0.5 mi from mouth and up creek	MI	6000 ft	*WA Archer 6871*	RENO, DUKE
T7N R29E	Cory Canyon	MI	7600 ft	*Ann Pinzl in 1974*	NSM
T8N R28E S28 W/4	N of Rattlesnake Canyon, NW aspect	MI	6820 ft	*Charlet 1959*	RENO
T9N R28E S35	Cottonwood Canyon	MI	–	*K Hawkins in 1981*	RENO
–	———	MI	6000 ft	*WA Archer 6871*	UC

[DAC: T12N R28E, T11N R28E R29E, T10N R28E R29E, T9N R29E, T8N R29E, T7N R28E R30E, T6N R30E]
[Little: T12N R27E]

Reported as present in the range by Little (1971).
DAC: this species dominates the pinyon-juniper woodlands of the western slope of the range and the pinyon woodlands of the eastern slope (1990–1995).

Weepah Hills (4)
Esmeralda County

Not reported as present in the range by Little (1971).
RWR: a small conifer stand at T1N R40E S18 observed with binoculars was probably *P. monophylla* (6/1993).

Wellington Hills (1)
1 record in 1 county (Douglas)

Towns/Range/Section	Location—Habitat	County	Elevation	Collector	Herbarium/a
T9N R23E S4	Long Dry Canyon, W side Wellington Hills	DO	–	*M Lavin SW112*	RENO

[DAC: T10N R23E, T9N R23E]

Reported as present in the range by Little (1971).
DAC: abundant in pinyon-juniper woodlands throughout the range (1988–1994).

West Humboldt Range (5)
Pershing County
[Little: T27N R33E]

Reported as present in the range by Little (1971).
DAC: 3 Nevada mountain ranges contain the name Humboldt, and all contribute to the Humboldt River drainage: the East Humboldt Range (north of the Ruby Mountains and south of Wells, Elko County), the Humboldt Range (northeast of Lovelock and southwest of Winnemucca, Pershing County), and the West Humboldt Range (east and south of Lovelock, Pershing County). The use of the name "Western Humboldt River Mountains" by the Pacific Railroad Survey when referring to both the Humboldt and the West Humboldt Ranges, and the use of the name "Clover Mountains" by the King Survey when referring to the East Humboldt Range, have caused much confusion. Modern herbarium labels still commonly refer to collections made in the Humboldt Range as the "West Humboldt Range" (for nomenclatural discussion, see McLane 1978). I have found no *Pinus monophylla* material in western herbaria from either the Humboldt or the West Humboldt Range, nor have I been able to verify its presence in the field, but 2 isolated occurrences are clearly resolved as the Humboldt Range and the West Humboldt Range on Little's map (1971). The only conifer species present in the West Humboldt Range is *Juniperus osteosperma* (1993–1994).

Whistler Range (4)
Eureka County

Possibly reported as present in the range by Little (1971), and likely to be there, but not verified by DAC or RWR.

White Mountains (1)
elevation record: 7200 ft
2 records in 2 counties (Mineral and Esmeralda)

Towns/Range/Section	Location—Habitat	County	Elevation	Collector	Herbarium/a
T1N R32E	in canyon W of Boundary Peak	MI	7200 ft	*Vasek 620928-7*	UCR
T1S R34E S31	mouth of Davis Creek	ES	—	*TS Elias 10377 & J Morefield*	RSA

[DAC: (MI) T1N R33E; (ES) T1S R33E R34E, T3S R34E]

Reported as present in the range by Little (1971).
DAC: abundant in extensive pinyon-juniper woodlands at middle to high (> 9000 ft) elevations (1993).

White Pine Range (1)
elevation record: 7600 ft
1 record in 1 county (Nye) and reports in another (White Pine)

Towns/Range/Section	Location—Habitat	County	Elevation	Collector	Herbarium/a
—	White Pine Range	NY	7600 ft	*Peter Herlan et al. in 1976*	NSM

[DAC: (WP) T18N R58E R59E, T17N R57E R58E R59E, T16N R57E R58E R59E, T15N R57E R58E R59E R60E, T14N R57E R58E R59E R60E, T13N R57E R58E R59E, T12N R58E R59E; (NY) T12N R58E R59E, T11N R58E R59E]

Reported as present in the range by Little (1971).
DAC: abundant in pinyon-juniper woodlands at low and middle elevations (1994).

White Rock Mountains (1)
elevation record: 6820 ft
1 record in 1 county (Lincoln)

Towns/Range/Section	Location—Habitat	County	Elevation	Collector	Herbarium/a
T5N R69E S36 NE/4	pinyon-juniper woodland near *P. ponderosa*	LI	6820 ft	*Charlet 2098*	RENO

[Little: T5N R70E, T4N R69E R70E, T3N R70E]

Reported as present in the range by Little (1971).
DAC: codominant with *Juniperus osteosperma* in the extensive pinyon-juniper woodlands (1994).

Wilson Creek Range (1)
elevation record: 7980 ft
1 record in 1 county (Lincoln)

Towns/Range/Section	Location—Habitat	County	Elevation	Collector	Herbarium/a
T5N R68E S22 W/4	pinyon-juniper-mountain-mahogany woodland	LI	7980 ft	*Charlet 2093*	RENO
[DAC: T7N R68E, T6N R68E R69E, T5N R66E R67E, T4N R68E R69E, T3N R69E, T2N R69E]					

Reported as present in the range by Little (1971).
DAC: abundant in pinyon-juniper woodlands throughout much of the range (1994).

Windermere Hills (1)
elevation record: 6980 ft
1 record in 1 county (Elko)

Towns/Range/Section	Location—Habitat	County	Elevation	Collector	Herbarium/a
T38N R63E	17.5 mi W of Pequop Summit	EL	6980 ft	*Arthur Rhoads in 1944*	UTC

Reported as present in the range by Little (1971).

Wood Hills (1)
elevation record: 6500 ft
2 records in 1 county (Elko)

Towns/Range/Section	Location—Habitat	County	Elevation	Collector	Herbarium/a
T37N R63E	"East Independence"; E Fork Humboldt River drainage, along US 40	EL	6500 ft	*L Benson 4649*	RSA
T37N R63E	"Independence Mtns"; 10 mi E of Wells, S of US 40	EL	—	*Leila Shultz 2381 & J Shultz*	UTC
[Little: T36N R63E]					

Reported as present in the range by Little (1971).

Worthington Mountains (2)
Lincoln County
[RWR: T1N R57E, T2S R57E]

Reported as present in the range by Little (1971).

OTHER FEATURES IN NEVADA, REPORTS AND RECORDS (4)

Douglas County
elevation record: 5000 ft
1 record

Towns/Range/Section	Location—Habitat	Elevation	Collector	Herbarium/a
T12N R20E	6 mi S of Gardnerville [E Fork Carson River]	5000 ft	*JW Vetter in 1956*	RENO

[DAC: T11N R21E, T10N R21E (both in Double Springs Flat area), T10N R22E, T9N R22E (both West Walker River drainage)]

Elko County
[RWR: T29N R61E R62E (Belcer Buttes)]

Eureka County
[Little: T20N R53E]

Lander County
elevation record: 6000 ft
1 record

Towns/Range/Section	Location—Habitat	Elevation	Collector	Herbarium/a
—	Tobe Roger's Ranch [Smokey Valley]	6000 ft	*EVA Murphy 481*	RENO, UTC

Lincoln County
[Little: T6N R65E, T5N R65E (Dutch John Mountain); T2S R71E, T3S R71E, T4S R70E R71E, T5S R70E R71E, T6S R69E R70E]

Lyon County
1 record

Towns/Range/Section	Location—Habitat	Elevation	Collector	Herbarium/a
T16N R21E	3 mi W of Dayton along Carson-Fallon Hwy	—	*Goodner 81 & Henning*	UTC

[DAC: T7N R25E]

Nye County
[Little: T10N R51E]

Unknown Location in Nevada
1 record

Towns/Range/Section	Location—Habitat	Elevation	Collector	Herbarium/a
—	Bubb's Creek	—	*Mrs Joseph Clemens in 1910*	RSA

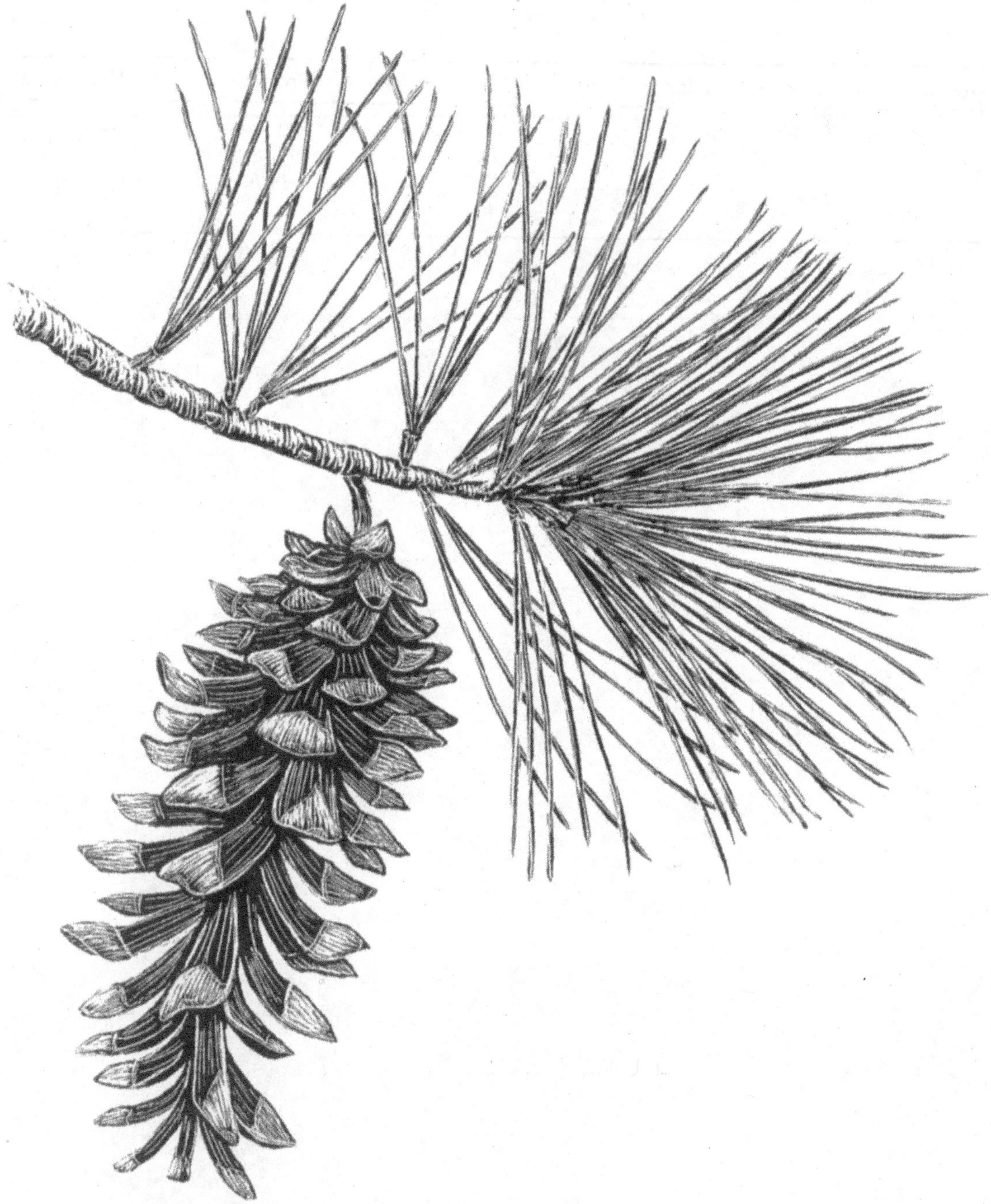

Pinus monticola Douglas ex D. Don

common name: western white pine
elevation: 5400–9200 ft (1645–2805 m)
zones: Canadian, Arid Transition, Upper Sonoran
27 Records for 4 counties (CC, DO, ST, WA) and 4 mountain ranges

Western white pine, an important commercial tree in the Pacific Northwest, is abundant and sometimes dominant in British Columbia, the northern Rocky Mountains, the Cascades, the Sierra Nevada, and the high mountains of the central Columbia Plateau. In the Great Basin, its presence is restricted to the California-Nevada border.

Pinus monticola is found in 4 Great Basin mountain ranges in far western Nevada, all west of the Lahontan Basin. The species is a major component of the Canadian Zone forests in the Carson Range of the Sierra Nevada and occurs sporadically in 3 other mountain ranges in western Nevada, and is present in 4 Nevada counties altogether (Carson City, Douglas, Storey, and Washoe). In the Virginia Range (Storey and Washoe Counties), the species is restricted to altered andesite soils surrounded by Upper Sonoran Zone pinyon-juniper woodlands. In the Pine Nut Range (Douglas County), western white pine occurs at high elevations (> 9000 ft) at the upper end of mountain brush communities in the Canadian Zone. Despite its narrow geographic range in Nevada, western white pine grows over a rather wide elevation range of 3800 ft (1160 m).

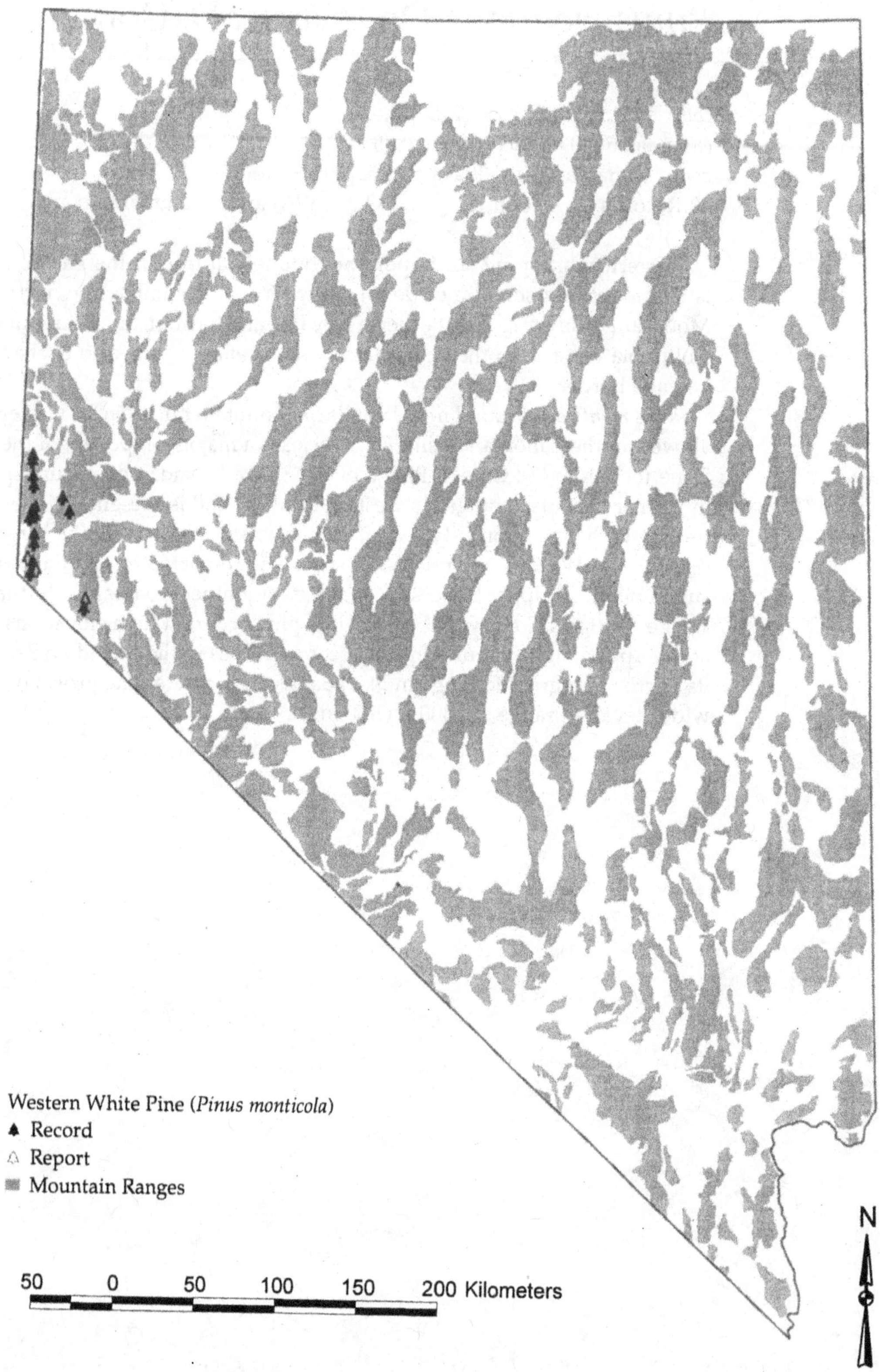
Western White Pine (*Pinus monticola*)
Record
Report
Mountain Ranges
50 0 50 100 150 200 Kilometers
N

Carson Range (1)
elevation record: 7200–9000 ft
13 records in 3 counties (Carson City, Douglas, and Washoe)

Towns/Range/Section	Location—Habitat	County	Elevation	Collector	Herbarium/a
T13N R18E S13 N/8	N of Castle Rock, in mixed conifer forest	DO	7700 ft	*Charlet 2169 & T Charlet*	DUKE, RENO
T14N R19E S30	Genoa Peak	DO	8700 ft	*Ann Pinzl in 1973*	NSM
T15N R19E S6 NW/4	N of Marlette Peak	WA	8550 ft	*Ann Pinzl 8417 Kimbraugh & Weaver*	NSM, BRY
T15N R19E S19	below Snow Valley Peak	CC	7660 ft	*Ann Pinzl 2541*	NSM
T15N R19E S19	Snow Valley Peak	CC	8500 ft	*Peter Herlan in 1971*	NSM
T16N R18E S2 NW/4	Slide Mountain, lower western slopes below Tahoe Meadows	WA	8000 ft	*Charlet 2061 & T Charlet*	DUKE, RENO
T17N R19E S18	Mount Rose, second-growth red fir forest, SE slopes	WA	8200 ft	*Billings 1439*	RENO
T17N R18E	Mount Rose	WA	8000 ft	*PA Lehenbauer in 1927*	RENO
T17N R18E S24	S slope Mount Rose, w/whitebark & lodgepole pines	WA	9000 ft	*Billings 1261*	RENO
T17N R19E S31	S side, Slide Mountain	WA	7500 ft	*AA Heller 10665*	RENO
T17N R19E S18	Galena Creek headwaters	WA	8000 ft	*WA Archer 6680*	RENO
T18N R18E S1	Hunter Creek canyon	WA	7200 ft	*AA Heller 10356*	RENO
T19N R18E	Hunter Creek, near sawmill	WA	—	*PA Lehenbauer in 1925*	RENO

[DAC: (DO) T14N R18E, T13N R19E]

Reported as present in the range by Little (1971).
DAC: common in upper montane forests with *Pinus jeffreyi* and *Abies concolor*, and in subalpine woodlands with *A. magnifica*, *P. contorta*, and/or *P. albicaulis*. In the mixed conifer forest between Spooner Summit and Daggett Pass (Douglas County), there are places where up to 90% of the *Abies magnifica* and *Pinus jeffreyi* have died, and *P. monticola* along with a few *P. contorta* are the only living conifers that remain (1971–1995).

Pah Rah Range (5)
Washoe County

Reported as present in the range by Little (1971).
DAC: this is almost certainly an erroneous report. I have seen no material in herbaria and have not seen the species on the mountain. See note below under "Virginia Range" (1975–1994).

Peavine Mountain (1)
elevation record: 7100–8270 ft
3 records in 1 county (Washoe)

Towns/Range/Section	Location—Habitat	County	Elevation	Collector	Herbarium/a
T20N R18E S23	Peavine Mountain	WA	7100 ft	*M Williams & Tiehm 75-81-6*	CAS
T20N R18E S23	N of lookout	WA	8200 ft	*M Williams 72-45-3*	RENO
T20N R18E S23	near peak	WA	8270 ft	*Billings 1452 & Wayne Roderik*	RENO

Not reported as present in the range by Little (1971).
DAC: rare; known only from a few individuals near the peak (1971–1995).

Pine Nut Mountains (1)
elevation record: 9000–9200 ft
4 records in 1 county (Douglas)

Towns/Range/Section	Location—Habitat	County	Elevation	Collector	Herbarium/a
T11N R22E	small, isolated grove in upper sagebrush zone, NE slopes Bald Mtn	DO	[9300 ft]	*Billings 1836*	RENO
T11N R22E S16	ridge NE of Bald Mountain Peak	DO	9000 ft	*Tiehm 7360 & Lavin*	CAS, UTC, RENO
T11N R22E S16	Bald Mountain	DO	9200 ft	*Lavin 4225 & Tiehm*	UTC
T11N R22E S16	Bald Mountain, NE slope in snowbank	DO	9000 ft	*RJ Bandoni in 1953*	DUKE
[DAC: T12N R22E S27 (Galena Peak)]					

Reported as present in the range by Little (1971).
DAC: a previously unreported stand of 5 individuals was seen near Galena Peak (T12N R22E) (1991), and the stand from which the above collections were made was observed with binoculars at Galena Peak. Bald Mountain is only 9205 ft tall, so the 9300-ft elevation reported must be in error (1989–1994).

Virginia Range (1)
elevation record: 5400–6800 ft
7 records in 2 counties (Storey and Washoe)

Towns/Range/Section	Location—Habitat	County	Elevation	Collector	Herbarium/a
T17N R21E S20	Christmas Tree Canyon	ST	6800 ft	*J Darling 4*	RENO
T17N R21E S20	in canyon 1 mi NW of Virginia City	ST	6800 ft	*Billings 1771*	RENO
T17N R21E S20	in canyon 1 mi NW of Virginia City	ST	6800 ft	*Billings 1773*	RENO
T17N R21E S20 NW/4	N slope of Cedar Hill in Cedar Hill Canyon NW of Virginia City	ST	6780 ft	*Jan Nachlinger 1494 & Billings*	DUKE
T18N R20E S26 NE/4	N of Geiger Grade lookout on altered andesite	WA	5400 ft	*Billings 1774*	RENO
T18N R20E S26 NE/4	in canyon NE of lookout, Geiger Grade (5700 ft)	WA	5400 ft	*VK Johnson in 1949*	RENO
T18N R20E S26 NE/4	2.2 mi ENE of Steamboat Spring (5500 ft)	WA	5400 ft	*TC Adams 206 in 1938*	RSA

Not reported as present in the range by Little (1971).
DAC: Perhaps *Pinus monticola* is missing from Little (1971) because "Virginia Range" was used for "Virginia Mountains" on the map, and the bubble was misplaced over the Pah Rah Range instead. The T18N R20E collections are all from the same location; 5400 ft is the correct elevation, as recorded on Billings's label (Billings, in litt. 1987) and verified by me (1973–1995).

var. *ponderosa*

Pinus ponderosa Douglas ex Lawson & C. Lawson

varieties in Nevada:
P. ponderosa var. *ponderosa* [Pacific ponderosa pine]
P. ponderosa var. *scopulorum* Engelmann [Rocky Mountain ponderosa pine]
common name: ponderosa pine, western yellow pine, yellow pine
elevation: 4200–9000 ft (1280–2745 m)
zones: Canadian, Arid Transition, Upper Sonoran
86 Records for 7 counties (CL, LI, MI, NY, ST, WA, WP)
Reports for 23 mountain ranges in Nevada, Records for 22 of these
New Reports for 3 mountain ranges, New Records for 2 of these
P. ponderosa var. *ponderosa* occurs in 14 mountain ranges; *P. ponderosa* var. *scopulorum* in 11

Next to Douglas-fir, ponderosa pine has the widest distribution of all western conifer trees. It occurs in Canada, 16 western states of the United States, and south to the mountains of tropical Mexico.

The typical variety and the Rocky Mountain variety (*Pinus ponderosa* var. *scopulorum*) occur in a total of 23 Nevada mountain ranges in 7 counties (Clark, Lincoln, Mineral, Nye, Storey, Washoe, and White Pine). In addition, Critchfield (1984b) recognizes a "North Plateau" race of ponderosa pine found in northern Washoe County near the Oregon border and extending north into Oregon, Washington, Idaho, and Montana, and in extreme northeastern California.

The Rocky Mountain variety of ponderosa pine (var. *scopulorum*) is more common; it occurs in eastern and southern Nevada, where it is can be found in canyons with perennial streams in many ranges. In western Nevada, var. *ponderosa* is a minor component of the Arid Transition Zone forests in the Carson Range, and small stands are present at many Great Basin locations within and west of the Lahontan Basin on altered andesite islands in the midst of pinyon-juniper woodland in the Upper Sonoran Zone. On the basis of long (12–18 cm) needles borne exclusively in threes, *Pinus ponderosa* var. *ponderosa* is also found in the Highland and Wilson Creek Ranges in Lincoln County, and the White Pine and Quinn Canyon Ranges in Nye County. I found only var. *ponderosa* in the White Pine and Highland Ranges, and while both varieties occur in the Quinn Canyon and Wilson Creek Ranges, the distinct varieties grow in different stands. The presence of var. *ponderosa* in Nye and Lincoln Counties has never before been reported, and the variety should be looked for in the Spring Mountains, Sheep Range, and Snake Range. I report new records for var. *ponderosa* in the Fort Sage Mountains and Junction House Range (Washoe County), and a new report for var. *scopulorum* in Irish Mountain (Lincoln County).

Planted ponderosa pines can be found in the Jarbidge, Painted Point, and Santa Rosa Ranges. Only the Jarbidge population is not yet producing cones. The species

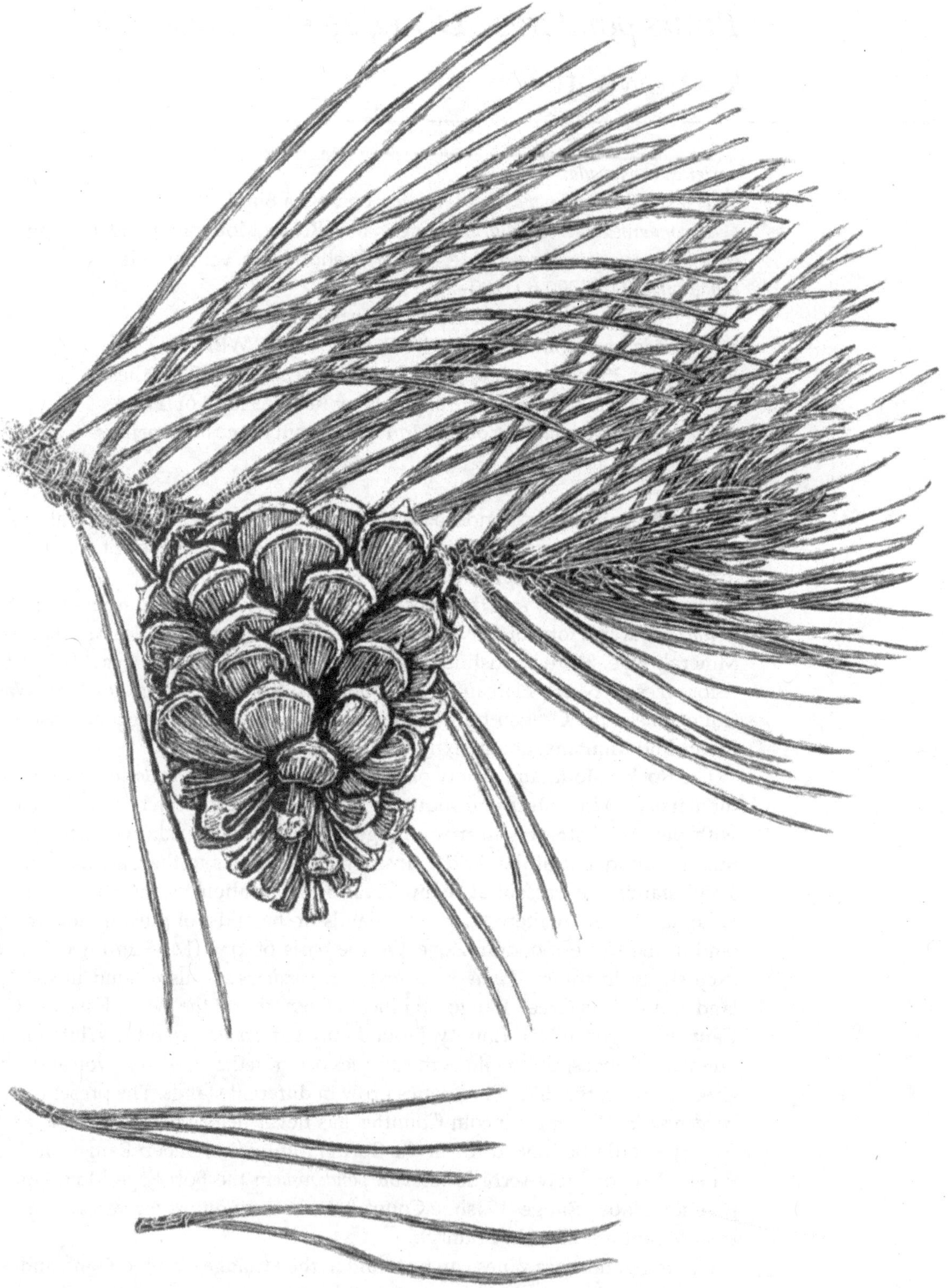

var. *scopulorum*

may escape and naturalize from these locations, but I have not seen any seedlings. The broad elevation range of 4800 ft (1465 m) reflects the broad geographic range of the species in Nevada.

Bodie Mountains (1)
elevation record: 6220–6760 ft
3 records in 1 county (Mineral)

Towns/Range/Section	Location—Habitat	County	Elevation	Collector	Herbarium/a
T6N R26E S3 SW/4	above Red Wash in altered andesite, lower end	MI	6420 ft	*Charlet 2198 & Ann Buell*	DUKE, RENO
T6N R26E S3 SW/4	above Red Wash in altered andesite, upper end	MI	6760 ft	*Charlet 2200 & Ann Buell*	DUKE, RENO, UC
T6N R26E S6 NW/4	above Sonoma Canyon, w/*P. jeffreyi* in altered andesite	MI	6220 ft	*Jan Nachlinger 1497 & Billings & Billings*	DUKE

Not reported as present in the range by Little (1971).
DAC: the taxon is var. *ponderosa.*
DAC: At least 2 reproductive individuals and 50 saplings occur in the Red Wash stand. Several other altered andesite sites with yellow pine occur in Red Wash, and *P. ponderosa* may occur in these (1991–1995).

Carson Range (1)
elevation record: 6000–7800 ft
6 records in 1 county (Washoe)

Towns/Range/Section	Location—Habitat	County	Elevation	Collector	Herbarium/a
T16N R18E S16	Carson Range, Incline Village	WA	6300 ft	*JT Howell 39791*	CAS
T16N R18E S18	3 mi E of Brockway	WA	6300 ft	*Nichols 679 & Lund*	RENO
T17N R19E S4	Jones Creek, near 1950s burn area	WA	6000 ft	*Charlet 1811 & T Charlet*	DUKE
T17N R19E S4	Jones Creek, near 1950s burn area	WA	6000 ft	*Charlet 1812 & T Charlet*	DUKE
T17N R19E S17	Galena Creek, on NW slope w/*P. washoensis* & *P. jeffreyi* about 100 yd NW of upper terminus of ski tow, N side of Mount Rose Rd, opposite Redmond Lodge; 2.8 mi by road SW of Galena Guard Station	WA	7000 ft	*JR Haller 10341-14*	IFGP
T18N R19E S18	head of Evans Creek	WA	7800 ft	*Nichols 641 & Lund*	RENO
[SB Vander Wall: (WA) T17N R19E S16 (S of Franktown)]					

Reported as present in the range by Little (1971).
DAC: var. *ponderosa* is uncommon but widespread at middle and lower elevations. Stephen Vander Wall reported the species along the lower east flank of the range, just south of Franktown (pers. comm. 1994) (1971–1995).

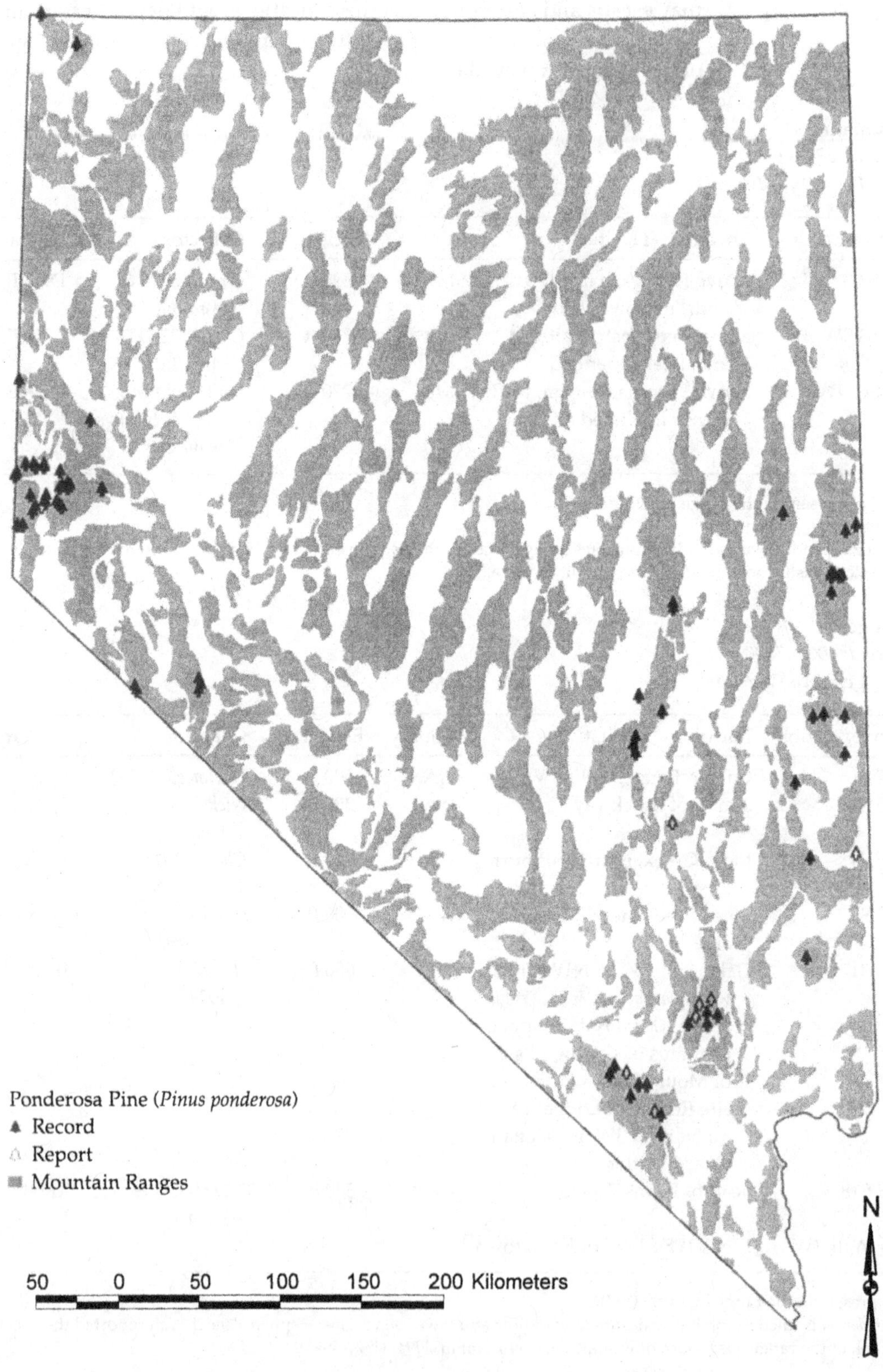
Ponderosa Pine (*Pinus ponderosa*)
Record
Report
Mountain Ranges
N
50 0 50 100 150 200 Kilometers

Clover Mountains (1)
elevation record: 5500–5900 ft
2 records in 1 county (Lincoln)

Towns/Range/Section	Location—Habitat	County	Elevation	Collector	Herbarium/a
T5S R67E S23	9 mi S of Caliente, near Cabin Pines campground	LI	5500 ft	*Williams 83-69-4 & Tiehm*	RENO
T5S R67E S23	Cabin Pines camp	LI	5900 ft	*Ann Pinzl 8626*	RSA, NSM
[RWR: T5S R71E (SW of Schroeder Lake, Beaver Dam State Park)]					

Reported as present in the range by Little (1971).
DAC: var. *scopulorum* is the taxon here.

****Fort Sage Mountains (1)**
elevation record: 4940–5180 ft
2 records in 1 county (Washoe)

Towns/Range/Section	Location—Habitat	County	Elevation	Collector	Herbarium/a
T25N R18E S8 NW/4	unnamed canyon, N slope of range	WA	5180 ft	*Charlet 2159*	DUKE, RENO, UC, BRY
T25N R18E S8 NW/4	unnamed canyon, N slope of range	WA	4940 ft	*Charlet 2160*	DUKE, RENO, UC

Not reported as present in the range by Little (1971).
DAC: a stand of more than 35 large reproductive individuals (var. *ponderosa*) grows in a steep canyon on the north side of the range, less than 1 km from the California border. Many saplings and seedlings are present in the stand (1993–1994).
New record for the species in the range.

Grant Range (1)
elevation record: 7440 ft
2 records in 1 county (Nye)

Towns/Range/Section	Location—Habitat	County	Elevation	Collector	Herbarium/a
T5N R58E S15	Scofield Canyon	NY	7440 ft	*Charlet 2013*	DUKE, RENO, UC
T6N R57E	Irwin Canyon, common	NY	—	*RS Peterson 63-328*	UTC
[RWR: T5N R58E S18 S20 (Scofield Canyon)]					

Reported as present in the range by Little (1971).
DAC: I found only var. *scopulorum*, which is locally common in both Irwin and Scofield Canyons; var. *ponderosa* is present in at least 2 drainages in the adjacent Quinn Canyon Range, and may yet be found here (1994).

Highland Range (1)
elevation record: 7160 ft
1 record in 1 county (Lincoln)

Towns/Range/Section	Location—Habitat	County	Elevation	Collector	Herbarium/a
T1N R66E S35 NW/4	stand of 20 reproductives surrounded by pinyon-juniper-oak woodlands	LI	7160 ft	*Charlet 1994*	DUKE, RENO, UC

Reported as present in the range by Little (1971).
DAC: these large, impressive trees have long needles (13–17 cm) almost exclusively in threes, and so are var. *ponderosa* (1994).

****Irish Mountain (3)**
Lincoln County
[McLane: T3S R59E]

Not reported as present in the range by Little (1971).
DAC: uncommon; small stands in canyons near the summit were reported by Alvin McLane (pers. comm. 1993).
New report for the species in the range.

Jarbidge Mountains (7)
Elko County

DAC: 3 trees, now 30–50 years old, were planted at the graveyard north of Jarbidge town, near the W Fork Jarbidge River (1988–1994).

****Junction House Range (1)**
elevation record: 4780–5085 ft
2 records in 1 county (Washoe)

Towns/Range/Section	Location—Habitat	County	Elevation	Collector	Herbarium/a
T20N R19E S13 S/4	N of radio towers, lower end of large stand, altered andesite	WA	4780 ft	*Charlet 1798*	DUKE
T20N R19E S24 SW/4	off Dandini Dr, W of Desert Research Inst, E of US 395	WA	5085 ft	*Charlet 1774*	DUKE

Not reported as present in the range by Little (1971).
DAC: var. *ponderosa* is common in 2 large stands in altered andesite at the southern end of the range (1989–1995).

Mormon Mountains (1)
elevation record: 6560 ft
1 record in 1 county (Clark)

Towns/Range/Section	Location—Habitat	County	Elevation	Collector	Herbarium/a
T11S R67E S21	Davies Spring, 11 mi E of Meadow Valley Wash summit; SE of Mormon Peak	CL	6560 ft	*F Landau et al. MM213*	UNLV

Not reported as present in the range by Little (1971).

Mosquito Mountains (1)
elevation record: 5000 ft
1 record in 1 county (Washoe)

Towns/Range/Section	Location—Habitat	County	Elevation	Collector	Herbarium/a
T47N R18E S15 SW/4	bottom of Twentymile Creek, 41°59.5'N, 119°38.5'W	WA	5000 ft	*Critchfield sn*	IFGP

Reported as present in the range by Critchfield & Allenbaugh (1969) and Little (1971).
DAC: these individuals of var. *ponderosa* and the stand in the Painted Point Range (Washoe County) are the only known members of the North Plateau race of ponderosa pine (as defined by Critchfield 1984b) in Nevada (1994).

Pah Rah Range (1)
elevation: 4950–5770 ft
2 records in 1 county (Washoe)

Towns/Range/Section	Location—Habitat	County	Elevation	Collector	Herbarium/a
T23N R22E S26	canyon N of Pah Rah Mountain, W of Nixon, S side Pyramid Lake [Tom Anderson Canyon]	WA	—	*LaRivers in 1948*	RENO
T23N R22E S26	"yellow pine cyn," altered rock, 60 ft tall, 30 in. dbh	WA	—	*Billings 2046*	RENO
[DAC: T23N R22E S27 NE/4, T23N R22E S23 SE/4]					

Not reported as present in the range by Little (1971).
DAC: *LaRivers in 1948* and *Billings 2046* are from a large stand (200+) of big reproductive trees (var. *ponderosa*) on the northern aspect of the canyon, immediately west of Tom Anderson Canyon, in altered andesite at an old mine site, at 4950–5770 ft. I found a woodrat midden at 5495 ft, < 50 m from the nearest ponderosa, and reported it to Peter Wigand at the Desert Research Institute, who later collected it (TAC260690PEW1) and found ponderosa pine needles in strata data at 1700 yr BP. The 2 other reported sites are nearby, and each contains a few trees (1975–1994).

Painted Point Range (1)
elevation record: 6500 ft
2 records in 1 county (Washoe)

Towns/Range/Section	Location—Habitat	County	Elevation	Collector	Herbarium/a
T45N R21E S8	Bald Mountain headquarters, Sheldon NWR, 41°50'N 119°38.5'W	WA	6500 ft	*Critchfield & Allenbaugh 1-9/3/65*	IFGP
T45N R21E S8 SW/4	Bald Mountain [Critchfield label (1965): "Tree 16 inch dbh, 35 ft tall. 50 annual rings @ dbh, est. total age to be 70yr. This stand of North Plateau–type ponderosa consists of 4 older trees and ca. 40 scrubby seedlings. Nearest ponderosa are in the Warner Mts."]	WA	6500 ft	*Brant Rogers 883 & Tiehm*	RENO

Reported as present in the range by Little (1971).
DAC: the stands here and in the Mosquito Mountains of Washoe County are the only native stands of the Northern Plateau race of var. *ponderosa* known in Nevada. There is also a population of about 15 reproductive-age individuals that were planted in the 1930s at the old Bald Mountain headquarters, Sheldon National Wildlife Refuge (1994).

Peavine Mountain (1)
elevation record: 4600–6200 ft
8 records in 1 county (Washoe)

Towns/Range/Section	Location—Habitat	County	Elevation	Collector	Herbarium/a
T19N R18E S7	Verdi fish hatchery (determined Billings)	WA	4600 ft	*R Sumner 109*	RENO
T19N R18E S7	summit, Dog Valley Grade near Verdi	WA	5500 ft	*JW Vetter in 1956*	RENO
T20N R18E S13	N slope along creek, altered andesite	WA	5800 ft	*Charlet 2164*	DUKE, RENO, UC
T20N R19E S16	1 mi S of Black Springs, 150 ft W of *2052*	WA	5450 ft	*Billings 2053*	RENO
T20N R19E S16	0.25 mi W of *2053*	WA	5450 ft	*Billings 2054*	RENO
T20N R19E S22 S/4	by North Virginia Street, E side of mountain	WA	4860 ft	*Charlet 1800*	DUKE
T20N R20E S36	Peavine Mountain, top of altered knob, S side of Peavine	WA	6200 ft	*Billings 2064*	RENO
—	Peavine Mountain	WA	5600 ft	*JT Howell 48640*	CAS

Reported as present in the range by Little (1971).
DAC: var. *ponderosa* is restricted to altered andesite soils here within the lower sagebrush–grass zone (1971–1995).

Quinn Canyon Range (1)
elevation record: 6380–6980 ft
4 records in 1 county (Nye)

Towns/Range/Section	Location—Habitat	County	Elevation	Collector	Herbarium/a
T3N R56E S14 SW/4	Sawmill Canyon	NY	6840 ft	*Charlet 2011*	DUKE, RENO, UC
T3N R56E S36 NW/4	Sawmill Canyon	NY	6380 ft	*Charlet 2012*	DUKE
T4N R56E S36	Little Cherry Creek	NY	6980 ft	*Charlet 2007*	DUKE, RENO
T4N R56E S36	Little Cherry Creek	NY	6980 ft	*Charlet 2008*	DUKE

Reported as present in the range by Little (1971) and Lewis (1975).
DAC: both var. *scopulorum* (*Charlet 2007, 2008*) and var. *ponderosa* are present but easily distinguished because var. *scopulorum* has short (7–12 cm), frequently double needles, while var. *ponderosa* has much longer (13–22 cm) needles exclusively in threes (1994).

Santa Rosa Range (7)
Humboldt County

DAC: about 10 trees planted near the head of Singas Creek have reached reproductive age, but no seedlings were seen. The species could naturalize from this location (6/29/1993).

Schell Creek Range (1)
elevation record: 7500 ft
1 record in 1 county (White Pine)

Towns/Range/Section	Location—Habitat	County	Elevation	Collector	Herbarium/a
T17N R66E	Taft Creek, 5 trees	WP	7500 ft	*Sandow in 1953*	RENO

Not reported as present in the range by Little (1971).
DAC: rare; var. *scopulorum* is apparently the only variety present (1994).

Seven Lakes Mountain (5)
Washoe County

Towns/Range/Section	Location—Habitat	County	Elevation	Collector	Herbarium/a
T24N R18E	400 yd E of major Red Rock formation, tuff slope, not fruiting	WA	6500 ft	*WA Bailey in 1951*	RENO

Not reported as present in the range by Little (1971).
DAC: I visited the locality on the label and saw some 20–25 yellow pine individuals, 2–10 m tall. I did not reach the yellow pines at 6500 ft near the top of the ridge and cannot comment on them, but all of the yellow pines I saw at the lower elevations were *P. jeffreyi*, not *P. ponderosa*. Based on this and on the distinct stomatal lines of this material, this collection is probably *P. jeffreyi*. Without visiting the yellow pines at 6500 ft, however, I cannot say for certain that the species does not occur here. It may yet be found (1990–1994).

Sheep Range (1)
elevation record: 6600–8040 ft
4 records in 1 county (Clark)

Towns/Range/Section	Location—Habitat	County	Elevation	Collector	Herbarium/a
T15S R61E S17	w/pinyon in saddle, Mormon Wells Rd	CL	7500 ft	*Beatley, Cronquist & Wells 1784*	CAS, RENO
T15S R60E S18	Wire Grass Spring	CL	8040 ft	*BH Banta in 1950*	RENO
T14S R61E S29	Sawmill Canyon	CL	8000 ft	*BH Banta in 1950*	RENO
T14S R61E S36	Mormon Wells Rd, Desert National Wildlife Range	CL	6600 ft	*Ann Pinzl 3836*	NSM
[A. McLane: T14S R60E S25 (Basin Spring drainage), T14S R61E S17 (Timber Canyon), T15S R60E S10 (Pine Spring)]					

Reported as present in the range by Little (1971).
DAC: Alvin McLane told me (pers. comm. 1994) that Mormon pioneers from Moapa cut many of the ponderosa pines on the east side of the range for construction, and so the species probably was much more common before the Moapa settlement was built. Only var. *scopulorum* has been collected here, but var. *ponderosa* may yet be found as it occurs north of the Sheep Range in the Quinn Canyon, Highland, Wilson Creek, and White Pine Ranges.

Snake Range (1)
elevation record: 6500–8500 ft
8 records in 1 county (White Pine)

Towns/Range/Section	Location—Habitat	County	Elevation	Collector	Herbarium/a
T12N R69E	10 mi above fish hatchery in Snake Canyon	WP	—	*I Ting 19374*	RENO
T13N R69E S8	along creek in pinyon-juniper, 2 mi above Lehman Caves	WP	8500 ft	*Billings 1896*	RENO
T13N R69E S15	Lehman Caves	WP	8500 ft	*PA Lehenbauer sn*	RENO
T13N R69E S7	Lehman Creek, E slope Snake Range	WP	8000 ft	*Percy Train 1160*	RENO
T13N R69E	Mount Wheeler [Baker Creek]	WP	6500 ft	*P Herlan in 1973*	NSM
T13N R69E S14	Baker Canyon, 5 mi SW of Baker	WP	6500 ft	*SJ Preece, Jr 1457*	UT
T16N R69E S24	E of Mount Moriah in Hendrys Creek Canyon	WP	8000 ft	*JC Lyman 878 et al.*	UCR
T16N R70E S10	Mount Moriah, Hampton Creek, abundant	WP	7300 ft	*Charlet 2111 & T Charlet*	DUKE, RENO, UC

Reported as present in the range by Little (1971).
DAC: var. *scopulorum* is common along streams and probably occurs at many additional locations in canyons along the east slope (1994).

Spring Mountains (1)
elevation record: 4200–9000 ft
9 records in 1 county (Clark)

Towns/Range/Section	Location—Habitat	County	Elevation	Collector	Herbarium/a
T18S R55E S3	upper Clark Canyon	CL	7000–7600 ft	*Beatley 10717 & Reveal*	RENO
T18S R55E S20	dominant near Trough Spring, S slope of NW Spring Mts	CL	8100 ft	*Beatley 11223 & Reveal*	RENO
T19S R56E S34	between summits of Charleston Peak	CL	9000 ft	*L Benson 15289*	RSA
T19S R57E S7	Deer Creek, 5-6 mi NE of Charleston Peak	CL	8500 ft	*LaRivers 501 & Hancock*	RENO, DUKE
T19S R57E S10 NE/4	——	CL	8700 ft	*AC Sanders 6877 et al.*	UCR
T21S R58E S16	Pine Creek	CL	4200 ft	*Ann Pinzl 4904*	NSM
T22S R58E S22	Middle Mud Springs	CL	7500 ft	*D Atwood 7795*	BRY
—	wash & hillsides	CL	7445 ft	*IW Clokey 5382*	RENO, UT
—	hillsides & bottom	CL	7445–7710 ft	*IW Clokey 7820*	UT, OGDF, BRY, RENO

[RWR: T21S R58E (Icebox Canyon)]
[DAC: T18S R56E (upper Lee Canyon)]

Reported as present in the range by Little (1971).
DAC: var. *scopulorum* is abundant along streams and on forested slopes with *Abies concolor. Pinzl 4904* is the low-elevation record and *Benson 15289* is the high-elevation record for the species in the state (1982 & 1993).

Toquima Range (7)
Nye County

Planted young trees are present in Jefferson Canyon (John Brack, Toiyabe NF, Tonopah District, pers. comm. July 1994).

Virginia Range (1)
elevation record: 4400–6400 ft
16 records in 2 counties (Washoe and Storey)

Towns/Range/Section	Location—Habitat	County	Elevation	Collector	Herbarium/a
—	Virginia Range	WA	6400 ft	*Billings 2019*	RENO
T17N R21E S6 N/4	above old Geiger Grade, along new Geiger Grade	WA	6240 ft	*Charlet 1781*	DUKE
T18N R20E S26 S/4	"White gold" stand, S aspect, altered andesite	WA	5300 ft	*Charlet 1813 & T Charlet*	DUKE, RENO, UC
T18N R20E S26 S/4	"White gold" stand, S aspect, altered andesite	WA	5300 ft	*Charlet 1815 & T Charlet*	DUKE
T18N R20E	[new] Geiger Grade	WA	—	*Peter Herlan sn*	NSM
T18N R20E	new Geiger Grade	WA	4700 ft	*WA Archer 5075*	RENO
T18N R20E S36	several trees in altered area in canyon N of Geiger Grade, N aspect	WA	—	*Billings 2057*	RENO
T19N R20E S35 NE/4	S of Hidden Valley Park, in large altered andesite stand	WA	5385 ft	*Charlet 2201 & James Morefield*	DUKE, RENO, UC
T19N R20E S36	in large altered andesite stand	ST	5220 ft	*Charlet 1803 & T Charlet*	DUKE
T19N R21E S21	Long Valley, lower Lago-marsino Canyon; altered andesite; farthest N	ST	4400 ft	*Charlet 1775*	DUKE
T19N R21E S30	top of ridge above Hidden Valley	ST	5660 ft	*Charlet 1809*	DUKE, RENO, UC
T19N R21E S27 S/4	Lagomarsino Canyon, 3.7 road miles S of Lockwood, S aspect, altered	ST	4520 ft	*Charlet 1776*	DUKE
T19N R21E S30	NW end of range, along powerline rd	ST	4960 ft	*Charlet 1807*	DUKE, RENO, UC, BRY
T19N R213 S32 NE/4	W of Washington, altered andesite	ST	5340 ft	*Charlet 1780*	DUKE, RENO, UC
T19N R21E S33 S/4	W of Washington, altered andesite	ST	5100 ft	*Charlet 1778*	DUKE, RENO, UC
T19N R23E S34	10 mi SE of Clarks Station, 0.5 mi N of Ramsey; extreme E ST County	ST	5700 ft	*Billings 1426*	RENO

Reported as present in the range by Little (1971).
DAC: var. *ponderosa*, the most abundant yellow pine in the Virginia Range, is uncommon but locally dominant and widespread here. It occurs exclusively on altered andesite within sagebrush-grass and pinyon-juniper zones. Billings (1950) and DeLucia & Schlesinger (1990) studied the physiology of this phenomenon (1973–1995).

Wassuk Range (1)
elevation record: 5875–6540 ft
6 records in 1 county (Mineral)

Towns/Range/Section	Location—Habitat	County	Elevation	Collector	Herbarium/a
T6N R30E S6	S arm of Alum Creek	MI	6540 ft	*Charlet 1970*	DUKE, RENO
T6N R30E S6	Alum Creek	MI	6520 ft	*Charlet 1969*	DUKE
T7N R30E S29 SW/4	Alum Creek	MI	5875 ft	*Ann Pinzl 9955*	NSM, BRY
T7N R30E S29 S/4	Alum Creek	MI	5980 ft	*Charlet 1967*	DUKE, RENO, UC
T7N R30E	Alum Creek campground	MI	6075 ft	*Ann Pinzl 4454*	NSM
T7N R30E S29 S/4	Alum Creek	MI	5980 ft	*Charlet 1968*	DUKE, RENO, UC, BRY

Not reported as present in the range by Little (1971).
DAC: var. *ponderosa* was abundant along Alum Creek, the only location known for the species in this range. Much of this stand was destroyed by a fire in 1995 (1990–1995).

White Pine Range (1)
elevation record: 6640–6880 ft
2 records in 1 county (Nye)

Towns/Range/Section	Location—Habitat	County	Elevation	Collector	Herbarium/a
T11N R59E S5	along stream surrounded by pinyon-juniper	NY	6640 ft	*Charlet 1986*	DUKE, RENO, UC, BRY
T12N R59E S29	along stream w/*J. scopulorum*	NY	6880 ft	*Charlet 1990*	DUKE, RENO

Reported as present in the range by Little (1971).
DAC: these collections are from the only stand I know in this range. All the individuals I have seen are var. *ponderosa,* determined on the basis of their long leaves (13–20 cm) exclusively in bundles of 3 (1994).

White Rock Mountains (1)
elevation record: 6820 ft
1 record in 1 county (Lincoln)

Towns/Range/Section	Location—Habitat	County	Elevation	Collector	Herbarium/a
T5N R69E S36 NE/4	small grove on white rock surrounded by pinyon-juniper	LI	6820 ft	*Charlet 2097*	DUKE, RENO, UC

Reported as present in the range by Little (1971).
DAC: var. *scopulorum* is between Camp Valley Creek and Upper Burnt Canyon Spring (1994).

Wilson Creek Range (1)
elevation record: 6000–9000 ft
3 records in 1 county (Lincoln)

Towns/Range/Section	Location—Habitat	County	Elevation	Collector	Herbarium/a
T5N R67E S36	Wilson Mountain, W slope (determined Billings)	LI	8600–9000 ft	*Percy Train 2559*	RENO
T5N R68E S27 W/4	grove of 20+ trees, also *Abies concolor* & *Arctostaphylos*	LI	8140 ft	*Charlet 2096*	DUKE
T2N R70E S7	Meadow Valley Wash, 4 mi E of Horsethief Mountain; red sandstone	LI	6000 ft	*Charlet 2100*	DUKE, RENO, UC
[RWR: T4N R68E S34]					

Reported as present in the range by Little (1971).
DAC: *Train 2559* and *Charlet 2096* are var. *scopulorum,* and *Charlet 2100* is var. *ponderosa* (1994).

OTHER LOCATION IN NEVADA, RECORD

Washoe County
elevation record: 5000 ft
1 record

Towns/Range/Section	Location—Habitat	Elevation	Collector	Herbarium/a
T18N R20E S32	Steamboat Springs, S end of Truckee Meadows	5000 ft	*Billings 1642*	RENO

DAC: var. *ponderosa* occurs here in a large stand on altered andesite, between the Carson Range and the Virginia Range (1971–1995).

Pinus washoensis Mason & Stockwell

common name: Washoe pine
elevation: 6240–8500 ft (1900–2590 m)
zone: Canadian
9 Records for 1 county (WA) and 2 mountain ranges
New Record for 1 mountain range
New Record for the Intermountain Region

Washoe pine was formerly known only from Mount Rose (Washoe County) in the Carson Range, immediately northeast of Lake Tahoe, where it was discovered in 1938 and described by Mason & Stockwell (1945). Since then, populations of Washoe pine have been found and collected at 4 northeastern California sites: Mount Lassen (Lanner 1984), Mount Babbit (Critchfield & Allenbaugh 1965), and in a sand dune at 5540 ft in Madelin Mesa, 6 miles west of the Nevada border (*Charlet 1924* at DUKE, RENO, BRY, UC) (all in Lassen County); a large population occurs in the Warner Mountains (Modoc County) (Haller 1961). The species has been reported to occur as far north as southern British Columbia (Haller, in Critchfield 1984b).

Washoe pine is one of the most restricted of all Nevada conifers, both ecologically and geographically, and there are certainly fewer individuals of this species than of any other conifer in the state. It is common in Nevada only from 7240 to 8500 ft on lateral moraines along Galena Creek on Mount Rose in the Carson Range (Washoe County), an area that was clear-cut in the 1860s. The species was excluded from *Intermountain Flora* (Cronquist et al. 1972) because it had not been found east of US 395 except in Modoc County in northeastern California, a region arguably excluded from the Intermountain Region as defined by Cronquist et al. (1972). I present the first record for the species within the Intermountain Region: a single very large individual found at 6240 ft in the Mosquito Mountains in northern Washoe County (*Charlet 1840* at DUKE; *Charlet 1932* at DUKE, RENO, UC, BRY, UNLV). The elevation range of Washoe pine in the state is 2260 ft (690 m). The species may yet be found at other locations on ridges at rocky openings in the Carson Range, and it should be looked for on altered andesite soils on the adjacent Peavine Mountain and the Virginia Range.

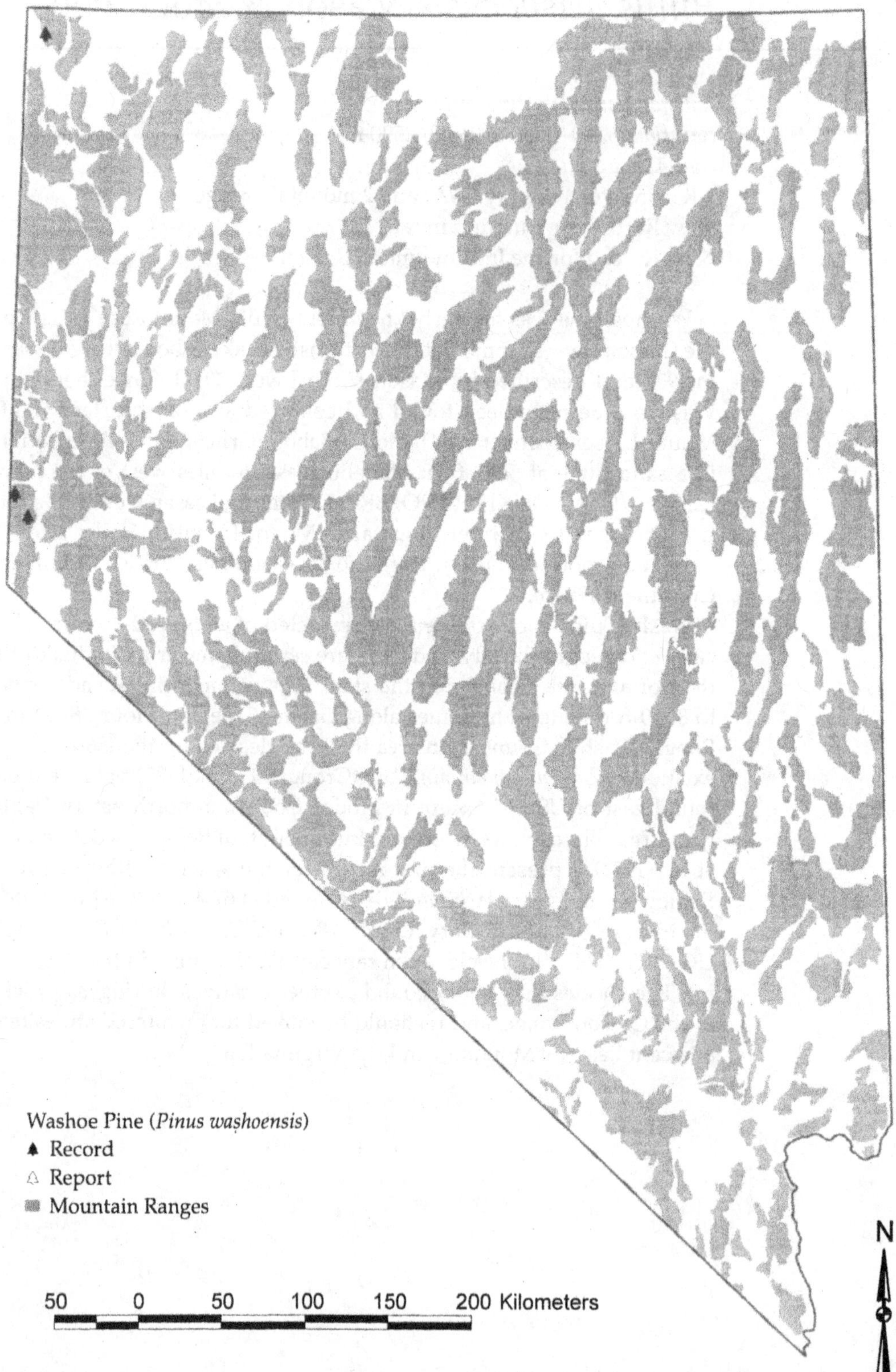

Washoe Pine (*Pinus washoensis*)
Record
Report
Mountain Ranges
50 0 50 100 150 200 Kilometers
N

Carson Range (1)
elevation record: 7240–8500 ft
7 records in 1 county (Washoe)

Towns/Range/Section	Location—Habitat	County	Elevation	Collector	Herbarium/a
T17N R19E S17	E slope of Mount Rose	WA	8000 ft	*Billings 1656*	RENO
T17N R19E S17	E side of Mount Rose (topotype)	WA	8500 ft	*Stockwell & Cumming 1455*	RSA, UTC
T17N R19E S17	E slope Mount Rose, from type tree	WA	7850 ft	*Billings in 1954*	DUKE
T17N R19E S17	Washoe pine forest w/*P. contorta* & *Abies concolor*	WA	7240 ft	*Charlet 1914 & T Charlet*	DUKE
T17N R19E S17	along Galena Creek	WA	7240 ft	*Charlet 1915 & T Charlet*	RENO, DUKE, UC, BRY
T17N R19E S20 N/8	Galena Creek	WA	7700 ft	*Charlet 2195 & T Charlet*	RENO
T18N R18E S4	———	WA	8200 ft	*K Genz 8558*	BRY

Reported as present in the range by Little (1971).
DAC: I have searched most of the Nevada portion of the Carson Range for this species but have found it only near the type locality. I have not visited the *Genz 8558* locality, which is in the vicinity of Deep Canyon, Cone Peak, and a radio facility. Old trees that were spared from logging in the 1860s remain along Galena Creek (1971–1995).

Madelin Mesa (4)
Washoe County

DAC: a stand of 50 reproductive individuals occurs 6 mi west of the Nevada-California border (*Charlet 1924* DUKE, RENO, BRY, UC) in a sand dune with *Purshia tridentata* and near *Populus tremuloides.* The species may occur in the Nevada portion of the range (1994).

****Mosquito Mountains (1)**
elevation record: 6240 ft
2 records in 1 county (Washoe)

Towns/Range/Section	Location—Habitat	County	Elevation	Collector	Herbarium/a
T45N R19E S4	ENE aspect, low edge aspen stand at spring; single large individual	WA	6240 ft	*Charlet 1840*	DUKE
T45N R19E S4	ENE aspect, low edge aspen stand at spring; single large individual	WA	6240 ft	*Charlet 1932*	DUKE, RENO, UC, BRY, UNLV

Not reported as present in the range by Critchfield & Allenbaugh (1969), Little (1971), or Critchfield (1984b).
DAC: both collections are from the same large (17 m tall, 1.2 m dbh) individual; *1840* is cones and bark only, and *1932* includes foliage, cones, and bark. This is the only individual of the species known to occur in the Intermountain Region as defined by Cronquist et al. (1972). I collected a core from this tree (84 years at 20 cm), and on this basis I estimate the age of the tree to be at least 250 years (1994).
New record for the species in the range.
New record for the Intermountain Region.

var. *menziesii*

Pseudotsuga menziesii (Mirbel) Franco

varieties in Nevada:
P. menziesii var. *menziesii* [Pacific Douglas-fir]
P. menziesii var. *glauca* (Beissner) Franco [Rocky Mountain Douglas-fir]
common name: Douglas-fir, Oregon pine
elevation: 6240–9800 ft (1905–2990 m)
24 Records for 4 counties (CL, DO, LI, WP) and 5 mountain ranges; possible twentieth-century EXTINCTION in 2 mountain ranges
P. menziesii var. *menziesii* occurs in 1 mountain range; *P. menziesii* var. *glauca* in 4

Douglas-fir is the most widely distributed conifer tree in North America, extending from latitude 55°N in Canada to latitude 18°N in southern Mexico, although it is far less common in Nevada than in the other 11 states in which it occurs. In Nevada, the typical variety occurs only in the Carson Range at Lake Tahoe (Douglas County) and is known only from collections made from the same tree, a spectacular individual 34 m tall and 1.35 m in diameter; its age at breast height (28.5-cm core) was determined to be 214 years, and so the tree is likely to be at least 400 years old. The species may have been much more widely distributed throughout the Lake Tahoe basin before these mountains were clear-cut to supply timbers for the Comstock mines (1860s). Rocky Mountain Douglas-fir (*P. menziesii* var. *glauca*) forms small stands in 4 mountain ranges in 3 counties (Clark, Lincoln, and White Pine) in eastern and southern Nevada.

There are only 2 records of Rocky Mountain Douglas-fir in the Jarbidge Mountain area (Elko County): *Nichols 373 & Lund* CAS & RENO (Cobb Creek, Bull Run Mountains) and *Nichols 453 & Lund* RENO (Merritt Mountain, Bruneau Range), both collected in 1937. Neither Loope (1969) nor I was able to verify the presence of the species at either location.

I have searched without success for Douglas-fir throughout the Jarbidge, Bull Run, and Independence Mountains and the intervening ranges for 7 years. The area was extensively logged in the early 1900s, and the species may have been widespread before that. Merritt Mountain is far from the mining activity, and so may have escaped logging, but it appears that it has not escaped fire, as the *Abies lasiocarpa* individuals present on the north slope of Merritt Mountain are quite young. A fire in the 1940s may have killed the last of the Douglas-fir trees in this area.

At least 2 populations of Douglas-fir have been planted in the wild in Nevada, at Kingston Canyon in the Toiyabe Range (Lander County) and at Hinkey Summit in the Santa Rosa Range (Humboldt County). Both populations are producing cones, but neither appears to be spreading. The elevation range of Douglas-fir in Nevada is 3560 ft (1085 m).

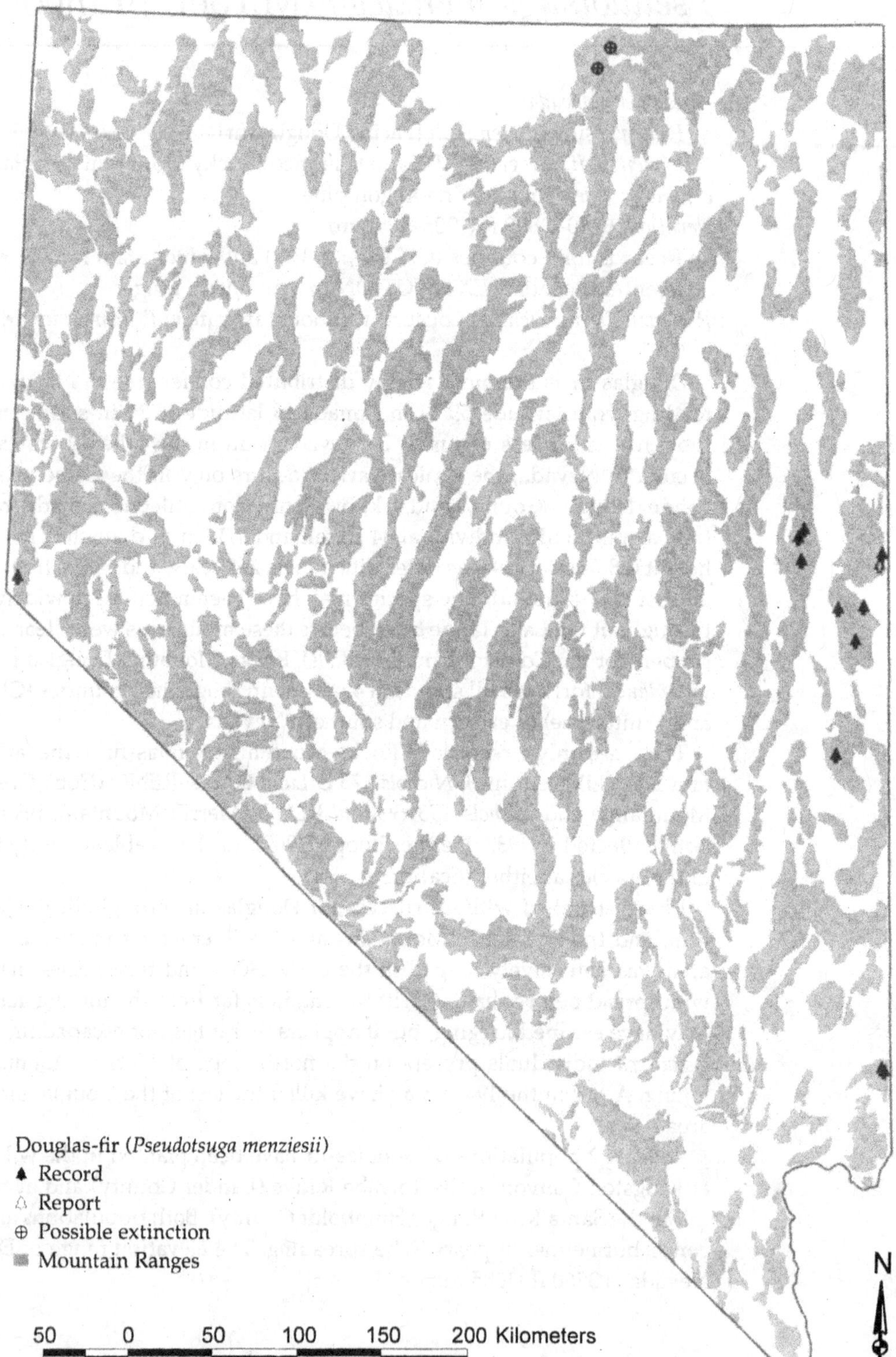

Douglas-fir (*Pseudotsuga menziesii*)
Record
Report
Possible extinction
Mountain Ranges
N
50 0 50 100 150 200 Kilometers

Bruneau Range (6)
elevation record: [6800 ft]
1 record in 1 county (Elko)

Towns/Range/Section	Location—Habitat	County	Elevation	Collector	Herbarium/a
T46N R54E	N slope Merritt Mountain, 15 mi NE of Mountain City	EL	6800 ft	*Nichols 453 & Lund in 1937*	RENO

Not reported as present in the range by Little (1971).
DAC: I have not visited all the possible sites on the mountain, so the species may still be here. From the 1910s to the 1930s a great deal of mining and homesteading occurred in the Jarbidge Mountains area. Sawmills processed much timber for the mines, and many cabins were built. If Douglas-fir was present in the Jarbidge Mountains proper, it may have been extirpated by the logging. However, little mining occurred in the vicinity of Merritt Mountain, and so a few individuals may remain here, particularly since they survived until 1937. If Douglas-fir does occur here, it certainly is not common. I visited conifer stands on the northwest slope of the mountain, and they were composed exclusively of *Abies lasiocarpa.* Loope (1969) visited Merritt Mountain and found *Abies lasiocarpa* to be the only conifer present. *Nichols 453 & Lund* was determined to be var. *glauca* by Billings (1988–1994).

Bull Run Mountains (6)
elevation record: [6200 ft]
1 record in 1 county (Elko)

Towns/Range/Section	Location—Habitat	County	Elevation	Collector	Herbarium/a
T45N R53E S29	Cobb Creek, 17 mi S of Mountain City, tree 20–40 ft tall	EL	6200 ft	*Nichols 373 & Lund in 1937*	CAS, RENO

Reported as present in the range by Little (1971).
DAC: probably extirpated; Little's report (1971) was apparently made on the basis of this collection. The locality was visited by Loope (1969), but he was unable to verify the presence of Douglas-fir. Both RWR and I attempted to verify it as well, but neither of us was able to locate the species. *Nichols 373 & Lund* was determined to be var. *glauca* by Billings (1990–1994).

Carson Range (1)
elevation record: 6240 ft
3 records in 1 county (Douglas)

Towns/Range/Section	Location—Habitat	County	Elevation	Collector	Herbarium/a
T14N R18E S22	1.5–2 mi N of Cave Rock, old road along shore, 85–100 ft tall	DO	[6228 ft]	*J Henrichs 552 in 1940*	RENO
T14N R18E S22	Cave Rock, US 50 lakeshore	DO	[6250 ft]	*Peter Herlan in 1972*	NSM
T14N R18E S22	1 air mile N of Cave Rock, 40 m from lake, w/Jeffrey pine	DO	6240 ft	*Charlet 2059 & T Charlet*	DUKE, RENO, BRY, UC

Reported as present in the range by Little (1971).
DAC: these collections are all from a single individual. In 1994, the tree was 34 m tall, with a dbh of 1.35 m. I obtained a core to a depth of 28.5 cm, and the age was 214 years at that depth. This is clearly a native tree, probably at least 400 years old. This individual was saved from logging probably as a result of its location below the logging road. Now known from only this location, the species is likely to have been more common before the Lake Tahoe basin was logged to provide timbers for the Comstock mines in the 1860s and 1870s. This is the only *Pseudotsuga* known from the Lake Tahoe basin, and the only member of var. *menziesii* known from the state of Nevada. The species should be looked for in Montreal Canyon of Glenbrook Creek and in Logan House Creek, streams that drain near the cove where *Charlet 2059* occurs (1971–1995).

Jarbidge Mountains (5)
Elko County

DAC: Neither Loope (1966) nor Lewis (1975) was able to find the species in either the Mountain City or Jarbidge USDA FS Ranger Districts. I presume that all reports of the species for the "Jarbidge Mountains area" (e.g., Billings 1954) refer to *Nichols & Lund 373* (CAS & RENO) from the Bull Run Mountains (see above commentary under "Bull Run Mountains") and/or *Nichols 453 & Lund* RENO from the Bruneau Range. In 7 years of looking, I have been unable to find Douglas-fir in the Jarbidge Mountains (1988–1994).

Ruby Mountains (7)
Elko County

DAC: reported by Kartesz (1988) to have been introduced here.

Santa Rosa Range (7)
Humboldt County

DAC: about 10 trees were planted at the Martin Creek Ranger Station, probably during the 1930s. They are of reproductive age, but I found no seedlings (1990–1994).

var. *glauca*

Schell Creek Range (1)
elevation record: 7380–9200 ft
3 records in 1 county (White Pine)

Towns/Range/Section	Location—Habitat	County	Elevation	Collector	Herbarium/a
T16N R65E S23	Kolchek Basin, limestone cliffs w/*Abies concolor*	WP	7380 ft	*Charlet 1999*	DUKE, RENO, UC
T17N R65E S10	2.5–4 mi above Berry Creek Ranger Station, N Fork Berry Creek	WP	7500–8000 ft	*Percy Train 1040*	CAS, RENO
T18N R65E S25	Timber Creek	WP	9200 ft	*Ann Pinzl 4358*	NSM

Reported as present in the range by Little (1971).
DAC: var. *glauca* is uncommon at moderately high elevations in a few locations (1994).

Snake Range (1)
elevation record: 7300–9800 ft
8 records in 1 county (White Pine)

Towns/Range/Section	Location—Habitat	County	Elevation	Collector	Herbarium/a
T11N R69E	S of Wheeler Peak, along trail to Lexington Arch	WP	9500 ft	*WH Blackburn 71*	BRY
T12N R69E S9	Snake Creek	WP	7900 ft	*Ann Pinzl 1783*	NSM
T13N R68E S2 SW/4	road to Stella Lake, Wheeler Peak area	WP	9800 ft	*Mozingo 74-43a*	BRY, RENO
T13N R68E	Mount Wheeler	WP	9500 ft	*Peter Herlan in 1973*	NSM
T13N R69E S15	Lehman Caves	WP	8500 ft	*PA Lehenbauer in 1926*	RENO
T16N R70E S10	Hampton Creek	WP	7300 ft	*Richard W. Rust 38*	UCR
T16N R70E S10	Hampton Creek	WP	7300 ft	*Charlet 2110 & T Charlet*	DUKE, RENO, UC, BRY, UNLV
—	Snake Range	WP	—	*Lehenbauer in 1926*	RENO

[Lyman & Sanders: T16N R70E (Hendrys Creek, S of Hampton Creek)]
[McLane: T12N R69E (N Fork Big Wash)]

Reported as present in the range by Little (1971).
DAC: common in spruce-fir forests at moderately high elevations. Alvin McLane reported (pers. comm. 1994) a very large tree (35 m tall) in the North Fork Big Wash (1994).

Toiyabe Range (7)
Lander County

Towns/Range/Section	Location—Habitat	County	Elevation	Collector	Herbarium/a
T16N R43E	Kingston Canyon [introduced]	LA	7400 ft	*M Williams 76-40-1 & Lott & Mozingo*	CAS, RENO, UTC

Not reported as present in the range by Little (1971).
DAC: the above collection was not marked "introduced" and was not in the introduced file at CAS or UTC, but the individuals in the canyon were introduced by the USFS (Goodrich 1981), probably in the 1940s. The trees are producing cones, but I have seen no seedlings (1991–1994).

Virginia Range (5)
Washoe County

DAC: erroneously reported as present in the range by Little (1971). I have not seen any material, nor have I seen the species on the range (1973–1995).

Virgin Mountains (1)
elevation record: 6900–7800 ft
6 records in 1 county (Clark)

Towns/Range/Section	Location—Habitat	County	Elevation	Collector	Herbarium/a
—	Virgin Mountains	CL	7600 ft	*JD Armstrong 901*	UNLV
T15S R70E S24	NE of Virgin Peak	CL	7000 ft	*Mozingo 71-5*	RENO
T15S R70E S25	———	CL	7500 ft	*Mozingo 75-23*	RENO
T15S R71E S20	Virgin Peak	CL	6900 ft	*JD Armstrong 119*	NSM
T15S R71E	N Virgin Peak	CL	7800 ft	*Ann Pinzl in 1974*	NSM
T15S R71E	E flank of Virgin Peak, in forest	CL	7300 ft	*Gary Baird 2862*	BRY

Not reported as present in the range by Little (1971).
DAC: var. *glauca* is the taxon here.

Wilson Creek Range (1)
elevation record: 8500–9500 ft
2 records in 1 county (Lincoln)

Towns/Range/Section	Location—Habitat	County	Elevation	Collector	Herbarium/a
T4N R68E S6	top of Mount Wilson	LI	9250 ft	*M Williams & Tiehm 81-51-8*	CAS, RENO
—	Wilson Mountain	LI	8500–9500 ft	*Percy Train 2561*	CAS, RENO

Reported as present in the range by Little (1971).
DAC: var. *glauca* is present in the mountain range (1994).

Tsuga mertensiana (Bongard) Carrière

common name: mountain hemlock
elevation: 7920–9650 ft (2415–2940 m)
zone: Hudsonian
15 Records in 3 counties (CC, DO, WA) and 2 mountain ranges

Mountain hemlock is found in the Pacific Mountain System along the Coast Ranges from above latitude 60°N in southern Alaska, through Canada, to the Olympic Mountains of Washington, throughout the Cascades, northwestern California, and south to the central Sierra Nevada. In the Intermontane Plateaus, it occurs only in the Wallowa Mountains of Oregon, on Glass Mountain in California, and on Peavine Mountain in Nevada, merely 20 km from the Sierran populations. It occurs in the northern Rocky Mountains of British Columbia south to northern Montana and Idaho, but does not occur south of latitude 45°N in the Rockies. It reaches its southern limits at latitude 36°38′N in the Sierra Nevada.

In Nevada, mountain hemlock is narrowly restricted to high elevations in the Hudsonian Zone of the Carson Range and a few isolated individuals growing at the summit of nearby Peavine Mountain. The narrow environmental tolerance of the species is reflected in its elevation amplitude of 1730 ft (525 m) in Nevada. Mountain hemlock is found in Washoe, Douglas, and Carson City Counties.

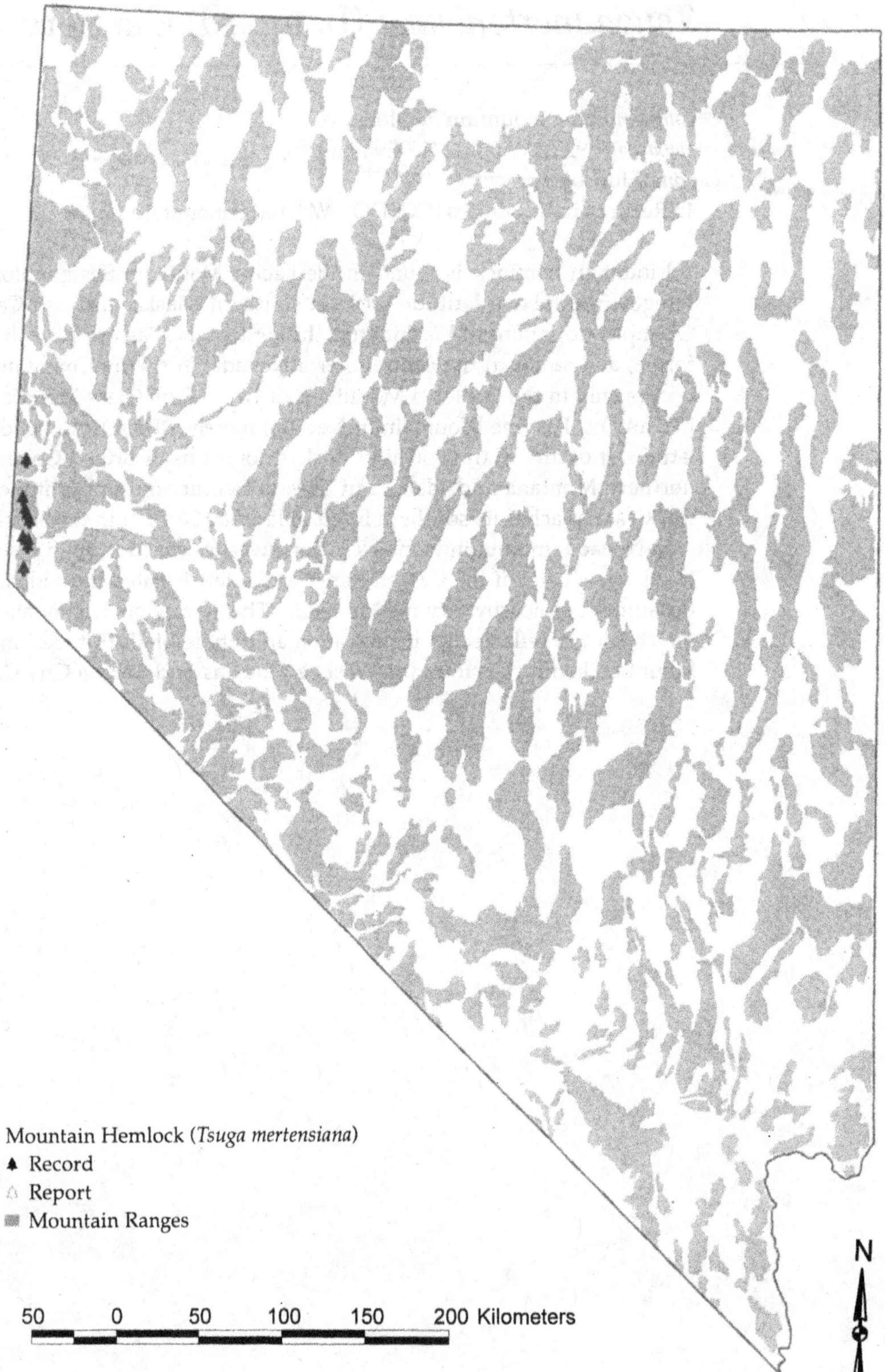
Mountain Hemlock (*Tsuga mertensiana*)
Record
Report
Mountain Ranges
50 0 50 100 150 200 Kilometers
N

Carson Range (1)
elevation record: 7920–9650 ft
13 records in 3 counties (Carson City, Douglas, and Washoe)

Towns/Range/Section	Location—Habitat	County	Elevation	Collector	Herbarium/a
T14N R19E S30 SE/4	Genoa Peak	DO	9140 ft	*Charlet 2172*	DUKE, RENO, UC
T14N R19E S30 SE/4	Genoa Peak	DO	9000 ft	*Ann Pinzl 10209*	NSM
T15N R19E S5 NW/4	N of Snow Valley Peak	WA	7920 ft	*Charlet 2063*	DUKE, RENO
T15N R19E S17	Snow Valley (determined Greene)	CC	8070 ft	*CF Baker 1437 in 1902*	RSA
T16N R18E S36 S/2	———	WA	8800 ft	*Ann Pinzl 8443 & Kimbraugh*	NSM
T17N R18E S12	Mount Rose, divide between Galena & White Creeks, E slope	WA	9000 ft	*Percy Train 4435*	RENO
T17N R18E	Mount Rose	WA	9650 ft	*AA Heller 9908*	RSA, RENO
T17N R18E	Mount Rose	WA	9000 ft	*PA Lehenbauer in 1926*	RENO
T17N R18E	near Mount Rose	WA	—	*NF Petersen 230*	RENO
T17N R18E	near Mount Rose	WA	—	*NF Petersen 514*	RENO
T17N R19E S19 N/4	Mount Rose	WA	8400 ft	*Charlet 2062 & T Charlet*	DUKE, RENO, UC, BRY
T17N R19E S32	Prices Lake	WA	8000 ft	*Nichols 722 & Lund*	RENO
T18N R18E S23	above Hunter Lake	WA	9500 ft	*Billings 1480*	RENO

Reported as present in the range by Little (1971).
DAC: occasionally codominant with *Pinus albicaulis* or *P. monticola* in the upper subalpine forests and woodlands of the Carson Range, especially on northern aspects (1971–1995).

Peavine Mountain (1)
elevation record: 8000 ft
2 records in 1 county (Washoe)

Towns/Range/Section	Location—Habitat	County	Elevation	Collector	Herbarium/a
T20N R18E	N of Peavine in a deep canyon	WA	8000 ft	*M Williams 72-89-11*	RENO
T20N R18E	Peavine Mountain	WA	8000 ft	*RJ Bandoni 3*	RENO

Reported as present in the range by Little (1971).
DAC: a few individuals are present at the summit and a few others at the head of a steep canyon north of the peak (1971–1995).

Appendix

Conifer Species Distributions and Native Diversity in Nevada Mountains

This appendix compiles all the distribution information concerning Nevada conifers as presented in the text. All Nevada mountain ranges are listed in alphabetical order (rows), and all conifer species are arranged in the order that they are presented in the text (columns). The first column following the mountain range name indicates the mountains that the author has collected conifers on (C) or has visited (V). Dashes in each row by column location indicates that the species is apparently absent, has not been collected or reported, and is not likely to be found in the mountain range. Numerals in each row by column location are codes that indicate the species has been collected, observed, reported, may occur, has been erroneously reported, is now extirpated, or has been introduced in the mountain range (see Introduction and following key). Numerals in the final column (species diversity) are the total number of naturally occurring native conifer species known to be present in the mountain range.

	Collected and/or visited	*Calocedrus decurrens*	*Juniperus californica*	*Juniperus communis*	*Juniperus occidentalis*	*Juniperus osteosperma*	*Juniperus scopulorum*	*Abies concolor*	*Abies lasiocarpa*	*Abies magnifica*	*Picea engelmannii*	*Pinus albicaulis*	*Pinus contorta*	*Pinus flexilis*	*Pinus jeffreyi*	*Pinus lambertiana*	*Pinus longaeva*	*Pinus monophylla*	*Pinus monticola*	*Pinus ponderosa*	*Pinus washoensis*	*Pseudotsuga menziesii*	*Tsuga mertensiana*	Species diversity
1. Adobe Range / EL	CV	–	–	–	–	1	–	–	–	–	–	–	–	–	–	–	–	–	–	–	–	–	–	1
2. Anchorite Hills / MI	CV	–	–	–	1	1	–	–	–	–	–	–	–	–	–	–	–	1	–	–	–	–	–	3
3. Antelope Range / PE, HU	V	–	–	–	–	–	–	–	–	–	–	–	–	–	–	–	–	–	–	–	–	–	–	0
4. Antelope Range / WP, EL	CV	–	–	–	–	1	4	1	–	–	–	–	–	1	–	–	–	2	–	–	–	–	–	4
5. Antelope Range / NY, EU	–	–	–	–	–	3	–	–	–	–	–	–	–	–	–	–	–	3	–	–	–	–	–	2
6. Ararat Hills / HU	V	–	–	–	–	–	–	–	–	–	–	–	–	–	–	–	–	–	–	–	–	–	–	0
7. Arrow Canyon Range / CL	–	–	–	–	–	–	–	–	–	–	–	–	–	–	–	–	–	–	–	–	–	–	–	0
8. Augusta Mountains / HU, LA, PE	CV	–	–	–	–	1	–	–	–	–	–	–	–	–	–	–	–	4	–	–	–	–	–	1
9. Badger Mountains / WA, HU	CV	–	–	–	1	5	–	–	–	–	–	–	–	–	–	–	–	–	–	–	–	–	–	1
10. Bare Mountain / NY	V	–	–	–	–	–	–	–	–	–	–	–	–	–	–	–	–	–	–	–	–	–	–	0
11. Barnett Hills / CH	–	–	–	–	–	–	–	–	–	–	–	–	–	–	–	–	–	–	–	–	–	–	–	0
12. Battle Mountain / LA, HU	CV	–	–	–	–	1	–	–	–	–	–	–	–	–	–	–	–	5	–	–	–	–	–	1
13. Belted Range / NY	–	–	–	–	–	1	–	–	–	–	–	–	–	–	–	–	–	1	–	–	–	–	–	2
14. Bilk Creek Mountains / HU, *OR*	CV	–	–	–	1	–	–	–	–	–	–	–	–	–	–	–	–	–	–	–	–	–	–	1
15. Bird Spring Range / CL	V	–	–	–	–	1	–	–	–	–	–	–	–	–	–	–	–	1	–	–	–	–	–	2
16. Black Canyon Range / LI	–	–	–	–	–	3	–	–	–	–	–	–	–	–	–	–	–	3	–	–	–	–	–	2

	Collected and/or visited	*Calocedrus decurrens*	*Juniperus californica*	*Juniperus communis*	*Juniperus occidentalis*	*Juniperus osteosperma*	*Juniperus scopulorum*	*Abies concolor*	*Abies lasiocarpa*	*Abies magnifica*	*Picea engelmannii*	*Pinus albicaulis*	*Pinus contorta*	*Pinus flexilis*	*Pinus jeffreyi*	*Pinus lambertiana*	*Pinus longaeva*	*Pinus monophylla*	*Pinus monticola*	*Pinus ponderosa*	*Pinus washoensis*	*Pseudotsuga menziesii*	*Tsuga mertensiana*	Species diversity
17. Black Mountains / CL, *AZ*	-	-	-	-	-	-	-	-	-	-	-	-	-	-	-	-	-	-	-	-	-	-	-	0
18. Black Rock Range / HU, PE	CV	-	-	-	1	-	-	-	-	-	-	-	-	-	-	-	-	-	-	-	-	-	-	1
19. Bloody Run Hills / HU	V	-	-	-	-	-	-	-	-	-	-	-	-	-	-	-	-	-	-	-	-	-	-	0
20. Blow Sand Mountain / CH	-	-	-	-	-	-	-	-	-	-	-	-	-	-	-	-	-	-	-	-	-	-	-	0
21. Blue Wing Mountains / PE	V	-	-	-	-	-	-	-	-	-	-	-	-	-	-	-	-	-	-	-	-	-	-	0
22. Bodie Mountains / MI, LY, *CA*	CV	-	-	-	1	2	-	-	-	-	-	-	1	1	1	-	-	1	-	1	-	-	-	7
23. Bone Mountains / EL	V	-	-	-	-	-	-	-	-	-	-	-	-	-	-	-	-	-	-	-	-	-	-	0
24. Bristol Range / LI	CV	-	-	-	-	1	-	-	-	-	-	-	-	-	-	-	-	1	-	-	-	-	-	2
25. Broken Hills / CH, MI	-	-	-	-	-	3	-	-	-	-	-	-	-	-	-	-	-	-	-	-	-	-	-	1
26. Bruneau Range / EL	CV	-	-	1	1	-	1	-	1	-	-	-	-	1	-	-	-	-	-	-	-	6	-	5
27. Buck Creek Mountains / EL	V	-	-	-	-	-	2	-	2	-	-	2	-	2	-	-	-	-	-	-	-	-	-	4
28. Buckskin Range / DO, LY	V	-	-	-	-	2	-	-	-	-	-	-	-	-	-	-	-	-	-	-	-	-	-	1
29. Buffalo Hills / WA	CV	-	-	-	1	1	-	-	-	-	-	-	-	-	-	-	-	-	-	-	-	-	-	2
30. Buffalo Mountain / HU	V	-	-	-	-	-	-	-	-	-	-	-	-	-	-	-	-	-	-	-	-	-	-	0
31. Bullfrog Hills / NY	V	-	-	-	-	-	-	-	-	-	-	-	-	-	-	-	-	-	-	-	-	-	-	0
32. Bull Run Mountains / EL	CV	-	-	-	1	5	1	-	2	-	-	1	-	1	-	-	-	-	-	-	-	6	-	5
33. Bunejug Mountains / CH	-	-	-	-	-	-	-	-	-	-	-	-	-	-	-	-	-	-	-	-	-	-	-	0
34. Buried Hills / LI	-	-	-	-	-	-	-	-	-	-	-	-	-	-	-	-	-	-	-	-	-	-	-	0
35. Burnt Spring Range / LI	-	-	-	-	-	1	-	-	-	-	-	-	-	-	-	-	-	3	-	-	-	-	-	2
36. Butte Mountains / WP	V	-	-	-	-	2	-	-	-	-	-	-	-	-	-	-	-	2	-	-	-	-	-	2
37. Cactus Range / NY	-	-	-	-	-	1	-	-	-	-	-	-	-	-	-	-	-	3	-	-	-	-	-	2
38. Calico Mountains / HU, PE	V	-	-	-	-	-	-	-	-	-	-	-	-	-	-	-	-	-	-	-	-	-	-	0
39. Calico Mountains / HU	V	-	-	-	-	-	-	-	-	-	-	-	-	-	-	-	-	-	-	-	-	-	-	0
40. Cambridge Hills / LY	V	-	-	-	-	-	-	-	-	-	-	-	-	-	-	-	-	-	-	-	-	-	-	0
41. Candelaria Hills / MI, ES	V	-	-	-	-	2	-	-	-	-	-	-	-	-	-	-	-	2	-	-	-	-	-	2
42. Carson Range / WA, DO, *CA*	CV	1	-	4	1	1	-	1	-	1	-	1	1	-	1	1	-	1	1	1	1	1	1	15
43. Castle Mountains / CL	-	-	-	-	-	-	-	-	-	-	-	-	-	-	-	-	-	-	-	-	-	-	-	0
44. Cedar Mountains / MI, ES, NY	-	-	-	-	-	1	-	-	-	-	-	-	-	-	-	-	-	3	-	-	-	-	-	2
45. Cedar Range / LI	-	-	-	-	-	1	-	-	-	-	-	-	-	-	-	-	-	1	-	-	-	-	-	2
46. Cherry Creek Range / WP	CV	-	-	-	-	1	1	1	-	-	-	-	-	1	-	-	2	2	-	-	-	-	-	6
47. Chief Range / LI	-	-	-	-	-	3	-	-	-	-	-	-	-	-	-	-	-	3	-	-	-	-	-	2
48. Clan Alpine Mountains / CH	CV	-	-	-	1	1	-	-	-	-	-	-	-	-	-	-	-	1	-	-	-	-	-	3
49. Clayton Ridge / ES	-	-	-	-	-	-	-	-	-	-	-	-	-	-	-	-	-	-	-	-	-	-	-	0
50. Clover Mountains / LI	-	-	-	-	-	1	3	-	-	-	-	-	-	-	-	-	-	1	-	1	-	-	-	4
51. Cocoon Mountains / CH	-	-	-	-	-	-	-	-	-	-	-	-	-	-	-	-	-	-	-	-	-	-	-	0
52. Copper Mountains / EL	CV	-	-	1	-	-	2	-	1	-	-	1	-	1	-	-	-	-	-	-	-	-	-	5
53. Corn Creek Range / CL	-	-	-	-	-	-	-	-	-	-	-	-	-	-	-	-	-	-	-	-	-	-	-	0
54. Cortez Mountains / EU, LA	CV	-	-	-	-	1	1	-	-	-	-	-	-	-	-	-	-	1	-	-	-	-	-	3
55. Coyote Spring Range / LI	-	-	-	-	-	-	-	-	-	-	-	-	-	-	-	-	-	-	-	-	-	-	-	0
56. Cucomungo Mountains / ES, *CA*	-	-	-	-	-	-	-	-	-	-	-	-	-	-	-	-	-	-	-	-	-	-	-	0

	Collected and/or visited	*Calocedrus decurrens*	*Juniperus californica*	*Juniperus communis*	*Juniperus occidentalis*	*Juniperus osteosperma*	*Juniperus scopulorum*	*Abies concolor*	*Abies lasiocarpa*	*Abies magnifica*	*Picea engelmannii*	*Pinus albicaulis*	*Pinus contorta*	*Pinus flexilis*	*Pinus jeffreyi*	*Pinus lambertiana*	*Pinus longaeva*	*Pinus monophylla*	*Pinus monticola*	*Pinus ponderosa*	*Pinus washoensis*	*Pseudotsuga menziesii*	*Tsuga mertensiana*	Species diversity
57. Cuprite Hills / ES, NY	V	–	–	–	–	–	–	–	–	–	–	–	–	–	–	–	–	–	–	–	–	–	–	0
58. Curnow Range / WA	V	–	–	–	–	1	–	–	–	–	–	–	–	–	–	–	–	–	–	–	–	–	–	1
59. Dead Camel Mountains / CH, LY	–	–	–	–	–	–	–	–	–	–	–	–	–	–	–	–	–	–	–	–	–	–	–	0
60. Deep Creek Mountains / WP, *UT*	V	–	–	–	–	2	–	–	–	–	–	–	–	–	–	–	–	2	–	–	–	–	–	2
61. De Lamar Mountains / LI	–	–	–	–	–	1	–	–	–	–	–	–	–	–	–	–	–	1	–	–	–	–	–	2
62. Delano Mountains / EL	V	–	–	–	–	2	–	–	–	–	–	–	–	–	–	–	–	–	–	–	–	–	–	1
63. Desatoya Mountains / CH, LA	CV	–	–	–	–	1	–	–	–	–	–	–	–	–	–	–	–	2	–	–	–	–	–	2
64. Desert Creek Mountains / LY	CV	–	–	–	1	1	–	–	–	–	–	–	–	–	1	–	–	2	–	–	–	–	–	4
65. Desert Hills / LI	–	–	–	–	–	–	–	–	–	–	–	–	–	–	–	–	–	–	–	–	–	–	–	0
66. Desert Mountains / LY, CH	V	–	–	–	–	–	–	–	–	–	–	–	–	–	–	–	–	–	–	–	–	–	–	0
67. Desert Range / CL, LI	–	–	–	–	–	1	–	–	–	–	–	–	–	–	–	–	–	–	–	–	–	–	–	1
68. Devils Hole Hills / NY	V	–	–	–	–	–	–	–	–	–	–	–	–	–	–	–	–	–	–	–	–	–	–	0
69. Diabase Hills / WA	V	–	–	–	–	–	–	–	–	–	–	–	–	–	–	–	–	–	–	–	–	–	–	0
70. Diamond Mountains / EU, WP	CV	–	–	–	–	1	3	–	–	–	–	–	–	3	–	–	5	1	–	–	–	–	–	4
71. Division Range / WA, HU	CV	–	–	–	1	–	–	–	–	–	–	–	–	–	–	–	–	–	–	–	–	–	–	1
72. Dixie Hills / EL	CV	–	–	–	–	1	–	–	–	–	–	–	–	–	–	–	–	–	–	–	–	–	–	1
73. Dogskin Mountain / WA	CV	–	–	–	1	1	–	–	–	–	–	–	–	–	–	–	–	–	–	–	–	–	–	2
74. Dolly Varden Mountains / EL	–	–	–	–	–	1	–	2	–	–	–	–	–	–	–	–	–	1	–	–	–	–	–	3
75. Double H Mountains / HU	V	–	–	–	–	–	–	–	–	–	–	–	–	–	–	–	–	–	–	–	–	–	–	0
76. Dry Hills / EU	–	–	–	–	–	–	–	–	–	–	–	–	–	–	–	–	–	–	–	–	–	–	–	0
77. Dry Lake Range / CL	–	–	–	–	–	–	–	–	–	–	–	–	–	–	–	–	–	–	–	–	–	–	–	0
78. Duck Creek Range / WP	CV	–	–	–	–	1	1	1	–	–	–	–	–	1	–	–	1	1	–	–	–	–	–	6
79. East Desert Range / CL, LI	–	–	–	–	–	–	–	–	–	–	–	–	–	–	–	–	–	–	–	–	–	–	–	0
80. East Gate Range / CH	V	–	–	–	–	1	–	–	–	–	–	–	–	–	–	–	–	1	–	–	–	–	–	2
81. East Humboldt Range / EL	CV	–	–	1	–	1	1	–	–	–	–	1	–	1	–	–	–	1	–	–	–	–	–	6
82. East Mormon Mountains / LI	–	–	–	–	–	–	–	–	–	–	–	–	–	–	–	–	–	–	–	–	–	–	–	0
83. East Pahranagat Range / LI	–	–	–	–	–	–	–	–	–	–	–	–	–	–	–	–	–	–	–	–	–	–	–	0
84. East Range / PE	CV	–	–	–	–	1	–	–	–	–	–	–	–	–	–	–	–	5	–	–	–	–	–	1
85. Edna Mountain / HU	V	–	–	–	–	–	–	–	–	–	–	–	–	–	–	–	–	–	–	–	–	–	–	0
86. Egan Range / WP, LI	CV	–	–	1	–	1	1	1	–	–	–	–	–	1	–	–	1	2	–	–	–	–	–	7
87. Elbow Range / CL	–	–	–	–	–	–	–	–	–	–	–	–	–	–	–	–	–	–	–	–	–	–	–	0
88. Eleana Range / NY	–	–	–	–	–	–	–	–	–	–	–	–	–	–	–	–	–	3	–	–	–	–	–	1
89. El Dorado Mountains / CL	–	–	–	–	–	–	–	–	–	–	–	–	–	–	–	–	–	–	–	–	–	–	–	0
90. Elk Mountains / EL	CV	–	–	–	–	–	3	–	1	–	–	1	–	3	–	–	–	–	–	–	–	–	–	4
91. Elko Mountains / EL	V	–	–	–	–	1	–	–	–	–	–	–	–	–	–	–	–	–	–	–	–	–	–	1
92. Ely Springs Range / LI	–	–	–	–	–	3	–	–	–	–	–	–	–	–	–	–	–	–	–	–	–	–	–	1
93. Eugene Mountains / PE	CV	–	–	–	–	1	–	–	–	–	–	–	–	–	–	–	–	–	–	–	–	–	–	1
94. Excelsior Mountains / MI	CV	–	–	–	1	1	–	–	–	–	–	–	–	–	–	–	–	2	–	–	–	–	–	3
95. Fairview Range / LI	CV	–	–	–	–	1	–	–	–	–	–	–	–	–	–	–	–	2	–	–	–	–	–	2
96. Fairview Range / CH	CV	–	–	–	–	1	–	–	–	–	–	–	–	–	–	–	–	1	–	–	–	–	–	2

	Collected and/or visited	*Calocedrus decurrens*	*Juniperus californica*	*Juniperus communis*	*Juniperus occidentalis*	*Juniperus osteosperma*	*Juniperus scopulorum*	*Abies concolor*	*Abies lasiocarpa*	*Abies magnifica*	*Picea engelmannii*	*Pinus albicaulis*	*Pinus contorta*	*Pinus flexilis*	*Pinus jeffreyi*	*Pinus lambertiana*	*Pinus longaeva*	*Pinus monophylla*	*Pinus monticola*	*Pinus ponderosa*	*Pinus washoensis*	*Pseudotsuga menziesii*	*Tsuga mertensiana*	Species diversity
97. Fernley Hills / LY, CH	V	-	-	-	-	-	-	-	-	-	-	-	-	-	-	-	-	-	-	-	-	-	-	0
98. Fish Creek Mountains / LA, PE	CV	-	-	-	1	1	-	-	-	-	-	-	-	-	-	-	-	5	-	-	-	-	-	2
99. Fish Creek Range / EU, NY	CV	-	-	-	-	1	1	-	-	-	-	-	-	1	-	-	1	2	-	-	-	-	-	5
100. Flowery Range / ST	CV	-	-	-	-	1	-	-	-	-	-	-	1	-	1	-	-	2	-	-	-	-	-	4
101. Fortification Range / LI, WP	V	-	-	-	-	1	-	-	-	-	-	-	-	-	-	-	-	3	-	-	-	-	-	2
102. Fort Sage Mountains / WA, *CA*	CV	-	-	-	1	1	-	-	-	-	-	-	-	-	1	-	-	-	-	1	-	-	-	4
103. Fox Creek Range / EL	CV	-	-	-	-	-	1	-	1	-	-	1	-	1	-	-	-	-	-	-	-	-	-	4
104. Fox Range / WA	CV	-	-	-	-	1	-	-	-	-	-	-	-	-	-	-	-	-	-	-	-	-	-	1
105. Frenchman Range / CL	CV	-	-	-	-	-	-	-	-	-	-	-	-	-	-	-	-	-	-	-	-	-	-	0
106. French Mountains / WA	CV	-	-	-	-	1	-	-	-	-	-	-	-	-	-	-	-	-	-	-	-	-	-	1
107. Gabbs Valley Range / MI	CV	-	-	-	-	1	-	-	-	-	-	-	-	-	-	-	-	1	-	-	-	-	-	2
108. Gap Mountains / NY	-	-	-	-	-	2	-	-	-	-	-	-	-	-	-	-	-	-	-	-	-	-	-	1
109. Garfield Hills / MI	V	-	-	-	-	-	-	-	-	-	-	-	-	-	-	-	-	-	-	-	-	-	-	0
110. General Thomas Hills / ES	-	-	-	-	-	-	-	-	-	-	-	-	-	-	-	-	-	-	-	-	-	-	-	0
111. Gillis Range / MI	V	-	-	-	-	-	-	-	-	-	-	-	-	-	-	-	-	-	-	-	-	-	-	0
112. Golden Gate Range / LI, NY	V	-	-	-	-	2	-	-	-	-	-	-	-	-	-	-	-	2	-	-	-	-	-	2
113. Goldfield Hills / ES, NY	CV	-	-	-	-	1	-	-	-	-	-	-	-	-	-	-	-	-	-	-	-	-	-	1
114. Gold Mountain Range / ES, *CA*	-	-	-	-	-	-	-	-	-	-	-	-	-	-	-	-	-	3	-	-	-	-	-	1
115. Goose Creek Mts / EL, *UT, ID*	V	-	-	-	-	2	-	-	-	-	-	-	-	-	-	-	-	-	-	-	-	-	-	1
116. Goshute Mountains / EL	-	-	-	1	-	1	1	1	-	-	-	-	-	1	-	-	1	2	-	-	-	-	-	7
117. Granite Range / EL	V	-	-	-	-	2	-	-	-	-	-	-	-	-	-	-	-	-	-	-	-	-	-	1
118. Granite Range / WA, PE	CV	-	-	-	1	1	-	-	-	-	-	-	-	-	-	-	-	-	-	-	-	-	-	2
119. Grant Range / NY	CV	-	-	-	1	1	2	1	-	-	-	-	-	1	-	-	1	1	-	1	-	-	-	8
120. Grapevine Mts / ES, NY, *CA*	-	-	-	-	-	1	-	-	-	-	-	-	-	1	-	-	-	1	-	-	-	-	-	3
121. Gray Hills / LY, MI	-	-	-	-	-	-	-	-	-	-	-	-	-	-	-	-	-	-	-	-	-	-	-	0
122. Groom Range / LI	-	-	-	-	-	1	-	1	-	-	-	-	-	1	-	-	-	3	-	-	-	-	-	4
123. Halfpint Range / NY	-	-	-	-	-	3	-	-	-	-	-	-	-	-	-	-	-	-	-	-	-	-	-	1
124. Hannan Range / PE	V	-	-	-	-	-	-	-	-	-	-	-	-	-	-	-	-	-	-	-	-	-	-	0
125. Hays Canyon Range / WA	CV	-	-	-	1	1	-	2	-	-	-	-	-	-	-	-	-	-	-	-	-	-	-	3
126. HD Range / EL	V	-	-	-	-	2	-	-	-	-	-	-	-	-	-	-	-	-	-	-	-	-	-	1
127. Highland Range / LI	CV	-	-	-	-	1	1	1	1	-	-	-	-	4	-	-	1	1	-	1	-	-	-	7
128. Highland Spring Range / CL	-	-	-	-	-	-	-	-	-	-	-	-	-	-	-	-	-	-	-	-	-	-	-	0
129. High Rock Canyon Hills / HU, WA	V	-	-	-	-	-	-	-	-	-	-	-	-	-	-	-	-	-	-	-	-	-	-	0
130. Hiko Range / LI	-	-	-	-	-	2	-	-	-	-	-	-	-	-	-	-	-	-	-	-	-	-	-	1
131. Hiller Mountains / CL	-	-	-	-	-	-	-	-	-	-	-	-	-	-	-	-	-	-	-	-	-	-	-	0
132. Hog Ranch Mountains / WA	V	-	-	-	-	-	-	-	-	-	-	-	-	-	-	-	-	-	-	-	-	-	-	0
133. Home Camp Range / WA	CV	-	-	-	1	-	-	-	-	-	-	-	-	-	-	-	-	-	-	-	-	-	-	1
134. Horse Range / NY	CV	-	-	-	1	2	-	-	-	-	-	-	-	-	-	-	-	1	-	-	-	-	-	3
135. Hot Creek Range / NY	CV	-	-	-	1	2	-	-	-	-	-	-	-	1	-	-	1	2	-	-	-	-	-	5
136. Hot Springs Mountains / CH	V	-	-	-	-	-	-	-	-	-	-	-	-	-	-	-	-	-	-	-	-	-	-	0

	Collected and/or visited	*Calocedrus decurrens*	*Juniperus californica*	*Juniperus communis*	*Juniperus occidentalis*	*Juniperus osteosperma*	*Juniperus scopulorum*	*Abies concolor*	*Abies lasiocarpa*	*Abies magnifica*	*Picea engelmannii*	*Pinus albicaulis*	*Pinus contorta*	*Pinus flexilis*	*Pinus jeffreyi*	*Pinus lambertiana*	*Pinus longaeva*	*Pinus monophylla*	*Pinus monticola*	*Pinus ponderosa*	*Pinus washoensis*	*Pseudotsuga menziesii*	*Tsuga mertensiana*	Species diversity
137. Hot Springs Range / HU	V	–	–	–	–	–	–	–	–	–	–	–	–	–	–	–	–	–	–	–	–	–	–	0
138. Humboldt Range / PE	CV	–	–	–	1	1	–	–	–	–	–	–	–	–	–	–	–	5	–	–	–	–	–	2
139. Hungry Range / WA	CV	–	–	–	–	1	–	–	–	–	–	–	–	–	–	–	–	–	–	–	–	–	–	1
140. Hunter Point Mountains / WP	CV	–	–	–	–	1	–	–	–	–	–	–	–	–	–	–	–	1	–	–	–	–	–	2
141. Huntoon Mountains / MI, *CA*	V	–	–	–	–	2	–	–	–	–	–	–	–	–	–	–	–	2	–	–	–	–	–	2
142. Ichabod Range / EL	CV	–	–	–	–	–	1	–	1	–	–	1	–	1	–	–	–	–	–	–	–	–	–	4
143. Idaho Canyon Range / HU	V	–	–	–	–	–	–	–	–	–	–	–	–	–	–	–	–	–	–	–	–	–	–	0
144. Independence Mountains / EL	CV	–	–	1	–	1	1	–	1	–	–	1	7	1	–	–	–	–	–	–	–	–	–	6
145. Irish Mountain / LI	–	–	–	–	–	1	–	–	–	–	–	–	–	–	–	–	3	2	–	3	–	–	–	4
146. Jackson Mountains / HU	CV	–	–	–	1	1	–	–	–	–	–	–	–	–	–	–	–	–	–	–	–	–	–	2
147. Jarbidge Mountains / EL	CV	–	–	1	–	5	1	–	1	–	–	1	7	1	–	–	–	–	–	7	–	5	–	5
148. Johnnie Range / NY	–	–	–	–	–	–	–	–	–	–	–	–	–	–	–	–	–	–	–	–	–	–	–	0
149. Jumbled Hills / LI	–	–	–	–	–	–	–	–	–	–	–	–	–	–	–	–	–	–	–	–	–	–	–	0
150. Junction House Range / WA	CV	–	–	–	1	1	–	–	–	–	–	–	–	–	–	–	–	1	–	1	–	–	–	4
151. Kamma Mountains / PE, HU	CV	–	–	–	1	1	–	–	–	–	–	–	–	–	–	–	–	–	–	–	–	–	–	2
152. Kawich Range / NY	CV	–	–	–	–	1	–	–	–	–	–	–	–	1	–	–	–	1	–	–	–	–	–	3
153. Kern Mountains / WP, *UT*	CV	–	–	–	–	1	1	1	–	–	–	–	–	1	–	–	–	1	–	–	–	–	–	5
154. Kinsley Mountains / EL	–	–	–	–	–	2	–	–	–	–	–	–	–	–	–	–	–	2	–	–	–	–	–	2
155. Lahontan Mountains / CH	V	–	–	–	–	–	–	–	–	–	–	–	–	–	–	–	–	–	–	–	–	–	–	0
156. Lake Range / WA	CV	–	–	–	1	1	–	–	–	–	–	–	–	–	–	–	–	–	–	–	–	–	–	2
157. Last Chance Range / NY	–	–	–	–	–	–	–	–	–	–	–	–	–	–	–	–	–	–	–	–	–	–	–	0
158. Las Vegas Range / CL	–	–	–	–	–	1	–	–	–	–	–	–	–	–	–	–	–	1	–	–	–	–	–	2
159. Leach Range / EL	–	–	–	–	–	3	–	–	–	–	–	–	–	–	–	–	–	3	–	–	–	–	–	2
160. Leppy Range / EL, *UT*	–	–	–	–	–	3	–	–	–	–	–	–	–	–	–	–	–	3	–	–	–	–	–	2
161. Limestone Hills / LI	–	–	–	–	–	3	–	–	–	–	–	–	–	–	–	–	–	–	–	–	–	–	–	1
162. Little High Rock Mountains / WA	V	–	–	–	–	–	–	–	–	–	–	–	–	–	–	–	–	–	–	–	–	–	–	0
163. Lodi Hills / NY	–	–	–	–	–	–	–	–	–	–	–	–	–	–	–	–	–	–	–	–	–	–	–	0
164. Lone Mountain / ES	–	–	–	–	–	1	–	–	–	–	–	–	–	–	–	–	–	2	–	–	–	–	–	2
165. Lost Creek Hills / WA	CV	–	–	–	1	–	–	–	–	–	–	–	–	–	–	–	–	–	–	–	–	–	–	1
166. Louderback Mountains / CH	V	–	–	–	–	–	–	–	–	–	–	–	–	–	–	–	–	–	–	–	–	–	–	0
167. Lucy Grey Range / CL	–	–	–	–	–	–	–	–	–	–	–	–	–	–	–	–	–	–	–	–	–	–	–	0
168. Madelin Mesa / WA, *CA*	CV	–	–	–	1	–	–	–	–	–	–	–	–	–	–	–	–	–	–	–	4	–	–	1
169. Magruder Mountain / ES	–	–	–	–	–	–	–	–	–	–	–	–	–	–	–	–	–	1	–	–	–	–	–	1
170. Mahogany Hills / EU	CV	–	–	–	–	1	–	–	–	–	–	–	–	–	–	–	–	2	–	–	–	–	–	2
171. Mahogany Mountains / LI, *UT*	–	–	–	–	–	3	–	–	–	–	–	–	–	–	–	–	–	1	–	–	–	–	–	2
172. Majuba Mountains / PE	CV	–	–	–	–	1	–	–	–	–	–	–	–	–	–	–	–	–	–	–	–	–	–	1
173. Mallard Hills / EL	V	–	–	–	–	–	–	–	–	–	–	–	–	–	–	–	–	–	–	–	–	–	–	0
174. Martin Creek Mountains / HU	V	–	–	–	–	–	–	–	–	–	–	–	–	–	–	–	–	–	–	–	–	–	–	0
175. Marys River Range / EL	V	–	–	1	4	–	1	–	1	–	–	1	–	1	–	–	–	–	–	–	–	–	–	5
176. Massacre Range / WA	CV	–	–	–	1	–	–	–	–	–	–	–	–	–	–	–	–	–	–	–	–	–	–	1

	Collected and/or visited	*Calocedrus decurrens*	*Juniperus californica*	*Juniperus communis*	*Juniperus occidentalis*	*Juniperus osteosperma*	*Juniperus scopulorum*	*Abies concolor*	*Abies lasiocarpa*	*Abies magnifica*	*Picea engelmannii*	*Pinus albicaulis*	*Pinus contorta*	*Pinus flexilis*	*Pinus jeffreyi*	*Pinus lambertiana*	*Pinus longaeva*	*Pinus monophylla*	*Pinus monticola*	*Pinus ponderosa*	*Pinus washoensis*	*Pseudotsuga menziesii*	*Tsuga mertensiana*	Species diversity
177. Maverick Springs Range / EL, WP	–	–	–	–	–	3	–	–	–	–	–	–	–	–	–	–	–	3	–	–	–	–	–	2
178. McCullough Range / CL	–	–	–	–	–	1	–	–	–	–	–	–	–	–	–	–	–	1	–	–	–	–	–	2
179. Meadow Valley Mts / LI, CL	–	–	–	–	–	1	–	–	–	–	–	–	–	–	–	–	–	–	–	–	–	–	–	1
180. Medicine Range / EL	CV	–	–	–	–	1	–	–	–	–	–	–	–	4	–	–	–	1	–	–	–	–	–	2
181. Monitor Hills / NY	–	–	–	–	–	–	–	–	–	–	–	–	–	–	–	–	–	–	–	–	–	–	–	0
182. Monitor Range / NY, EU, LA	CV	–	–	3	–	1	3	–	–	–	–	–	–	1	–	–	1	1	–	–	–	–	–	6
183. Montana Mountains / HU	V	–	–	–	–	–	–	–	–	–	–	–	–	–	–	–	–	–	–	–	–	–	–	0
184. Monte Christo Mts / NY, MI, CH	–	–	–	–	–	–	–	–	–	–	–	–	–	–	–	–	–	–	–	–	–	–	–	0
185. Monte Cristo Range / ES	V	–	–	–	–	1	–	–	–	–	–	–	–	–	–	–	–	–	–	–	–	–	–	1
186. Montezuma Range / ES	V	–	–	–	–	–	–	–	–	–	–	–	–	–	–	–	–	2	–	–	–	–	–	1
187. Mormon Mountains / LI, CL	V	–	–	–	–	1	–	–	5	–	5	–	–	–	–	–	–	1	–	1	–	–	–	3
188. Mosquito Mountains / WA	CV	–	–	–	1	–	–	1	–	–	–	–	–	–	–	–	–	–	–	1	1	–	–	4
189. Mountain Boy Range / EU	CV	–	–	–	–	1	–	–	–	–	–	–	–	–	–	–	–	1	–	–	–	–	–	2
190. Muddy Mountains / CL	–	–	–	–	–	–	–	–	–	–	–	–	–	–	–	–	–	–	–	–	–	–	–	0
191. Needle Range / NY	–	–	–	–	–	1	–	–	–	–	–	–	–	–	–	–	–	3	–	–	–	–	–	2
192. Newberry Mountains / CL	–	–	1	–	–	1	–	–	–	–	–	–	–	–	–	–	–	1	–	–	–	–	–	3
193. New Pass Range / CH, LA	CV	–	–	–	–	1	–	–	–	–	–	–	–	–	–	–	–	2	–	–	–	–	–	2
194. New York Mountains / CL, *CA*	–	–	–	–	–	1	–	–	–	–	–	–	–	–	–	–	–	–	–	–	–	–	–	1
195. Nightingill Mountains / PE	–	–	–	–	–	3	–	–	–	–	–	–	–	–	–	–	–	–	–	–	–	–	–	1
196. North Muddy Mountains / CL	–	–	–	–	–	–	–	–	–	–	–	–	–	–	–	–	–	–	–	–	–	–	–	0
197. North Pahroc Range / LI	–	–	–	–	–	2	–	–	–	–	–	–	–	–	–	–	–	2	–	–	–	–	–	2
198. Opal Mountains / CL	–	–	–	–	–	–	–	–	–	–	–	–	–	–	–	–	–	–	–	–	–	–	–	0
199. Osgood Mountains / HU	V	–	–	–	–	–	–	–	–	–	–	–	–	–	–	–	–	–	–	–	–	–	–	0
200. Owyhee Bluffs / HU, EL	V	–	–	–	–	–	–	–	–	–	–	–	–	–	–	–	–	–	–	–	–	–	–	0
201. Pah Rah Range / WA	CV	–	–	–	1	1	–	–	–	–	–	–	–	–	–	–	–	1	5	1	–	–	–	4
202. Pahranagat Range / LI	–	–	–	–	–	2	–	–	–	–	–	–	–	–	–	–	–	2	–	–	–	–	–	2
203. Pahsupp Mountains / PE	V	–	–	–	–	–	–	–	–	–	–	–	–	–	–	–	–	–	–	–	–	–	–	0
204. Pahute Mesa / NY	–	–	–	–	–	–	–	–	–	–	–	–	–	–	–	–	–	3	–	–	–	–	–	1
205. Painted Point Range / WA	CV	–	–	–	1	–	7	–	–	–	7	–	7	–	–	–	–	–	–	1	–	–	–	2
206. Palmetto Mountains / ES	–	–	–	–	–	2	–	–	–	–	–	–	–	–	–	–	–	1	–	–	–	–	–	2
207. Pancake Range / NY, WP	CV	–	–	–	–	1	2	–	–	–	–	–	–	–	–	–	–	2	–	–	–	–	–	3
208. Papoose Range / LI	–	–	–	–	–	1	–	–	–	–	–	–	–	–	–	–	–	–	–	–	–	–	–	1
209. Paradise Range / NY	CV	–	–	–	1	1	–	–	–	–	–	–	–	3	–	–	–	1	–	–	–	–	–	4
210. Park Range / NY	–	–	–	–	–	3	–	–	–	–	–	–	–	–	–	–	–	3	–	–	–	–	–	2
211. Paymaster Ridge / ES	–	–	–	–	–	–	–	–	–	–	–	–	–	–	–	–	–	–	–	–	–	–	–	0
212. Peavine Mountain / WA, *CA*	CV	1	–	–	1	1	–	1	–	–	–	1	1	–	1	1	–	1	1	1	–	–	1	12
213. Peko Hills / EL	V	–	–	–	–	–	–	–	–	–	–	–	–	–	–	–	–	–	–	–	–	–	–	0
214. Pequop Mountains / EL	–	–	–	–	–	1	2	1	1	–	–	–	–	1	–	–	1	1	–	–	–	–	–	7
215. Petersen Mountain / WA, *CA*	CV	–	–	–	1	1	–	3	–	–	–	–	–	–	1	–	–	–	–	–	–	–	–	4
216. Pilot Mountains / MI	V	–	–	–	–	1	–	–	–	–	–	–	–	–	–	–	–	5	–	–	–	–	–	1

	Collected and/or visited	*Calocedrus decurrens*	*Juniperus californica*	*Juniperus communis*	*Juniperus occidentalis*	*Juniperus osteosperma*	*Juniperus scopulorum*	*Abies concolor*	*Abies lasiocarpa*	*Abies magnifica*	*Picea engelmannii*	*Pinus albicaulis*	*Pinus contorta*	*Pinus flexilis*	*Pinus jeffreyi*	*Pinus lambertiana*	*Pinus longaeva*	*Pinus monophylla*	*Pinus monticola*	*Pinus ponderosa*	*Pinus washoensis*	*Pseudotsuga menziesii*	*Tsuga mertensiana*	Species diversity
217. Pilot Range / EL *UT*	–	–	–	–	–	1	1	1	–	–	1	–	–	3	–	–	–	1	–	–	–	–	–	6
218. Pine Forest Range / HU	CV	–	–	–	1	1	–	–	–	–	–	1	–	1	–	–	–	–	–	–	–	–	–	4
219. Pine Grove Hills / LY	CV	–	–	–	1	1	–	–	–	–	–	–	1	1	1	–	–	1	–	–	–	–	–	6
220. Pine Nut Mountains / LY, DO	CV	–	–	–	1	1	–	–	–	–	–	–	5	–	1	–	–	1	1	–	–	–	–	5
221. Piñon Range / EL	V	–	–	–	–	2	–	–	–	–	–	–	–	–	–	–	–	2	–	–	–	–	–	2
222. Pinto Peak Range / WA	CV	–	–	–	1	–	–	–	–	–	–	–	–	–	–	–	–	–	–	–	–	–	–	1
223. Pintwater Range / LI, CL	–	–	–	–	–	–	–	–	–	–	–	–	–	–	–	–	–	–	–	–	–	–	–	0
224. Pioche Hills / LI	V	–	–	–	–	1	–	–	–	–	7	–	–	–	–	–	–	1	–	–	–	–	–	2
225. Piute Range / CL, *CA*	–	–	–	–	–	–	–	–	–	–	–	–	–	–	–	–	–	–	–	–	–	–	–	0
226. Poker Brown Mountains / PE	V	–	–	–	–	–	–	–	–	–	–	–	–	–	–	–	–	–	–	–	–	–	–	0
227. Pueblo Mountains / HU, *OR*	CV	–	–	–	5	–	–	–	–	–	–	–	–	–	–	–	–	–	–	–	–	–	–	0
228. Quinn Canyon Range / NY, LI	CV	–	–	–	–	1	1	1	1	–	–	–	–	3	–	–	1	2	–	1	–	–	–	8
229. Ranger Mountains / CL, NY	–	–	–	–	–	–	–	–	–	–	–	–	–	–	–	–	–	–	–	–	–	–	–	0
230. Rawhide Hills / MI	–	–	–	–	–	–	–	–	–	–	–	–	–	–	–	–	–	–	–	–	–	–	–	0
231. Resting Spring Range / NY	–	–	–	–	–	–	–	–	–	–	–	–	–	–	–	–	–	–	–	–	–	–	–	0
232. Reveille Range / NY	–	–	–	–	–	3	–	–	–	–	–	–	–	–	–	–	–	1	–	–	–	–	–	2
233. River Mountains / CL	–	–	–	–	–	–	–	–	–	–	–	–	–	–	–	–	–	–	–	–	–	–	–	0
234. Roberts Mountains / EU	CV	–	–	3	1	1	1	–	–	–	–	–	–	1	–	–	–	1	–	–	–	–	–	6
235. Royston Hills / NY, ES	–	–	–	–	–	3	–	–	–	–	–	–	–	–	–	–	–	–	–	–	–	–	–	1
236. Ruby Mountains / EL, WP	CV	–	–	1	–	1	1	1	–	–	1	1	–	1	–	–	1	1	–	–	–	7	–	9
237. Sahwave Mountains / PE	CV	–	–	–	–	3	–	–	–	–	–	–	–	–	–	–	–	–	–	–	–	–	–	1
238. Salmon River Range / EL, *ID*	CV	–	–	–	–	–	–	–	–	–	–	–	–	3	–	–	–	–	–	–	–	–	–	1
239. San Antonio Mountains / NY, ES	V	–	–	–	–	3	–	–	–	–	–	–	–	–	–	–	–	3	–	–	–	–	–	2
240. Sand Range / WA	V	–	–	–	–	–	–	–	–	–	–	–	–	–	–	–	–	–	–	–	–	–	–	0
241. Sand Springs Range / CH, MI	V	–	–	–	–	–	–	–	–	–	–	–	–	–	–	–	–	–	–	–	–	–	–	0
242. Santa Renia Mountains / EL, LA	V	–	–	–	–	–	–	–	–	–	–	–	–	–	–	–	–	–	–	–	–	–	–	0
243. Santa Rosa Range / HU	CV	–	–	–	1	–	–	–	–	–	–	–	7	1	–	–	–	–	–	7	–	7	–	2
244. Schell Creek Range / LI, WP	CV	–	–	1	–	1	1	1	–	–	1	–	–	1	–	–	1	1	–	1	–	1	–	10
245. Seaman Range / LI, NY	–	–	–	–	–	2	4	4	–	–	–	–	–	4	–	–	4	2	–	–	–	–	–	2
246. Selenite Range / PE	CV	–	–	–	1	1	–	–	–	–	–	–	–	–	–	–	–	–	–	–	–	–	–	2
247. Sentinel Hills / HU	V	–	–	–	–	–	–	–	–	–	–	–	–	–	–	–	–	–	–	–	–	–	–	0
248. Seven Lakes Mountain / WA, *CA*	CV	–	–	–	–	1	–	–	–	–	–	–	–	–	1	–	–	–	–	5	–	–	–	2
249. Seven Troughs Range / PE	CV	–	–	–	1	1	–	–	–	–	–	–	–	–	–	–	–	–	–	–	–	–	–	2
250. Sheep Creek Range / EU, EL, LA	–	–	–	–	–	–	–	–	–	–	–	–	–	–	–	–	–	–	–	–	–	–	–	0
251. Sheephead Mountains / PE	V	–	–	–	–	–	–	–	–	–	–	–	–	–	–	–	–	–	–	–	–	–	–	0
252. Sheep Range / CL, LI	–	–	–	1	–	1	1	1	–	–	–	–	–	1	–	–	1	1	–	1	–	–	–	8
253. Shoshone Mountain / NY	–	–	–	–	–	3	–	–	–	–	–	–	–	–	–	–	–	3	–	–	–	–	–	2
254. Shoshone Mountains / NY, LA	CV	–	–	–	–	1	–	–	–	–	–	–	–	1	–	–	–	2	–	–	–	–	–	3
255. Shoshone Range / LA, EU	CV	–	–	–	1	1	–	–	–	–	–	–	–	–	–	–	–	1	–	–	–	–	–	3
256. Silver Peak Range / ES	V	–	–	–	–	1	–	–	–	–	–	–	–	–	–	–	3	1	–	–	–	–	–	3

	Collected and/or visited	*Calocedrus decurrens*	*Juniperus californica*	*Juniperus communis*	*Juniperus occidentalis*	*Juniperus osteosperma*	*Juniperus scopulorum*	*Abies concolor*	*Abies lasiocarpa*	*Abies magnifica*	*Picea engelmannii*	*Pinus albicaulis*	*Pinus contorta*	*Pinus flexilis*	*Pinus jeffreyi*	*Pinus lambertiana*	*Pinus longaeva*	*Pinus monophylla*	*Pinus monticola*	*Pinus ponderosa*	*Pinus washoensis*	*Pseudotsuga menziesii*	*Tsuga mertensiana*	Species diversity
257. Simpson Park Range / LA, EU	CV	–	–	–	1	1	–	–	–	–	–	–	–	3	–	–	–	1	–	–	–	–	–	4
258. Singatse Range / LY	V	–	–	–	–	1	–	–	–	–	–	–	–	–	–	–	–	3	–	–	–	–	–	2
259. Sinkavata Hills / MI	–	–	–	–	–	–	–	–	–	–	–	–	–	–	–	–	–	–	–	–	–	–	–	0
260. Skull Mountain / NY	–	–	–	–	–	–	–	–	–	–	–	–	–	–	–	–	–	–	–	–	–	–	–	0
261. Slate Ridge / ES	V	–	–	–	–	–	–	–	–	–	–	–	–	–	–	–	–	–	–	–	–	–	–	0
262. Slumbering Hills / HU	V	–	–	–	–	–	–	–	–	–	–	–	–	–	–	–	–	–	–	–	–	–	–	0
263. Smoke Creek Mountains / WA, *CA*	V	–	–	–	–	–	–	–	–	–	–	–	–	–	–	–	–	–	–	–	–	–	–	0
264. Snake Mountains / EL	CV	–	–	–	1	2	–	–	–	–	–	–	–	3	–	–	–	–	–	–	–	–	–	3
265. Snake Range / WP	CV	–	–	1	–	1	1	1	1	–	1	–	–	1	–	–	1	1	–	1	–	1	–	11
266. Snowstorm Mountains / EL, HU	V	–	–	–	–	–	–	–	–	–	–	–	–	–	–	–	–	–	–	–	–	–	–	0
267. Sonoma Range / HU, PE	CV	–	–	–	–	1	–	–	–	–	–	–	–	–	–	–	–	–	–	–	–	–	–	1
268. South Pahroc Range / LI	–	–	–	–	–	–	–	–	–	–	–	–	–	–	–	–	–	3	–	–	–	–	–	1
269. South Virgin Mountains / CL	–	–	–	–	–	3	–	–	–	–	–	–	–	–	–	–	–	1	–	–	–	–	–	2
270. Specter Range / NY	V	–	–	–	–	–	–	–	–	–	–	–	–	–	–	–	–	–	–	–	–	–	–	0
271. Spotted Range / CL	–	–	–	–	–	–	–	–	–	–	–	–	–	–	–	–	–	–	–	–	–	–	–	0
272. Spring Mountains / CL, NY	V	–	–	1	–	1	1	1	–	–	5	–	–	1	–	–	1	1	–	1	–	–	–	8
273. Stillwater Range / CH, PE	CV	–	–	–	–	1	–	–	–	–	–	–	–	–	–	–	–	2	–	–	–	–	–	2
274. Stonewall Mountain / NY	–	–	–	–	–	1	–	–	–	–	–	–	–	–	–	–	–	1	–	–	–	–	–	2
275. Sulphur Spring Range / EL, EU	CV	–	–	–	–	1	–	–	–	–	–	–	–	–	–	–	–	1	–	–	–	–	–	2
276. Sweetwater Mts / LY, DO, *CA*	V	–	–	–	4	2	–	1	–	–	–	2	1	3	1	–	–	2	–	–	–	–	–	7
277. Sylvania Mountains / ES, *CA*	–	–	–	–	–	2	–	–	–	–	–	–	–	–	–	–	–	2	–	–	–	–	–	2
278. Terraced Hills / WA	V	–	–	–	–	–	–	–	–	–	–	–	–	–	–	–	–	–	–	–	–	–	–	0
279. Terril Mountains / CH, MI, LY	–	–	–	–	–	–	–	–	–	–	–	–	–	–	–	–	–	–	–	–	–	–	–	0
280. Tikaboo Range / LI	–	–	–	–	–	–	–	–	–	–	–	–	–	–	–	–	–	–	–	–	–	–	–	0
281. Timber Mountain / NY	–	–	–	–	–	3	–	–	–	–	–	–	–	–	–	–	–	3	–	–	–	–	–	2
282. Timpahute Range / LI	–	–	–	–	–	2	–	–	–	–	–	–	–	–	–	–	–	2	–	–	–	–	–	2
283. Toana John Mountains / EL	CV	–	–	–	–	1	–	–	–	–	–	–	–	–	–	–	–	–	–	–	–	–	–	1
284. Toana Range / EL	–	–	–	–	–	1	–	3	–	–	–	–	–	3	–	–	–	1	–	–	–	–	–	4
285. Tobin Range / PE	CV	–	–	–	1	1	–	–	–	–	–	–	–	–	–	–	–	–	–	–	–	–	–	2
286. Toiyabe Range / NY, LA	CV	–	–	–	1	1	1	–	–	–	–	–	–	1	7	–	–	1	–	–	–	7	–	5
287. Toquima Range / NY	CV	–	–	–	1	1	–	–	–	–	–	–	–	1	–	–	–	1	–	7	–	–	–	4
288. Trinity Range / PE	CV	–	–	–	–	1	–	–	–	–	–	–	–	–	–	–	–	–	–	–	–	–	–	1
289. Trout Creek Mountains / HU, *OR*	V	–	–	–	5	–	–	–	–	–	–	–	–	–	–	–	–	–	–	–	–	–	–	0
290. Truckee Range / WA	V	–	–	–	–	–	–	–	–	–	–	–	–	–	–	–	–	–	–	–	–	–	–	0
291. Tule Springs Hills / LI	–	–	–	–	–	–	–	–	–	–	–	–	–	–	–	–	–	–	–	–	–	–	–	0
292. Tuscarora Mountains / EL, EU	V	–	–	–	–	–	–	–	–	–	–	–	–	–	–	–	–	–	–	–	–	–	–	0
293. Virginia Mountains / WA	CV	–	–	–	3	1	–	–	–	–	–	–	–	–	–	–	–	3	–	–	–	–	–	3
294. Virginia Range / WA, CC, LY, ST	CV	–	–	–	1	1	–	1	–	–	–	1	1	–	1	5	–	1	1	1	–	5	–	9
295. Virgin Mountains / CL, *UT*	–	–	–	–	–	1	1	1	–	–	–	–	–	–	–	–	–	1	–	–	–	1	–	5
296. Volcanic Hills / ES	V	–	–	–	–	–	–	–	–	–	–	–	–	–	–	–	–	–	–	–	–	–	–	0

	Collected and/or visited	*Calocedrus decurrens*	*Juniperus californica*	*Juniperus communis*	*Juniperus occidentalis*	*Juniperus osteosperma*	*Juniperus scopulorum*	*Abies concolor*	*Abies lasiocarpa*	*Abies magnifica*	*Picea engelmannii*	*Pinus albicaulis*	*Pinus contorta*	*Pinus flexilis*	*Pinus jeffreyi*	*Pinus lambertiana*	*Pinus longaeva*	*Pinus monophylla*	*Pinus monticola*	*Pinus ponderosa*	*Pinus washoensis*	*Pseudotsuga menziesii*	*Tsuga mertensiana*	Species diversity
297. Wassuk Range / MI, LY	CV	–	–	–	1	1	–	–	–	–	–	1	1	1	1	–	–	1	–	1	–	–	–	8
298. Weepah Hills / ES	–	–	–	–	–	–	–	–	–	–	–	–	–	–	–	–	–	4	–	–	–	–	–	0
299. Wellington Hills / DO	V	–	–	–	1	2	–	–	–	–	–	–	–	–	–	–	–	1	–	–	–	–	–	3
300. West Gate Range / CH	V	–	–	–	–	–	–	–	–	–	–	–	–	–	–	–	–	–	–	–	–	–	–	0
301. West Humboldt Range / PE	CV	–	–	–	–	1	–	–	–	–	–	–	–	–	–	–	–	5	–	–	–	–	–	1
302. West Range / LI	V	–	–	–	–	3	–	–	–	–	–	–	–	–	–	–	–	–	–	–	–	–	–	1
303. Whistler Range / EU	CV	–	–	–	–	1	–	–	–	–	–	–	–	–	–	–	–	4	–	–	–	–	–	1
304. White Mountains/ES, MI, *CA*	CV	–	–	–	4	1	–	–	–	–	–	4	3	1	4	–	1	1	–	–	–	–	–	5
305. White Pine Range/WP, NY	CV	–	–	1	–	1	1	1	–	–	–	–	–	1	–	–	1	1	–	1	–	–	–	8
306. White River Range/WP, NY	CV	–	–	–	–	1	–	–	–	–	–	–	–	–	–	–	–	–	–	–	–	–	–	1
307. White Rock Mountains/LI, *UT*	CV	–	–	–	–	1	2	2	–	–	–	–	–	–	–	–	–	1	–	1	–	–	–	5
308. White Throne Mountains / CH	–	–	–	–	–	–	–	–	–	–	–	–	–	–	–	–	–	–	–	–	–	–	–	0
309. Wild Horse Range / EL	CV	–	–	–	–	–	1	–	1	–	–	–	–	–	–	–	–	–	–	–	–	–	–	2
310. Wilson Creek Range / LI	CV	–	–	–	–	1	1	1	–	–	–	–	–	–	–	–	–	1	–	1	–	1	–	6
311. Windermere Hills / EL	CV	–	–	–	–	1	–	–	–	–	–	–	–	–	–	–	–	1	–	–	–	–	–	2
312. Wood Hills / EL	V	–	–	–	–	3	–	–	–	–	–	–	–	3	–	–	–	1	–	–	–	–	–	3
313. Worthington Mountains / LI	–	–	–	–	–	2	–	–	–	–	–	–	–	–	–	–	–	2	–	–	–	–	–	2
314. Yucca Mountain / NY	–	–	–	–	–	–	–	–	–	–	–	–	–	–	–	–	–	–	–	–	–	–	–	0

Totals

65 reports not verified but counted as occurrences (3)
78 reports verified, counted as occurrences, but no collections (2)
471 reports of occurrences in mountain ranges with accompanying records (1)
207 ranges with conifers
107 ranges with no conifers
202 mountain ranges visited by DAC
152 ranges with conifers visited by DAC
118 mountain ranges on which DAC collected conifers
57 ranges with no conifers not visited
50 ranges with no conifers visited
55 ranges with conifers not visited

Abbreviations

number preceding the mountain range refers to the corresponding number that appears on Map 1
C: mountain range in which conifer collections were made
V: mountain range visited by DAC

Presence

– = absent from the range
1 = collection seen at AUSTIN, BRY, CAS, DUKE, IFGP, IRSR, NSM, OGDF, RENO, RSA, UC, UCR, UNLV, UT, and/or UTC
2 = taxon observed in field by DAC and/or RWR
3 = taxon reported in reliable checklist or map, not verified by DAC or RWR
4 = taxon likely to occur in the range
5 = questionable or erroneous report, attempts to verify it were unsuccessful
6 = possible local extinction
7 = taxon introduced to the range

Counties

CC = Carson City; CH = Churchill; CL = Clark; DO = Douglas; EL = Elko; ES = Esmeralda; EU = Eureka; HU = Humboldt; LA = Lander; LI = Lincoln; LY = Lyon; MI = Mineral; NY = Nye; PE = Pershing; ST = Storey; WA = Washoe; WP = White Pine.

States

AZ = Arizona; *CA* = California; *ID* = Idaho; *OR* = Oregon; *UT* = Utah.

Literature Cited

Alverson WS, Kuhlmann W, Waller DM. 1994. Wild forests: conservation biology and public policy. Washington: Island Press. 300 p.

Axelrod DI. 1976. History of the coniferous forests, California and Nevada. Berkeley: University of California Publications in Botany 70.

Axelrod DI. 1995. The Miocene Purple Mountain flora of western Nevada. Geological Sciences 139:1–62.

Axelrod DI, Raven PH. 1985. Origins of the Cordilleran flora. Journal of Biogeography 12:21–47.

Beatley JC. 1976. Vascular plants of the Nevada Test Site and central-southern Nevada: ecologic and geographic distributions. Oak Ridge, Tenn.: Technical Information Center, Office of Technical Information, Energy Research and Development Administration, TID-26881. 308 p.

Billings WD. 1950. Vegetation and plant growth as affected by chemically altered rocks in the western Great Basin. Ecology 31:62–74.

Billings WD. 1951. Vegetational zonation in the Great Basin of western North America. In: Les bases écologiques de la régénération de la végétation des zones arides. Paris: International Union of Biological Sciences. Ser. B, no. 9, p 101–22.

Billings WD. 1954. Nevada trees. 2d ed. Agricultural Extension Service Bulletin 94. Reno: University of Nevada.

Billings WD. 1992. Islands of Sierran plants on the arid slopes of Peavine Mountain. Mentzelia 6(1):32–39.

Burns RM, Honkala BH. 1990. Silvics of North America. Volume 1, Conifers. Washington: USDA, Forest Service, Agriculture Handbook 654.

Clokey IW. 1951. Flora of the Charleston Mountains, Clark County, Nevada. Berkeley: University of California Press.

Critchfield WB. 1984a. Impact of the Pleistocene on the genetic structure of North American conifers. In: Lanner RL, editor. Proceedings of the Eighth North American Forest Biology Workshop, Logan, Utah, July 30–August 1, 1984. Logan: Utah State University. p 70–118.

Critchfield WB. 1984b. Crossability and relationships of Washoe pine. Madroño 31(3):144–70.

Critchfield WB, Allenbaugh GL. 1965. Washoe pine on the Bald Mountain Range, California. Madroño 18:63–64.

Critchfield WB, Allenbaugh GL. 1969. The distribution of Pinaceae in and near northern Nevada. Madroño 20(1):12–26.

Cronquist A, Holmgren AH, Holmgren NH, Reveal JL, editors. 1972. Intermountain flora, volume 1. New York: Hafner.

DeLucia EH, Schlesinger WH. 1990. Ecophysiology of Great Basin and Sierra

Nevada vegetation on contrasting soils. In: Osmond CB, Pitelka LF, Hidy GM, editors. Plant biology of the Basin and Range. New York: Springer-Verlag. p 143–78.

Elliot-Fisk DL, Peterson AM. 1991. Trees. In: Hall CA Jr., editor. Natural history of the White-Inyo Range, eastern California. Berkeley: University of California Press. p 87–107.

Goodrich S. 1981. A floristic study of central Nevada [MS thesis]. Provo (UT): Brigham Young University. 344 p.

Grayson D. 1993. The desert's past: a natural prehistory of the Great Basin. Washington: Smithsonian Institution Press. 356 p.

Greeley WB. 1925. The relation of geography to timber supply. Economic Geography 1(1):1–14.

Hall CA, editor. 1991. Natural history of the White-Inyo Range, eastern California. California Natural History Guides 55. Berkeley: University of California Press. 536 p.

Haller JR. 1961. Some recent observations on ponderosa, Jeffrey and Washoe pines in northeastern California. Madroño 16:126–32.

Harper KT, Freeman DC, Ostler WK, Klikoff LC. 1978. The flora of Great Basin mountain ranges: diversity, sources and dispersal ecology. Great Basin Naturalist Memoirs 2:81–103.

Hunt CB. 1967. Physiography of the United States. San Francisco: WH Freeman. 480 p.

Kartesz JT. 1988. A flora of Nevada [dissertation]. Reno: University of Nevada. 1729 p.

Kartesz JT. 1994. A synonymized checklist of the vascular plants of North America. Portland (OR): Timber Press.

Lanner RM. 1984. Trees of the Great Basin. Reno: University of Nevada Press. 215 p.

Lavin M. 1981. The floristics of the headwaters of the Walker River, California and Nevada [MS thesis]. Reno: University of Nevada. 141 p.

Lewis ME. 1971. Flora and major plant communities of the Ruby–East Humboldt Mountains with special emphasis on Lamoille Canyon. Unpublished report for the Intermountain Region, Humboldt National Forest Region 4. Ogden (UT): USDA, Forest Service. 62 p.

Lewis ME. 1973. Wheeler Peak area: species list. Unpublished report for the Intermountain Region. Ogden (UT): USDA, Forest Service. 17 p.

Lewis ME. 1974. Mount Moriah: species list. Unpublished report for the Intermountain Region. Ogden (UT): USDA, Forest Service.

Lewis ME. 1975a. Plant communities of the Jarbidge Mountain complex, Humboldt National Forest. Unpublished report for the Intermountain Region. Ogden (UT): USDA, Forest Service.

Lewis ME. 1975b. Quinn Canyon Division plant checklist, Humboldt National Forest. Unpublished report for the Intermountain Region. Ogden (UT): USDA, Forest Service.

Lewis ME. 1976. Flora of the Santa Rosa Mountains, Humboldt National Forest.

Unpublished report for the Intermountain Region. Ogden (UT): USDA, Forest Service.

Linsdale MA, Howell JT, Linsdale JM. 1952. Plants of the Toiyabe Mountains area, Nevada. Wasmann Journal of Biology 10:129–200.

Little EL Jr. 1956. Pinaceae, Rutaceae, Meliaceae, Anacardiaceae, Tamaricaceae, Cornaceae, Oleaceae, Bignoniaceae of Nevada. Contributions toward a Flora of Nevada no. 40. Washington: USDA. 81 p.

Little EL Jr. 1971. Atlas of United States trees. Volume 1, Conifers and important hardwoods. Washington: USDA, Forest Service, Miscellaneous Publication 1146.

Lloyd RM, Mitchell RS. 1973. A flora of the White Mountains, California and Nevada. Berkeley: University of California Press. 202 p.

Loope LL. 1969. Subalpine and alpine vegetation of northeastern Nevada [dissertation]. Durham (NC): Duke University. 292 p.

Lyman JC, Sanders AC. 1993. Mount Moriah botanical survey: Mount Moriah Wilderness, Humboldt National Forest, White Pine County, Nevada. Report prepared for USDA, Forest Service, Humboldt National Forest, Ely, NV.

Mason HL, Stockwell WP. 1945. A new pine from Mount Rose, Nevada. Madroño 8:61–63.

McLane AR. 1978. Silent Cordilleras. Reno (NV): Camp Nevada.

Merriam CH. 1898. Life zones and crop zones of the United States. Washington: USDA, Biological Survey Bulletin 10.

Morefield JD, Taylor DW, DeDecker M. 1988. Vascular flora of the White Mountains of California and Nevada: An updated, synonymized, working checklist. In: Hall CA Jr., Doyle-Jones V, editors. Plant biology of Eastern California. Los Angeles: White Mountain Research Station, University of California. p. 310–64.

Morin NR. 1993. Flora of North America. Volume 2, Conifers. Oxford: Oxford University Press.

Nachlinger JL, Peterson FF, Williams MJ. 1992a. Peavine Mountain, Nevada. Part 1: The character of Peavine Mountain. Mentzelia 6(1):1–64.

Nachlinger JL, Peterson FF, Williams MJ. 1992b. Peavine Mountain, Nevada. Part 2: Vascular plants of Peavine Mountain. Mentzelia 6(2):1–84.

Nevada Bureau of Mines and Geology. 1972. Topographic map of Nevada. Reno: University of Nevada.

Nowak CL, Nowak RS, Tausch RJ, Wigand PE. 1994. A 30,000 year record of vegetation dynamics at a semi-arid locale in the Great Basin. Journal of Vegetation Science 5:579–90.

Reveal JL. 1979. Biogeography of the Intermountain Region: A speculative appraisal. Menztzelia 4:1–92.

Schopmeyer CS. 1974. Seeds of woody plants in the United States. Washington: USDA, Forest Service, Agriculture Handbook 450.

Sudworth GB. 1915. The cypress and juniper trees of the Rocky Mountain region. Washington: USDA Bulletin 207.

Vasek FC. 1966. The distribution and taxonomy of three western junipers. Brittonia 18(4):350–72.

Wells PV. 1983. Paleobiogeography of montane islands in the Great Basin since the last glaciopluvial. Ecological Monographs 53(4):341–82.

Williams MJ, Howell JT, True GH, Tiehm A. 1992. A catalogue of vascular plants on Peavine Mountain. Mentzelia 6(2):3–84.

Young JA, Young CG. 1992. Seeds of woody plants of North America. Rev. & enl. ed. Portland (OR): Discorides Press.